Progress in Probability
Volume 52

Seminar on Stochastic Analysis, Random Fields and Applications III

Centro Stefano Franscini, Ascona, September 1999

Robert C. Dalang
Marco Dozzi
Francesco Russo
Editors

Springer Basel AG

Editors' addresses:

Robert C. Dalang
Département de Mathématiques
Ecole Polytechnique Fédérale
CH-1005 Lausanne
Switzerland
e-mail: robert.dalang@epfl.ch

Marco Dozzi
Institut Elie Cartan
Université Henri Poincaré
BP 239
F-54506 Vandoeuvre-les-
Nancy Cedex
France
e-mail: dozzi@iecn.u-nancy.fr

Francesco Russo
Département de Mathématiques
Institut Galilée
Université Paris 13
F-95430 Villetaneuse
France
e-mail: russo@math.univ-paris13.fr

2000 Mathematics Subject Classification: 60G, 60H, 60K; 35R60, 47D07, 91B28

A CIP catalogue record for this book is available from the Library of Congress, Washington D.C., USA

Deutsche Bibliothek Cataloging-in-Publication Data

Seminar on Stochastic Analysis, Random Fields and Applications <3, 1999, Ascona>:
Seminar on Stochastic Analysis, Random Fields and Applications :
Centro Stefano Franscini, Ascona September 1999 / Robert C. Dalang ... ed. –
Basel ; Boston ; Berlin : Birkhäuser, 2002
(Progress in probability ; Vol. 52)

ISBN 978-3-0348-9474-6 **ISBN 978-3-0348-8209-5 (eBook)**
DOI 10.1007/978-3-0348-8209-5

Originally published by Birkhäuser Verlag, Basel in 2002

Printed on acid-free paper produced from chlorine-free pulp. TCF ∞

9 8 7 6 5 4 3 2 1

www.birkhäuser-science.com

To Sergio Albeverio

on the occasion of his sixtieth birthday

Preface

This volume contains the Proceedings of the five-day *Third Seminar on Stochastic Analysis, Random Fields and Applications*, which took place at the Centro Stefano Franscini (Monte Verità) in Ascona (Ticino), Switzerland, from Monday, September 20 to Friday, September 24, 1999. The first two editions of this conference occured in 1993 and 1996. The Seminar focused on three topics: fundamental aspects of stochastic analysis, physical modeling, and applications to financial engineering. As in 1993 and 1996, the third topic was the subject of the *Third Minisymposium on Stochastic Methods in Financial Models.*

A major topic within *Stochastic Analysis* is the area of stochastic partial differential equations. The state of the art of a large part of this subject was presented in several lectures, that covered equations driven by correlated Gaussian noise in high space dimension, the relationship with super-Brownian motion and population dynamics, Malliavin calculus and asymptotics of law densities, stochastic algorithms and control theory. While of fundamental nature, several of these stochastic equations have also recently been used in the study of interest rates in finance.

In *Physical Modeling*, recent developments in quantum field theory, kinetic theory and magnetic fields were presented. This area was explored more extensively than in the two previous editions during a special session on Tuesday afternoon dedicated, on the occasion of his sixtieth birthday, to Professor Sergio Albeverio. Some aspects of the work of this renowned Swiss expert are described by Philippe Blanchard in this volume. This session was organized with the collaboration of the Centro di Ricerca in Fisica e Matematica (CERFIM), Locarno: Professor Albeverio was born in Ticino and he is still very actively involved in scientific activities in his home region. Lectures in honor of Professor Albeverio were given by Ph. Blanchard (Bielefeld), M. Röckner (Bielefeld), S. Scarlatti (Roma), Yu. Kondratiev (Bonn), R. Gielerak (Zielona Gora), R. Vilela Mendes (Lisbon), G. Losa and D. Merlini (CERFIM).

Financial Engineering is a rapidly developing theoretical and computational research area which requires a high degree of mathematical sophistication to meet the needs of the worldwide banking and insurance systems. Current areas of intense activity include: the modeling of the term structure of interest rates, via infinite dimensional stochastic processes; the analysis of extreme events, whose importance is highlighted by recent instability in financial markets; study of efficient replacements of the Black-Scholes model, in order to take into account the fact that statistical observations of the evolution of stock prices and indices seem to justify the use of Lévy processes with jumps; the study of credit risk and the pricing of possibly defaulting bonds, which concerns not only banks and corporations but also public authorities; weather derivatives, which are products created to hedge the risks associated with weather changes and which make use of random volatility models.

The Minisymposium on Stochastic Methods in Financial Models was this time co-organized with the Centro Studi Bancari in Lugano, directed by Professor René Chopard. We thank him and Dr. Mauro Guerra for their very competent collaboration. Thursday afternoon, four lectures directed towards a less-specialized audience were organized at the Centro Studi Bancari and were also widely attended by practitioners from the local banking community. The speakers and titles of their talks were:

- Wolfgang Runggaldier (Università di Padova), *On shortfall risk minimization under model uncertainty;*
- Paul Embrechts (ETH-Zürich), *Risk management after LTCM: extremes matter;*
- Nicole El Karoui (Ecole Polytechnique, Palaiseau), *Portfolio insurance with continuous constraints;*
- Marek Musiela (University of South Wales and Paribas London), *Infinite dimensional diffusions, Kolmogorov equations and interest rate models.*

Financial support for this meeting was provided by the Fonds National Suisse pour la Recherche Scientifique (Berne), the Centro Stefano Franscini (ETH-Zürich), the Centro Studi Bancari (Lugano), and the Ecole Polytechnique Fédérale de Lausanne (EPFL). We take this opportunity to thank these institutions.
We are honored to *dedicate this volume to Professor Sergio Albeverio* in recognition of his numerous high quality scientific contributions. On a more personal level, F. Russo would like to express his gratitude to his friend and director of his postdoctoral studies, who strongly motivated him to continue his research activities.

May 2001

Robert C. Dalang
Marco Dozzi
Francesco Russo

List of Participants

S. Albeverio,	Bochum, Germany
E. Alos,	Barcelona, Spain
V. Bally,	Paris 6, France
O. E. Barndorff-Nielson,	Aarhus, Denmark
G. Barone-Adesi,	Lugano, Switzerland
G. Ben Arous,	EPF Lausanne, Switzerland
P. Blanchard,	Bielefeld, Germany
R. Blei,	Connecticut, U.S.A.
S. Bonaccorsi,	Trento, Italy
R. Buckdahn,	Brest, France
E. Buffet,	Dublin, Ireland
E. Carlen,	Georgia, U.S.A.
R. Carmona,	Princeton, U.S.A.
C. Carvalho,	Lisbon, Portugal
S. Cerrai,	Florence, Italy
G. Chazal,	BNP Paris, France
P. Cheridito,	ETH Zurich, Switzerland
K. H. Cho,	EPF Lausanne, Switzerland
R. Chopard,	CSB Vezia, Switzerland
R. Cont,	Ecole Polytechnique, France
G. Da Prato,	SNS Pisa, Italy
R. C. Dalang,	EPF Lausanne, Switzerland
M. Deaconu,	INRIA Lorraine, France
A. Debussche,	Paris Sud, France
F. Delbaen,	ETH Zurich, Switzerland
R. Dell'Aquila,	CSB Vezia, Switzerland
A. Dermoune,	Lille, France
M. N. Dietsch,	Metz, France
M. Dozzi,	Nancy II, France
N. El Karoui,	Ecole Polytechnique, France
P. Embrechts,	ETH Zurich, Switzerland
M. L. Esquivel,	Lisbon, Portugal
A. Filinkov,	Adelaide, Australia
D. Filipović,	ETH Zurich, Switzerland
K. Fleischmann,	WIAS Berlin, Germany
P. Florchinger,	Metz, France
J. P. Fouque,	North Carolina, U.S.A.
S. Galluccio,	Parisbas London, U.K.
J. Garnier,	Ecole Polytechnique, France
D. Gatarek,	BRE Bank SA., Poland
H. Geman,	ESSEC and Dauphine, France

R. Gielerak,	Zielona Gora, Poland
F. Gozzi,	Roma 1, Italy
M. Gradinaru,	Nancy I, France
M. Grothaus,	Bonn, Germany
G. Guatteri,	Trento, Italy
M. Guerra,	Banca del Gottardo Lugano, Switzerland
A. Guidotti,	EPF Lausanne, Switzerland
Z. Haba,	Wroclaw, Poland
U. G. Haussmann,	British Columbia, Canada
M.-O. Hongler,	EPF Lausanne, Switzerland
P. Imkeller,	Humboldt Berlin, Germany
Y. Kondratiev,	Bonn, Germany
T. Kuna,	Bonn, Germany
B. Leblanc,	BNP Paris, France
O. Lévèque,	EPF Lausanne, Switzerland
M. Loffredo,	Siena, Italy
G. Losa,	CERFIM Locarno, Switzerland
A. Marazzi,	CSB Vezia, Switzerland
A. Martin,	Munich, Germany
C. Martini,	INRIA Rocquencourt, France
S. Méléard,	Paris X, France
D. Merlini,	CERFIM Locarno, Switzerland
L. Mesnager,	Paris X, France
A. Millet,	Paris X, France
P. L. Morien,	Paris X, France
C. Mueller,	Rochester, U.S.A.
M. Musiela,	New South Wales, Australia
B. Øksendal,	Oslo, Norway
M. J. Oliveira,	Lisbon, Portugal
G. Pagès,	Paris 12, France
E. Pardoux,	Aix-Marseille 1, France
E. Platen,	Sidney, Australia
M. Pontier,	Toulouse, France
A. Posilicano,	Como, Italy
N. Privault,	La Rochelle, France
M. C. Quenez-Kammerer,	Marne-la-Vallée, France
M. Röckner,	Bielefeld, Germany
A. Roncoroni,	ESSEC, France
B. Roynette,	Nancy I, France
B. L. Rozovskii,	Southern California, U.S.A.
W. J. Runggaldier,	Padova, Italy
F. Russo,	Paris 13, France
M. Sales,	Princeton, U.S.A.
M. Sanz-Solé,	Barcelona, Spain

S. Scarlatti,	Roma, Italy
W. R. Schneider,	ABB Baden, Switzerland
C. Stricker,	Franche-Comté, France
A. Sulem,	INRIA Rocquencourt, France
G. Tessitore,	Genova, Italy
M. Thieullen,	Paris 6, France
L. Tubaro,	Trento, Italy
F. Utzet,	Barcelona, Spain
P. Vallois,	Nancy I, France
F. Viens,	North Texas, U.S.A.
R. Vilela-Mendes,	Lisbon, Portugal
M. Zähle,	Jena, Germany
J. C. Zambrini,	Lisbon, Portugal
J. Zani,	Torino, Italy
Y. Zufferey,	EPF Lausanne, Switzerland

Table of Contents

Progress in Probability, Vol. 52, xv–xvii

SERGIO ALBEVERIO

by Philippe Blanchard

Sergio Albeverio was born on January 17, 1939, in Lugano, Switzerland. He is the only son of Olivetta and Luigi Albeverio. Sergio was admitted to the Eidgenössische Technische Hochschule (ETH) in Zürich in 1958, where he entered the program in mathematics and physics. This is where I first met him almost forty years ago.

Sergio acquired there a deep knowledge of mathematics and of the foundations of modern physics. He graduated in 1962 with a "Diploma" thesis under the supervision of D. Ruelle and M. Fierz, titled "An explicit solution of a generalized Ising model in two dimensions." His Ph.D. dissertation in 1966 concerned an idealized quantum mechanical stripping process and the diffraction of a plane electromagnetic wave by a wedge with an angle of $\frac{\pi}{4}$. Using a theory of difference equations with periodic coefficients developed by Sergio's "Doktorvater" Res Jost, he obtained an explicit analytical solution of the problem in terms of the Sommerfeld integral. The model can be interpreted as describing two particles forming a bound state via a δ-potential moving in one dimension and scattered against a wall which reflects one of the particles perfectly.

Sergio's curriculum vitae is impressive. He continues to publish countless articles and scientific contributions. Here, I will restrict myself to the books he has written.

During his time at the Hochstraße 60, Sergio started pioneering work in the field of singular interactions in quantum physics and Schrödinger operator theory which he has pursued until now. Sergio's book (with F. Gesztesy, Raphael Høegh-Krohn and Helge Holden) on "Solvable Models in Quantum Mechanics" appeared in 1988. It is one of the standard references on the subject as well as one of the most inspiring approaches in this field. The main topic of the book is the mathematical study of perturbations of the Laplace operator in $\mathbb{R}^d$ by potentials with support on a discrete finite or infinite set of point sources (crystal lattice, random solid, ...). Many of these models are solvable: indeed, the eigenvalues, the eigenfunctions and the scattering matrix can be computed explicitly. These calculations display interesting and complicated phenomena and give one a feeling of how to handle more realistic and complex situations.

Sergio's second book in this area (with P. Kurasov), titled "Singular Perturbations of Differential Operators," appeared in 2000 and offers a fine presentation of the enormous amount of material that has accumulated in this field over the last few decades. The bibliography contains 976 references! The aim of this book is to describe finite rank perturbations of self-adjoint operators, generalized singular perturbation and the scattering theory for finite rank operators. The study of this type of model began in 1931 with the model of Kronig-Penney that describes the quantum mechanical motion of an electron in a crystal. Today, this approach can be used to create models of nanoelectronic devices.

Sergio Albeverio is one of the most well-known people in mathematical physics. To his many colleagues, he inspires respect and affection for his passionate interests, not only in physics and mathematics, but also in philosophy and the history of science. His name is best known in connection with the formulation of Feynman's path integral as an oscillatory Fresnel integral, published jointly with his friend R. Høegh-Krohn in the book "Mathematical theory of Feynman path integrals," which was published in 1976.

Endowed with a profound knowledge of mathematics, Sergio can move freely from one field to another, discarding non-essentials in order to reach a deeper understanding of the problems. He continues to amass, without interruption, a truly impressive collection of papers and books encompassing stochastic analysis, infinite dimensional analysis, functional analysis, non-standard analysis, representation theory, mathematical physics, mathematical modeling, p-adic analysis and string theory. In all these fields, there are many important contributions associated with his name. For example, as early as 1975, Sergio together with Raphael Høegh-Krohn became interested in the theory of Dirichlet forms and diffusion processes on rigged Hilbert spaces. Using this formalism, Sergio made great contributions, together with various coauthors, in working out important results in Euclidean quantum field theory, stochastic quantization and ergodic theory.

In Halmos' classification, Sergio is a mathophysicist proving that "pure mathematics can be practically useful and applied mathematics artistically elegant." The innumerable contributions of the groups he has formed and inspired in Bielefeld, Bochum and Bonn, that draw in mathematicians and theoretical physicists from many nations, are the best proof of the "amazingly unified intellectual structure of mathematics, despite its many subdivisions and enormous rate of growth," as emphasized by Paul Halmos.

Using his thorough knowledge of stochastic analysis, field-theory and statistical physics techniques, Sergio Albeverio initiated more than twenty years ago some original work in the field of non-standard analysis, which he pursued in collaboration with J. Fenstad, R. Høegh-Krohn and T. Lindstrom for several years. This effort culminated in the book "Nonstandard methods in stochastic analysis and mathematical physics," published in 1986.

In the last eight years, two books have appeared on different aspects of modern quantum field theory. Sergio contributed greatly to completing these two projects. The first one, by Sergio, R. Høegh-Krohn, J. Marion, D. Testard and B. Torresani, describes "Non commutative distributions – Unitary representations of gauge groups and algebras" and appeared in 1993.

Many years ago, Sergio became interested in string theory. Carlo Rovelli's characterization of the last century as the "century of the incomplete revolution" suggests that there may be a greater revolution to come, greater than the "relativity revolution" of Albert Einstein and the "quantum revolution" which began with the quantum assumption in Max Planck's talk in Berlin on December 14, 1900. Superstring theory is an attempt to unify general relativity and quantum mechanics. In this approach, the point-like particles we observe are string-like loops existing in a ten dimensional space-time. These ten dimensions include the ordinary four-dimensional space-time plus six supplementary dimensions, which will only become apparent in experiments at extremely small scales. Sergio's book proposing "A mathematical introduction to string theory – Variational problems, geometric and probabilistic models" (with J. Jost, S. Paycha and S. Scarlatti) was published in 1997.

All those who know Sergio closely appreciate his warm sensitivity and the friendly criticism he uses to coordinate his team of numerous co-workers. Sergio's scientific policy can be summarized in the following way: avoid a too narrow specialization, submit all results to criticism and encourage communication among physicists and mathematicians through personal contacts, seminars, schools and conferences. His influence extends far beyond the limits of his Institute in Bonn. Sergio has a passion to share and knows that knowledge has to be transmitted from one generation to the next by constructing a solid chain that links past and present.

Now, and that will be all, let me just say how happy I am to write this brief foreword touching only upon a few highlights in the brilliant career of a very dear long-time friend and colleague.

Bibliography

ALBEVERIO, S., HŒGH-KROHN, R. (1976) *Mathematical theory of Feynman path integrals.* Springer Verlag.

ALBEVERIO, S., FENSTAD, J., HŒGH-KROHN, R., LINDSTROM, T. (1986) *Nonstandard methods in stochastic analysis and mathematical physics.* Academic Press.

ALBEVERIO, S., GESZTESY, F., HŒGH-KROHN, R., HOLDEN, H. (1988) *Solvable Models in Quantum Mechanics.* Springer Verlag.

ALBEVERIO, S., HŒGH-KROHN, R., MARION J., TESTARD, D., TORRESANI, B. (1993) *Non commutative distribution – Unitary representations of gauge groups and algebras.* Marcel Dekker, Inc.

ALBEVERIO, S., JOST, J., PAYCHA, S., SCARLATTI, S. (1997) *A mathematical introduction to string theory – Variational problems, geometric and probabilistic models.* Cambridge University Press.

ALBEVERIO, S., KURASOV, P. (2000) *Singular Perturbations of Differential Operators.* Cambridge University Press.

Progress in Probability, Vol. 52, 1–18

Light, Atoms, and Singularities

O. E. Barndorff-Nielsen, F. E. Benth and J. L. Jensen

Abstract. Motivated by questions concerning cooling and trapping of atoms using counterpropagating laser beams we consider continuous time Markovian jump processes with singular jump intensities and some related stochastic processes, modeling the atomic momentum. The asymptotic behaviour of these processes as time goes to infinity is studied, in particular drawing on methods from renewal theory. Results from this type of study have been instrumental for the understanding and improvement of the efficiency of the cooling schemes.

1. Introduction

One of the exciting areas of present day Physics is the study of the interaction of light and particles, in particular the cooling of atoms and molecules, and investigations flowing from this.

The technique(s) of cooling and trapping of clouds of atoms were developed over a period of about ten years, from the mid Eighties to the mid Nineties, resulting in the award of the 1997 Nobel price in Physics to the three pioneer physicists in the area, Steven Chu, William Phillips and Claude Cohen-Tannoudji.

As will be indicated below, a detailed study of the stochastic elements in the cooling processes has been essential for the understanding of a key element of the physics and has led to dramatic improvements of the cooling techniques, so that at present the most efficient techniques are capable of achieving temperatures at a staggeringly low level, at the order of nano Kelvins. The study was carried out by Cohen-Tannoudji and some of his collaborators in a tour de force mathematical analysis in which they reinvented and extended parts of the classical renewal theory of Probability. See Bardou, Bouchaud, Emile, Aspect and Cohen-Tannoudji [5], Bardou [6], Bardou and Castin [7] and in particular the extensive survey paper by Bardou, Bouchaud, Aspect and Cohen-Tannoudji [2].

A compact review of these achievements, as viewed from Stochastics and with some further work, has been given in Barndorff-Nielsen and Benth [3] and Barndorff-Nielsen, Benth and Jensen [4]. A focal point in that further work was

2000 *Mathematics Subject Classification.* Primary 60J75, Secondary: 60K05, 81V80, 37N20.
Key words and phrases. Laser cooling, Markov jump processes, renewal theory, Bessel processes.

a discussion of continuous time Markov jump processes with singularities in the jump intensities.

The present paper outlines the above-mentioned developments and adds some results. We start in Section 2 by describing the physical background for our study. Section 3 deals with continuous time jump models where the waiting times are mixtures or sums of exponential distributions, generalizing the results in Barndorff-Nielsen et al. [4]. The relation between laser cooling and Bessel processes is discussed in Section 4, and, finally, in Section 5, some other ramifications, concerning time-dependent and supersingular jump rates, are considered.

2. Background

The cooling techniques mentioned in the Introduction have opened the way for a multitude of fundamental studies in physics concerning the interaction of light and matter and for a variety of high tech applications, for instance to atom optics, atom lithography, atomic clocks, atomic lasers, Bose-Einstein condensation (see Townsend, Ketterle and Stringari [18], Ketterle [14] and Burnett, Edwards and Clark [9]), and slowing the speed of light (Hau, Harris, Dutton and Behroozi [12][1]; see also Marangos [15]).

Briefly, the cooling is achieved by directing three pairs of counterpropagating laser beams towards a chosen point in space where a cloud of the atoms is initially trapped by means of an inhomogeneous magnetic field (see for instance Aspect and Dalibard [1]). The operating effect of the lasers, which have to be suitably tuned, can only be properly understood at a basic quantum physical level.

However, the stochastic phenomenon that explains a key part of the efficiency of the methods can to a reasonable degree of realism be described as follows[2]. The momentum (or velocity) of the atom, viewed as a vector in one, two or three dimensions, behaves over time as a Markov jump process with a singularity at the origin. In other words, when the atom arrives at a position x in momentum space it remains there for an exponentially distributed time with a mean value $\lambda(x)^{-1}$, the jump rate $\lambda(x)$ being a continuous function with $\lambda(0) = 0$ and $\lambda(x) > 0$ for $x \neq 0$. Furthermore, when the atom shifts momentum the shift vector is, at least in the neighbourhood of the origin, stochastically independent of position[3]. Experimentally important cases are of the form $\lambda(x) = const.|x|^\gamma$, with $\gamma = 2$ or 4 in one dimension. In such cases atoms arriving near the origin will, due to the singularity, stay there for a time whose distribution is heavy tailed and belongs to the domain of attraction of a stable law with index $\alpha = d/\gamma$ where d denotes dimension. As is well known, the maximum of n positive, independent

---

[1]The front page of *Nature* referred to this result by the title "*Cycling at the speed of light*". The techniques have now been improved to velocities of light around 2km/h.

[2]For more details, see Bardou, Bouchaud, Aspect and Cohen-Tannoudji [2] or, for a brief account, Barndorff-Nielsen and Benth [3].

[3]The shifts occur when the atom emits a photon, previously absorbed from one of the laser beams.

and identically distributed random stable variables is of the same magnitude as the sum of the variables, implying that the longer the experiment lasts the greater the number of atoms with a very low velocity.

In Barndorff-Nielsen, Benth and Jensen [4] continuous state Markov jump processes x_t with a singularity, in the sense indicated above, are studied in general[4], the main topic being the behaviour of x_t near the origin for large t. In particular, results are derived concerning the limiting distributions, after suitable normalisations, of the momentum distribution near the origin and of the time spent in present state.

3. Generalization of the waiting time distribution

The models discussed in Barndorff-Nielsen et al. [4] capture the main aspects of the cooling process. However, as mentioned in Saubaméa, Leduc and Cohen-Tannoudji [17], the description of the waiting time distribution as an exponential distribution with mean $\lambda(x)^{-1}$ is only an approximation. We will in this section consider a model that allows for a more general waiting time distribution. We do this by extending the state space B of the Markov process x_t of the momentum by a variable having only a finite number of values. In this way we can accomodate a waiting time distribution that is either a mixture of exponential distributions (Subsection 3.1) or a sum of exponentially distributed terms (Subsection 3.2). From Kolmogorov's forward equation we derive a renewal type equation for a transform of the density of x_t, which in turn allows us to find the asymptotic form of the density for large t.

We consider a Markov process (x_t, I_t) with state space $B \times \{1, 2, \ldots, K\}$, where B is a region in $\mathbb{R}^d$. The intensity of leaving the state (x, j) is $\lambda(x, j)$. We assume that there exist positive constants c_1, c_2 such that

$$\begin{aligned} c_1 \;\leq\; \lambda(x,j) &\leq c_2, \qquad x \in B, \qquad j \in \{1,..,K-1\} \\ \lambda(x,K) &\leq c_2, \qquad\qquad\quad x \in B, \end{aligned}$$

and where we have in mind that $\lambda(0, K) = 0$. Given that we jump from state $(x_{t-}, I_{t-}) = (y, l)$ let $p(\cdot|y, l)$ be the probability density function of the new position. Furthermore, given also that the new position is x let $\alpha_j(x, y, l)$ be the probability that the new value of I_t is j. In the present setup if we want the waiting time distribution to be a sum of exponentials this can be achieved by letting some of the jumps of (x_t, I_t) involve a change of I_t only. We will therefore assume that there exists $\tilde{K} \in \{0, 1, \ldots, K-1\}$ such that if $I_t \leq \tilde{K}$ the next jump will change I_t only. The special case $\tilde{K} = 0$ means that all jumps will change both x_t and I_t and the waiting time distribution is a mixture of exponentials. When $l \leq \tilde{K}$ we use the notation $\alpha_j(y, l)$ instead of $\alpha_j(y, y, l)$. Finally, let $p(x, j; t)$ be the probability density of (x_t, I_t) with initial density $a(x, j) = p(x, j; 0)$, and let $p_t(x, j; t)$ be derivative of $p(x, j; t)$ with respect to t.

[4] Entailing extensions of classical renewal theory.

Kolmogorov's forward equation is

$$p_t(x,j;t) = -\lambda(x,j)p(x,j;t) + \sum_{l=1}^{\tilde{K}} \lambda(x,l)\alpha_j(x,l)p(x,l;t) + \sum_{l=\tilde{K}+1}^{K} \int_B \lambda(y,l)p(x|y,l)\alpha_j(x,y,l)p(y,l;t)dy. \quad (1)$$

Define

$$h_j(x,t) = \sum_{l=\tilde{K}+1}^{K} \int_B \lambda(y,l)p(x|y,l)\alpha_j(x,y,l)p(y,l;t)dy.$$

We will not analyze (1) in its full generality, but only consider the two cases $\tilde{K} = 0$ and $\tilde{K} = K - 1$.

3.1. The case $\tilde{K} = 0$

We now assume that $\tilde{K} = 0$, that is, all jumps change the position x_t. This models the situation where the waiting time for a jump is a mixture of exponential waiting times. We will also assume that

$$p(y|x,l)\alpha_j(x,y,l) \le k_1, \quad \forall\, x,y \in B,\ \forall\, l,j, \quad (2)$$

for some constant k_1. We first derive a general renewal type equation for h_j and then specialize to a simple model, corresponding to the 'simple model' in Barndorff-Nielsen et al. [4].

When the functions h_j are known the solution to (1) can be written

$$p(x,j;t) = a(x,j)\mathrm{e}^{-t\lambda(x,j)} + \int_0^t h_j(x,\tau)\mathrm{e}^{-(t-\tau)\lambda(x,j)}d\tau.$$

From this we derive

$$\begin{aligned} h_j(x,t) &= \sum_{l=1}^{K} \int_B \lambda(y,l)p(x|y,l)\alpha_j(x,y,l) \\ &\qquad \times \left[a(y,l)\mathrm{e}^{-t\lambda(y,l)} + \int_0^t h_l(y,\tau)\mathrm{e}^{-(t-\tau)\lambda(y,l)}d\tau\right] dy, \\ &= v_j(x,t) + \sum_{l=1}^{K} \int_B \int_0^t h_l(y,t-w)f_{lj}(x,y,w)dwdy, \end{aligned} \quad (3)$$

with

$$f_{lj}(x,y,s) = p(x|y,l)\alpha_j(x,y,l)\lambda(y,l)\mathrm{e}^{-s\lambda(y,l)}$$

and

$$v_j(x,t) = \sum_{l=1}^{K} \int_B a(y,l)f_{lj}(x,y,t)dy.$$

We next define the n-fold convolution

$$f_{ab}^{n*}(x,y,t) = \sum_{l} \int_B \int_0^t f_{al}^{(n-1)*}(z,y,t-w) f_{lb}(x,z,w)\,dw\,dz.$$

One can now repeat the proof of Proposition 2.1 in Barndorff-Nielsen et al. [4] using the bound

$$\begin{aligned}
&f_{l_n l_0}^{n*}(z_0,z_n,w_n)\\
&= \sum_{l_{n-1},\ldots,l_1} \int_B \cdots \int_B \int_0^{w_n} \cdots \int_0^{w_2} \prod_{i=1}^{n} f_{l_i,l_{i-1}}(z_{i-1},z_i,w_i-w_{i-1})\,dz_{n-1}\cdots dz_1\\
&\qquad \times dw_{n-1}\cdots dw_1,\\
&\le \frac{c_2^n t^{n-1}}{(n-1)!} \sum_{l_{n-1},\ldots,l_1} \int_B \cdots \int_B \prod_{i=1}^{n} [p(z_i|z_{i-1},l_{i-1})\alpha_{l_i}(z_i,z_{i-1},l_{i-1})]\,dz_{n-1}\cdots dz_1,\\
&\le \frac{c_2^n t^{n-1}}{(n-1)!} k_1
\end{aligned}$$

for $w_n \le t$. We are therefore able to express the solution to (3) as an integral of v_j times the renewal density $\sum_{n=1}^{\infty} f_{ab}^{n*}(x,y,t)$.

We now make some further assumptions to get back to some more simple equations. We first assume that the new state j only depends on the new position x,

$$\alpha_j(x,y,l) = \alpha_j(x). \tag{4}$$

Then $h_j(x,t) = \alpha_j(x)h(x,t)$ with

$$h(x,t) = \sum_{l=1}^{K} \int_B p(x|y,l)\lambda(y,l)p(y,l;t)\,dy.$$

Instead of (3) we get

$$h(x,t) = v(x,t) + \int_B \int_0^t h(y,t-w) f(x,y,w)\,dw\,dy \tag{5}$$

with

$$f(x,y,s) = \sum_{l=1}^{K} \alpha_l(y)p(x|y,l)\lambda(y,l)\mathrm{e}^{-s\lambda(y,l)} \tag{6}$$

and

$$v(x,t) = \sum_{l=1}^{K} \int_B a(y,l)p(x|y,l)\lambda(y,l)\mathrm{e}^{-t\lambda(y,l)}\,dy.$$

If we make the further assumption that the new position x is independent of the previous position y and previous state l,

$$p(x|y,l) = b(x) \tag{7}$$

we find $h(x,t) = b(x)g(t)$ with

$$g(t) = \sum_{l=1}^{K} \int_B \lambda(y,l)p(y,l;t)dy.$$

Instead of (5) we obtain

$$g(t) = v(t) + \int_0^t g(t-w)u(w)dw$$

with

$$u(s) = \sum_{l=1}^{K} \int_B b(y)\alpha_l(y)\lambda(y,l)\mathrm{e}^{-s\lambda(y,l)}dy$$

and

$$v(t) = \sum_{l=1}^{K} \int_B a(y,l)\lambda(y,l)\mathrm{e}^{-t\lambda(y,l)}dy.$$

Under the conditions (4) the only difference to the general case in Barndorff-Nielsen et al. [4] is that f in (6) now consists of a sum of terms instead of one term only. The results from Barndorff-Nielsen et al. [4] therefore carry over, especially, Proposition 2.1 of that paper is still valid. The Proposition states that the solution to (5) is given as an integral of v with respect to a generalized renewal density obtained from f. Similarly, under the conditions (4) and (7) the situation is as for the "simple model" in Barndorff-Nielsen et al. [4] where we have only one state and the new position is independent of the previous position. Thus from the form of $\lambda(y,l)$ for y close to zero we find the asymptotic form of $u(t)$ for $t \to \infty$. This in turn gives us the asymptotic form of $g(t)$ and finally also the asymptotic form of $p(x,j;t)$. As a concrete example assume that the $\lambda(y,l)$'s are such that $u(t) \sim ct^{-(1+\xi)}$ with $\frac{1}{2} < \xi < 1$. Then $g(t) \sim \xi\{c\Gamma(\xi)\Gamma(1-\xi)\}^{-1}t^{-(1-\xi)}$ for $t \to \infty$. Since $h_j(x,t) = \alpha_j(x)b(x)g(t)$ we get from (2) and Proposition 3.2 in Barndorff-Nielsen et al. [4] that

$$p(x,j;t) = t^{\xi}\frac{\xi\alpha_j(x)b(x)}{c\Gamma(\xi)\Gamma(1-\xi)}\Psi(t\lambda(x,j),\xi) + \frac{t^{\xi}}{1+t\lambda(x,j)}o(1),$$

where $\Psi(z,\xi) = \int_0^1 (1-s)^{\xi-1}\mathrm{e}^{-sz}ds$.

3.2. The case $\tilde{K} = K - 1$

We now consider the case where $\tilde{K} = K - 1$ and when we jump from the state $l < K$ we jump to $l+1$ and when we jump from $l = K$ we jump to $l = 1$. This models the situation where the waiting time is a sum of exponential waiting times.

The Kolmogorov forward equation (1) becomes

$$p_t(x,j;t) = \begin{cases} -\lambda(x,j)p(x,j;t) + \lambda(x,j-1)p(x,j-1;t) & j > 1 \\ -\lambda(x,1)p(x,1;t) + \int_B \lambda(y,K)p(x|y,K)p(y,K;t)dy & j = 1. \end{cases} \tag{8}$$

Defining

$$h(x,t) = \int_B p(x|y,K)\lambda(y,K)p(y,K;t)dy \tag{9}$$

we find from (8) with $j = 1$

$$p(x,1,t) = a(x,1)\mathrm{e}^{-t\lambda(x,1)} + \int_0^t h(x,\tau_0)\mathrm{e}^{-(t-\tau_0)\lambda(x,1)}d\tau_0. \tag{10}$$

We may now use this together with (8) with $j = 2$ to find $p(x,2,t)$,

$$\begin{aligned} &p(x,2,t)\\ &= a(x,2)\mathrm{e}^{-t\lambda(x,2)}\\ &\quad + \int_0^t \lambda(x,1)\left[a(x,1)\mathrm{e}^{-\tau_1\lambda(x,1)} + \int_0^{\tau_1} h(x,\tau_0)\mathrm{e}^{-(\tau_1-\tau_0)\lambda(x,1)}d\tau_0\right]\mathrm{e}^{-(t-\tau_1)\lambda(x,2)}d\tau_1\\ &= a(x,2)\mathrm{e}^{-t\lambda(x,2)} + \lambda(x,1)a(x,1)\int_0^t \mathrm{e}^{-\tau_1\lambda(x,1)-(t-\tau_1)\lambda(x,2)}d\tau_1\\ &\quad +\lambda(x,1)\int_0^t\int_0^{\tau_1} h(x,\tau_0)\mathrm{e}^{-(\tau_1-\tau_0)\lambda(x,1)-(t-\tau_1)\lambda(x,2)}d\tau_0 d\tau_1. \end{aligned} \tag{11}$$

Proceeding in this way we find with $\tau_j = t$

$$\begin{aligned} &p(x,j,t)\\ &= a(x,j)\mathrm{e}^{-t\lambda(x,j)}\\ &\quad +\sum_{r=1}^{j-1} a(x,r)\prod_{s=r}^{j-1}\lambda(x,s)\int_0^{\tau_j}\cdots\int_0^{\tau_{r+1}} \mathrm{e}^{-\tau_r\lambda(x,r)-\sum_{s=r+1}^{j}(\tau_s-\tau_{s-1})\lambda(x,s)}\prod_{s=r}^{j-1}d\tau_s\\ &\quad +\prod_{s=1}^{j-1}\lambda(x,s)\int_0^{\tau_j}\cdots\int_0^{\tau_1} h(x,\tau_0)\mathrm{e}^{-\sum_{s=1}^{j}(\tau_s-\tau_{s-1})\lambda(x,s)}\prod_{s=0}^{j-1}d\tau_s. \end{aligned} \tag{12}$$

Inserting (12) with $j = K$ in (9) we obtain a renewal equation for $h(x,t)$. In the special case with $K = 2$ this equation becomes

$$\begin{aligned} &h(x,t) = \\ &\int_B p(x|y,2)\lambda(y,2)\left[a(y,2) + \lambda(y,1)a(y,1)\frac{1-\mathrm{e}^{-t(\lambda(y,1)-\lambda(y,2))}}{\lambda(y,1)-\lambda(y,2)}\right]\mathrm{e}^{-t\lambda(y,2)}dy\\ &+\int_B\int_0^t h(y,\tau_0)p(x|y,2)\lambda(y,2)\lambda(y,1)\mathrm{e}^{-(t-\omega)\lambda(y,2)}\frac{1-\mathrm{e}^{-(t-\tau_0)(\lambda(y,1)-\lambda(y,2))}}{\lambda(y,1)-\lambda(y,2)}d\tau_0 dy. \end{aligned} \tag{13}$$

In the simple model with $p(x|y,2) = b(x)$ we have $h(x,t) = b(x)g(t)$ and instead of (13) we find

$$g(t) = \int_B \lambda(y,2)\left[a(y,2) + \lambda(y,1)a(y,1)\frac{1-\mathrm{e}^{-t(\lambda(y,1)-\lambda(y,2))}}{\lambda(y,1)-\lambda(y,2)}\right]\mathrm{e}^{-t\lambda(y,2)}dy + \int_0^t g(t-\tau_0)\left[\int_B b(y)\lambda(y,2)\lambda(y,1)\mathrm{e}^{-\tau_0\lambda(y,2)}\frac{1-\mathrm{e}^{-\tau_0(\lambda(y,1)-\lambda(y,2))}}{\lambda(y,1)-\lambda(y,2)}dy\right]d\tau_0.$$

This renewal equation for g can be analyzed as in Barndorff-Nielsen et al. [4]. We can then establish the asymptotic form of $g(t)$ for $t \to \infty$, and using this in (10) and (11) we find the asymptotic form of $p(x,j,t)$ for $t \to \infty$. The arguments are as in Barndorff-Nielsen et al. [4]. As an example assume that $\lambda(y,2) \leq \lambda(y,1) \leq c_1$, $\lambda(y,1) \geq c_2 > 0$ and that $\lambda(y,2) = |y|^\delta$ for small values of y with $\frac{1}{2} < d/\delta < 1$. Then

$$\begin{aligned} u(t) &= \int_B b(y)\lambda(y,2)\lambda(y,1)\mathrm{e}^{-t\lambda(y,2)}\frac{1-\mathrm{e}^{-t(\lambda(y,1)-\lambda(y,2))}}{\lambda(y,1)-\lambda(y,2)}dy \\ &\sim t^{-(1+d/\delta)}b(0)\frac{C_d\Gamma(1+d/\delta)}{\delta}, \end{aligned}$$

and from Proposition 3.1 in Barndorff-Nielsen et al. [4] we get $g(t) \sim c_3 t^{-(1-d/\delta)}$. From Proposition 3.2 in Barndorff-Nielsen et al. [4] and (10) we see that

$$p(x,1;t) = t^\xi c_4 b(x)\Psi(t\lambda(x,1),\xi) + \frac{t^\xi}{1+t\lambda(x,1)}o(1)$$

and from (11)

$$p(x,2;t) = t^\xi \frac{c_5 b(x)}{\lambda(x,1)-\lambda(x,2)}\Psi(t\lambda(x,2),\xi) + \frac{t^\xi}{1+t\lambda(x,2)}o(1),$$

with $\xi = d/\delta$.

4. Relation to Bessel processes

In this section we consider the different laser cooling schemes presented in Bardou et al. [2] in the context of continuous-time stochastic processes. Many of the different cooling schemes can be modelled, at least approximately, within the framework of diffusions. As we shall see, one is naturally lead to Langevin diffusions and Bessel processes, which capture many of the important features of laser cooling. We restrict our consideration to one space dimension throughout the section.

In laser cooling mechanisms based on friction there are two effects, which together results in atomic temperatures down to the level of the recoil limit[5]. Friction forces lead to a drift of the atomic momentum towards zero, while spontaneous

[5]The recoil limit is a temperature of the order of 1 microKelvin, and is the limit of the cooling schemes known as "Doppler" and "Sisyphus". These rely on different kinds of friction mechanisms.

and random emission of fluorescence photons introduces a "diffusion" of momentum. In the discussion of Bardou et al. [2] the authors say that the competition between friction forces and diffusion leads to a steady state momentum distribution, where the effective temperature of the cooling process can be expressed in terms of the half-width (or the standard deviation) of the stationary distribution. Let us now consider the standard cooling schemes in the framework of continuous-time stochastic processes. A natural model in light of the above presentation is the Langevin diffusion:

$$dX_t = -aX_t\,dt + c\,dB_t,$$

where X_t is the atomic momentum at time t and the constants a and c are positive and B_t is a standard Brownian motion. This model has a friction proportional to the momentum and a spontaneous emission of photons with rate c. It is well known that X_t has a normal distribution with expectation $X_0 \exp(-at)$ and variance $\frac{c^2}{2a}(1-\exp(-2at))$. Hence, X_t will reach a normally distributed steady state with expectation zero and variance $c^2/2a$. The "half-width" of the stationary distribution is therefore $c/\sqrt{2a}$, which we can relate to the effective temperature of the cooling process described by the Langevin diffusion.

As Bardou et al. [2, Sect. 2.2.1] describe, it is possible to cool atoms without friction. Instead of pushing the atomic momentum towards zero with a friction force, one resorts to cooling principles where the spontaneous rate of emission of photons[6] depends on the momentum and vanishes at zero. Such cooling schemes are in fact more effective than the standard ones, since one may circumvent the recoil limit[7]. In such cooling mechanisms the random walk in momentum space has an inhomogeneous diffusion coefficient dependent on the current momentum and vanishing at zero. If we model the emission rate at momentum x by $\lambda(x) := c^2x^\delta$, for positive constants c and δ (cf. Section 2), a diffusion approximation of the subrecoil cooling scheme may be

$$dX_t = cX_t^{\delta/2}\,dB_t \tag{14}$$

where X_t is the atomic momentum at time t. The invariant measure of X_t is $\lambda^{-1}(x) = c^{-2}x^{-\delta}$, the inverse of the fluorescence rate. Observe that the invariant measure is not a probability density when $\delta \geq 1$. We will investigate the asymptotic behaviour of such diffusions and compare with results for subrecoil cooling schemes like Raman cooling (where $\delta = 4$, see Bardou et al. [2]).

First, notice that X_t given as in (14) will never cross zero. If it reaches zero, the process will remain trapped ever after[8]. Recall that this is not in agreement with the actual laser cooling experiment, where discrete jumps in momentum across zero may take place.

[6]The spontaneous rate of emission of photons is called the fluorescence rate.

[7]The schemes are therefore called subrecoil laser cooling.

[8]The probability of hitting zero will be discussed further below.

The study of (14) naturally leads us to Bessel processes. In order to establish the connection, we consider the following class of models with $\delta > 0$

$$dX_t = \operatorname{sgn}(2-\delta)cX_t^{\delta/2}\,dB_t$$

where c is a positive constant and we use the convention $\operatorname{sgn}(0) = 1$. The reasons for introducing $\operatorname{sgn}(2-\delta)$ are purely technical and has nothing to do with the physical model. Note that when $\delta > 2$, we can always change the Brownian motion by $\tilde{B}_t = -B_t$ in order to remove the negative sign from the model. Shifting to $\tilde{B}_t$ will define the same stochastic process (in distribution) since $\tilde{B}_t$ again is a Brownian motion. Assume from now on that $X_0 = a > 0$.

Let Y_t be the pathwise unique solution of the stochastic differential equation

$$dY_t = 2\sqrt{Y_t^+}\,dB_t + \frac{2(1-\delta)}{2-\delta}\,dt$$

where $Y^+ = \max(y, 0)$. The process Y_t is called a $\mathrm{BES}^2(\alpha)$-process with $\alpha = \frac{2(1-\delta)}{2-\delta}$ in the notation of Rogers and Williams [16]. Furthermore, $R_t := \sqrt{Y_t}$ is known as an α-dimensional Bessel process. We need to have $\alpha > 0$, which is achieved if either $\delta \in [0, 1)$ or $\delta > 2$. In the latter case we know from [16] that $Y_t > 0$ for all $t \geq 0$ since $\alpha > 2$. When $\delta \in [0, 1)$, on the other hand, Y_t may become zero in finite time. Note that Y_t is non-explosive in the sense that it does not reach infinity in finite time with a positive probability. Define

$$X_t = \left(\frac{1}{4}c^2(2-\delta)^2 Y_t\right)^{1/(2-\delta)}$$

and let $Y_0 = \frac{4a^{2-\delta}}{c^2(2-\delta)^2} > 0$. When $\delta > 2$, X_t is well defined (does not explode) since Y_t is positive for all times. An easy application of Itô's Formula shows that when $\delta \in [0, 1)$,

$$dX_t = cX_t^{\delta/2}\,dB_t$$

and when $\delta > 2$,

$$dX_t = -cX_t^{\delta/2}\,dB_t$$

In the latter case, we have that X_t is non-explosive and positive for all times due to the properties of Y_t, while for $\delta \in [0, 1)$, there is a positive probability for X_t of eventually being trapped at zero in finite time. Recall again that the minus sign in the latter model is introduced for technical reasons (see the discussion above).

We prove pathwise uniqueness of solutions of (14) for a given Brownian motion B_t. Consider the case $\delta > 2$, the other is similar: Let X_t and $\hat{X}_t$ be two solutions of

$$dX_t = -cX_t^{\delta/2}\,dB_t$$

which are positive and non-explosive. Using Itô's Formula, we have that $Y_t = g(X_t)$ and $\hat{Y}_t = g(\hat{X}_t)$ with $g(x) = (4/c^2(2-\delta)^2)x^{2-\delta}$ are both positive solutions of

$$dY_t = 2\sqrt{Y_t}\,dB_t + \frac{2(1-\delta)}{(2-\delta)}\,dt$$

with the same initial condition. By pathwise uniqueness of this equation (see e.g. Rogers and Williams [16]), $X_t^{2-\delta} = \hat{X}_t^{2-\delta}$ and uniqueness thus follows.

When $\frac{2(1-\delta)}{(2-\delta)} = n \in \mathbb{N}$ the solution Y_t can be explicitly constructed when B_t is given in a special manner. Let $W(t)$ be an n-dimensional Brownian motion with $|W(0)|^2 = b > 0$. Define

$$dB_t = |W(t)|^{-1} W(t) \cdot dW(t)$$

which is seen to be a Brownian motion on $\mathbb{R}$ by Lévy's Theorem (see e.g. Rogers and Williams [16]). Moreover, $Y_t = |W(t)|^2$ is a solution to

$$dY_t = 2\sqrt{Y_t}\, dB_t + n\, dt,$$

i.e., for such δ we can produce explicit representations of X_t as a function of the modulus of an n-dimensional Brownian motion. In the example below we demonstrate this for Raman cooling, where $\delta = 4$.

Example 4.1. Choose $\delta = 4$ which gives $n = 3$. Hence, let $W(t)$ be a 3-dimensional Brownian motion and $Y_t = |W(t)|^2$. With $|W(0)| = (ac)^{-1}$ we have that

$$X_t = \frac{1}{c}|W(t)|^{-1}$$

is a solution to

$$dX_t = -cX_t^2\, dB_t$$

and $X_0 = a$.

We turn to the study of asymptotic properties of X_t when t goes to infinity. Only the case $\delta > 2$ will be treated, since this is the most interesting from the laser cooling point of view. Let, for a natural number $n \geq 3$, $\frac{2(\delta-1)}{\delta-2} = n$ or $\delta(n) := \delta = \frac{2(n-1)}{(n-2)}$. Letting n run through the natural numbers, we see that $\delta(n)$ decreases from $\delta(3) = 4$ to $\delta(\infty) = 2$. $\delta(3) = 4$ will correspond to Raman cooling and $\delta(\infty) = 2$ to VSCPT cooling[9], since $\lambda(x) = c^2x^4$ and $\lambda(x) = c^2x^2$, resp. For a given $n \geq 3$, we have that

$$Y_t = W_1(t)^2 + \ldots + W_n^2(t).$$

To simplify our considerations, we assume from now on that the Brownian motion $W(t)$ starts at the origin[10]. It is well known that

$$Y_t \sim f_{\frac{1}{2t},\frac{n}{2}}(x) \qquad \text{where} \qquad f_{\gamma,\nu} = \frac{\gamma^\nu}{\Gamma(\nu)} x^{\nu-1} e^{-\gamma x}.$$

This implies that

$$X_t = \left(\frac{1}{4}c^2(2-\delta(n))^2 Y_t\right)^{-1/(\delta(n)-2)}$$

---

[9] *V*elocity *S*elective *C*oherent *P*opulation *T*rapping, a subrecoil laser cooling scheme. See Bardou et al. [2]

[10] Of course, the transformation to X_t is then not well defined for all $t \geq 0$, since it implies that $X_0 = \infty$. However, we may form X_t from Y_t when $t > 0$, and the assumption simplifies the asymptotic considerations considerably.

has a probability distribution with density $p(x,t)$ given by

$$\begin{aligned} p(x,t) &= -f_{\frac{1}{2t},\frac{n}{2}}(k^{-2}x^{-(\delta(n)-2)}) \cdot k^{-2}(-1)(\delta(n)-2)x^{-(\delta(n)-2)-1} \\ &= k^{-n}\frac{\delta(n)-2}{2^{n/2}\Gamma(\frac{n}{2})}t^{-n/2}x^{-\frac{n}{2}\delta(n)+n-1}\exp\left(-(2tk^2x^{\delta(n)-2})^{-1}\right) \\ &= \frac{2^{1-n/2}}{k^n(n-2)\Gamma(\frac{n}{2})}t^{-\frac{n}{2}}x^{-2\frac{n-1}{n-2}}\exp\left(-(2k^2tx^{\frac{2}{n-2}})^{-1}\right) \end{aligned}$$

where $k = \frac{1}{4}c^2(2-\delta(n))^2$. Hence, when $t \to \infty$,

$$t^{\frac{n}{2}}p(x,t) \sim \frac{2^{1-\frac{n}{2}}}{k^n(n-2)\Gamma(\frac{n}{2})} \cdot x^{-2\frac{n-1}{n-2}}.$$

We conclude that the renormalized probability distribution of X_t has power law tails for big t, with power $\delta(n) \in (2,4], n \geq 3$. It is interesting to compare this with the asymptotic results of Bardou et al. [2]. Even though they do not consider the same fluorescence rate as here, there are close connections in the asymptotics. In Sect. 6.3.2 they show that for large times the momentum distribution has tails which are proportional to $t^{-1+1/\delta}x^{-\delta}$ (see Eq. (6.35) in [2]). The Bessel model has the same tail behaviour. The rate of t is, on the other hand, quite different. From the above considerations it is seen to be $t^{-1-1/(\delta-2)}$ in the Bessel model. However, our fluorescence function is unbounded outside the trapping region, while [2] use a flat or confined function.

We can easily calculate the moments of X_t for different n: A straightforward integration shows, for $m \geq 1$,

$$\begin{aligned} \mathrm{E}[X_t^m] &= \frac{2^{1-\frac{n}{2}}}{k^n(n-2)\Gamma(\frac{n}{2})}t^{-\frac{n}{2}}\int_0^\infty x^{m-2\frac{n-1}{n-2}}\exp\left(-(2k^2tx^{\frac{2}{n-2}})^{-1}\right)dx \\ &= \frac{2^{1+m(1-\frac{n}{2})}}{k^{m(n-2)}(n-2)\Gamma(\frac{n}{2})} \cdot t^{m(1-\frac{n}{2})}\int_0^\infty u^{(1-\frac{1}{2}(n-2)(m-1))-1}e^{-u}\,du \end{aligned}$$

which is finite if and only if $1 - \frac{1}{2}(n-2)(m-1) > 0$, or

$$m < 1 + \frac{2}{n-2}.$$

From this we see that X_t has finite expectation and variance when $n = 3$, while for $n > 3$, X_t has only finite expectation. More precisely,

$$t^{\frac{n}{2}-1}\mathrm{E}[X_t] = \frac{2^{2-\frac{n}{2}}}{k^{n-2}(n-2)\Gamma(\frac{n}{2})}$$

while for $n = 3$,

$$t\mathrm{E}[X_t^2] = \frac{2}{k^2}.$$

Recall that $n = 3$ corresponds to $\delta = 4$ (Raman cooling). For this case we see that X_t has finite expectation and variance, converging to zero at the rate $t^{-1/2}$ and t^{-1}, respectivly, when t tends to infinity. Hence, X_t will eventually be "trapped" at the origin almost surely. On the other hand, when $n > 3$ (i.e. $\delta \in (2,4)$) we have only finite expectation. X_t converges to zero in expectation at the rate $t^{-(n/2-1)}$, while its higher order moments are infinite.

In this section we have seen that diffusion processes in connection with laser cooling show many of the characteristics of Lévy statistics. Momentum distributions with power tails and non-existence of higher moments, signifying laser cooling schemes, may be reproduced in a diffusion framework using Bessel processes.

5. Some other ramifications

5.1. Time dependent λ

In Barndorff-Nielsen et al. [4] we considered jump intensities $\lambda(x)$ dependent on the current position x only. We will here generalize to Markov processes where the jump intensity depends on time as well and derive the corresponding renewal-type equations. The idea is to model situations where the experimental conditions are changed as the experiment is carried out.

Assume the state space B to be a measurable set of $\mathbb{R}^d$ and let $a(x)$ be the initial probability distribution of the Markov process. The probability of jumping from $y \in B$ to $x \in B$ is denoted $p(x|y)$. We assume that $\lambda(t,x) \leq \Lambda$ on $[0,T] \times B$ for some constant Λ, where T may be either finite of infinite.

5.1.1. General model The Kolmogorov backward equation is,

$$p_t(x,t) = -\lambda(t,x)p(x,t) + \int \lambda(t,y)p(x|y)p(y,t)\,dy.$$

Define as before

$$h(x,t) = \int \lambda(t,y)p(x|y)p(y,t)\,dy \tag{15}$$

and obtain, by using the backward equation, the following expression for $p(x,t)$:

$$p(x,t) = a(x)\mathrm{e}^{-\int_0^t \lambda(s,x)\,ds} + \int_0^t h(x,s)\mathrm{e}^{-\int_s^t \lambda(u,x)\,du}\,ds.$$

By inserting this relation into (15) we get a renewal type equation for $h(x,t)$,

$$\begin{aligned} h(x,t) &= \int \lambda(t,y)p(x|y)\Big\{a(y)\mathrm{e}^{-\int_0^t \lambda(s,y)\,ds} + \int_0^t h(y,s)\mathrm{e}^{-\int_s^t \lambda(u,y)\,du}\,ds\Big\}\,dy \\ &= \int a(y)p(x|y)\lambda(t,y)\mathrm{e}^{-\int_0^t \lambda(s,y)\,ds}\,dy \\ &\quad + \int_0^t \int h(y,s)p(x|y)\lambda(t,y)\mathrm{e}^{-\int_s^t \lambda(u,y)\,du}\,dy\,. \end{aligned}$$

Observe that it is not possible to write the last integral as a time convolution. We can therefore not resort to Laplace transformation techniques to study this renewal-type equation in further detail.

5.1.2. Simple model To simplify matters, consider $p(x|y) = b(x)$, i.e. the jump distribution is independent of the current state of the process. The Kolmogorov backward equation now becomes,

$$p_t(x,t) = -\lambda(t,x)p(x,t) + b(x)\int \lambda(t,y)p(y,t)\,dy.$$

Define

$$g(t) = \int \lambda(t,y)p(y,t)\,dy$$

and obtain

$$p(x,t) = a(x)\mathrm{e}^{-\int_0^t \lambda(s,x)\,ds} + b(x)\int_0^t g(s)\mathrm{e}^{-\int_s^t \lambda(u,x)\,du}\,ds.$$

This gives a relation for g:

$$\begin{aligned} g(t) &= \int \lambda(t,y)\Big\{a(y)\mathrm{e}^{-\int_0^t \lambda(s,y)\,ds} + b(y)\int_0^t g(s)\mathrm{e}^{-\int_s^t \lambda(u,y)\,du}\Big\}\,dy \\ &= \int a(y)\lambda(t,y)\mathrm{e}^{-\int_0^t \lambda(s,y)\,ds}\,dy + \int_0^t g(s)\Big\{\int b(y)\lambda(t,y)\mathrm{e}^{-\int_s^t \lambda(u,y)\,du}\,dy\Big\}\,ds. \end{aligned}$$

Letting

$$v(t) = \int a(y)\lambda(t,y)\mathrm{e}^{-\int_0^t \lambda(s,y)\,ds}\,dy \quad \text{and} \quad u(t,s) = \int b(y)\lambda(t,y)\mathrm{e}^{-\int_s^t \lambda(u,y)\,du}\,dy$$

we get a renewal type equation for $g(t)$:

$$g(t) = v(t) + \int_0^t g(s)u(t,s)\,ds. \tag{16}$$

In the terminology of Gripenberg, Londen and Staffans [11] equation (16) is a non-convolution Volterra equation of the second kind. They provide a rather extensive treatment of existence and uniqueness results for such problems under various integrability hypotheses on u and v. With the boundedness condition on λ, an explicit solution can be constructed in the following manner: Let $u^{(1)}(t,s) = u(t,s)$ and define the functions $u^{(n)}(t,s)$ for $n = 2,3,4,\dots$ inductively by

$$u^{(n)}(t,s) = \int_s^t u^{(n-1)}(t,\tau)u(\tau,s)\,d\tau. \tag{17}$$

We have the following proposition:

Proposition 5.1. *The solution of* (16) *is*

$$g(t) = v(t) + \int_0^t v(s)Q_t(ds) \tag{18}$$

where $Q_t(ds) = \sum_{n=1}^{\infty} u^{(n)}(t,s)\,ds$ and the $u^{(n)}(t,s)$ are defined in (17)*. Moreover, $g(t)$ is bounded on every compact subset of $[0,T]$. The solution is unique in the class of non-negative functions being bounded on every compact subset of $[0,T]$.*

Proof. We prove the result directly instead of appealing to the general theory of Gripenberg et al. [11]. Our argument is a simplified version of the proof of Proposition 2.1 in Barndorff-Nielsen et al. [4]:

We start by showing that $g(t)$ is well defined: By iterating the definition of $u^{(n)}(t,s)$ we get the following bound,

$$\begin{aligned}
u^{(n)}(t,s) &= \int_s^t \int_{\tau_1}^t \cdots \int_{\tau_{n-2}}^t u(t,\tau_{n-1})u(\tau_{n-1},\tau_{n-2})\cdots u(\tau_1,s)\,d\tau_{n-1}\cdots d\tau_1 \\
&= \int_s^t \int_{\tau_1}^t \cdots \int_{\tau_{n-2}}^t \Big(\int_B b(y)\lambda(t,y)e^{-\int_{\tau_{n-1}}^t \lambda(\theta,y)\,d\theta}\,dy\Big)\cdots \\
&\qquad \Big(\int_B b(y)\lambda(\tau_1,y)e^{-\int_{\tau_1}^s \lambda(\theta,y)\,d\theta}\,dy\Big)\,d\tau_{n-1}\cdots d\tau_1 \\
&\le \Lambda^n \int_s^t \int_{\tau_1}^t \cdots \int_{\tau_{n-2}}^t d\tau_{n-1}\cdots d\tau_1 \\
&\le \frac{\Lambda^n (t-s)^{n-1}}{(n-1)!}.
\end{aligned}$$

Hence, $\sum_{n=1}^{\infty} u^{(n)}(t,s) \le \Lambda e^{\Lambda(t-s)}$ for all $t > 0$. We can then bound the integral on the right-hand side of (18) by,

$$\int_0^t v(s)Q_t(ds) \le \Lambda \int_0^t v(s)e^{\Lambda(t-s)}\,ds \le \Lambda e^{\Lambda t}\,.$$

This means that $g(t)$ in (18) is well defined. Moreover, $\sup_{s\le t} g(s) \le \Lambda + \Lambda e^{\Lambda t}$ which implies that $g(t)$ is bounded on every compact subset of $[0,T]$.

It is a straightforward calculation to show that $g(t)$ solves (16). We turn to uniqueness of the solution: Assume $\tilde{g}(t)$ is a non-negative solution of (18) which is bounded on every compact subset of $[0,T]$. Then,

$$g(t) - \tilde{g}(t) = \int_0^t \big(g(s) - \tilde{g}(s)\big)u(t,s)\,ds.$$

Iterating this expression gives

$$g(t) - \tilde{g}(t) = \int_0^t \big(g(s) - \tilde{g}(s)\big)u^{(n)}(t,s)\,ds.$$

Hence,

$$\begin{aligned}
|g(t) - \tilde{g}(t)| &\le \frac{\Lambda^n}{(n-1)!}\int_0^t |g(s) - \tilde{g}(s)|(t-s)^{n-1}\,ds \\
&\le \big(\sup_{s\le t} g(s) + \sup_{s\le t} \tilde{g}(s)\big)\frac{\Lambda^n t^n}{n!}.
\end{aligned}$$

Letting $n \to \infty$ yields $g(t) = \tilde{g}(t)$. □

Proposition 5.1 tells us that Equation (16) has a solution that formally looks like the solution to an ordinary renewal equation. However, it is not clear how to extend these results to also get the asymptotic form of $g(t)$ as $t \to \infty$, a result that is needed to find the asymptotic form of the density $p(x,t)$.

5.2. Supersingular jump rate

In this subsection we return to the setting of Barndorff-Nielsen et al. [4] where the jump intensity λ depends on the current state x only. In Barndorff-Nielsen et al. [4] all the possibilities considered had $\lambda(x) = |x|^\alpha$ for x small and with $\alpha > 0$. Here we will consider an example outside this class. We take the state space to be $B = [-1, 1]$ and consider $\lambda(x) = \exp(-|x|^{-1})$. The transition density is $p(y|x) = b(y)$ with b a continuous function.

The density of a typical waiting time is

$$u(t) = \int_B b(y)\lambda(y)\mathrm{e}^{-t\lambda(y)}dy = \int_{-1}^{1} b(y)\mathrm{e}^{-|y|^{-1}}\exp(-t\mathrm{e}^{-|y|^{-1}})dy.$$

For $t \to \infty$ we find

$$\begin{aligned} u(t) &\sim 2b(0)\int_0^{\mathrm{e}^{-1}} (\log v)^{-2}\mathrm{e}^{-tv}dv \\ &= 2b(0)\int_0^{\mathrm{e}^{-1}} \mathrm{e}^{-tv}dK(v), \end{aligned} \tag{19}$$

where

$$\begin{aligned} K(v) &= \int_0^v (\log w)^{-2}dw = \frac{v}{(\log v)^2}\int_0^1 \left(1 + \frac{\log z}{\log v}\right)^{-2} dz \\ &\sim v(\log v)^{-2} \quad \text{for } v \to 0. \end{aligned} \tag{20}$$

Comparing (19) and (20) we find, by Tauberian theory (cf. Feller [10, Theorem 3, p. 445])

$$u(t) \sim 2b(0)(\log t)^{-2}t^{-1} \quad \text{for } t \to \infty. \tag{21}$$

For the cases considered in Barndorff-Nielsen and Benth [4] one had $u(t) \sim ct^{-(1+\alpha)}$ with $\alpha > 0$. The case (21) is therefore a limiting case corresponding to $\alpha = 0$. Referring, for instance, to Loéve [13, p. 334] we find that the distribution with density u is not in the domain of attraction of any (stable) law.

We can now proceed with the analysis as in Section 3 in Barndorff-Nielsen et al. [4]. From (21) we see that the tail probability is

$$1 - U(t) = \int_t^\infty u(s)ds \sim 2b(0)(\log t)^{-1}, \quad \text{for } t \to \infty,$$

and this implies (see Bingham, Goldie and Teugels [8, Cor. 8.1.7])

$$1 - \hat{U}(s) \sim 2b(0)(-\log s)^{-1}, \quad s \to 0.$$

As in Barndorff-Nielsen et al. [4] we then get

$$\hat{G}(s) \sim \frac{-\log s}{2b(0)} \quad \text{and} \quad G(t) \sim \frac{\log t}{2b(0)}, \tag{22}$$

where G appears in

$$p(x,t) = a(x)\mathrm{e}^{-t\lambda(x)} + b(x)\left\{G(t) - \int_0^t G(\tau)\lambda(x)\mathrm{e}^{-(t-\tau)\lambda(x)}d\tau\right\}. \tag{23}$$

To analyze the asymptotic form of (23) we rewrite (22) as

$$G(t) = \frac{1}{2b(0)}\log(1+t)(1+m(t)), \quad m(t) \to 0 \text{ for } t \to \infty.$$

In the asymptotic analysis we assume that $\xi = t\lambda(x)$ is bounded. From

$$\begin{aligned}&\int_0^t \log(1+\tau)\lambda(x)\mathrm{e}^{-(t-\tau)\lambda(x)}d\tau \\ &= \xi\int_0^1 [\log(t) + \log(\frac{1}{t} + 1 - u)]\mathrm{e}^{-u\xi}du \\ &= \log(t)(1-\mathrm{e}^{-\xi}) + O(\xi),\end{aligned}$$

we find

$$p(x,t) = \frac{1}{2}\log(t)\{\mathrm{e}^{-\xi} + o(1)\} + O\left(\int_0^t \log(1+\tau)m(\tau)\lambda(x)\mathrm{e}^{-(t-\tau)\lambda(x)}d\tau\right).$$

We split the last integral into the integral from 0 to s and from s to t. The first part is of order $O(\frac{s}{t}\log(t))$, and the second part is of order $O(m(s)\log(t))$. We therefore take $s = \sqrt{t}$ so that both terms are $o(\log(t))$ and we have

$$p(x,t) = \frac{1}{2}\log(t)\{\mathrm{e}^{-t\lambda(x)} + o(1)\}.$$

As compared with the results in Section 3 of Barndorff-Nielsen et al. [4] we see that the scaling here is with $\log t$ instead of t^α for some $\alpha > 0$. Also, the function $\Psi(\xi,\alpha)$ has here been replaced by $\mathrm{e}^{-\xi}$.

References

[1] A. Aspect and J. Dalibard, *Le refroidissement des atomes par laser*, La Recherche, **261** (1994), 30–37.

[2] F. Bardou, J.-P. Bochaud, A. Aspect and C. Cohen-Tannoudji, *Non-ergodic cooling: subrecoil laser cooling and Lévy statistics*, manuscript (2000).

[3] O. E. Barndorff-Nielsen and F. E. Benth, *Laser cooling and stochastics*, in M. C. M. de Gunst, C. A. J. Klaassen and A. W. van der Vaart (eds.), *State of the Art in Probability and Statistics; Festschrift for Willem R. van Zwet*, Lecture Notes – Monograph Series, Institute of Mathematical Statistics, to appear.

[4] O. E. Barndorff-Nielsen, F. E. Benth, and J. L. Jensen, *Markov processes with a singularity*, Adv. Appl. Probability (2000), to appear.

[5] F. Bardou, J.-P. Bouchaud, O. Emile, A. Aspect and C. Cohen-Tannoudji, *Subrecoil laser cooling and Lévy flights*, Physical Review Letters, **72** (1994), 203–206.

[6] F. Bardou, *Refroidissement laser sub-recul par résonances noires*, Ph.D. Thesis, **1995**.

[7] F. Bardou and Y. Castin, *Le refroidissement laser subrecul: fonctions d'onde stochastiques et vols de Lévy*, Images de la Physique, Revue du CNRS (1998), 7–14.

[8] N. H. Bingham, C. M. Goldie and J. L. Teugels, *Regular variation*, Encyclopedia of Mathematics and its Applications, Cambridge University Press (1987).

[9] K. Burnett, M. Edwards and C. W. Clark, *The theory of Bose-Einstein condensation of dilute gases*, Physics Today, December 1999, 37–42.

[10] W. Feller, *An Introduction to Probability Theory and Its Applications, Vol II*, second edition, J. Wiley & Sons New York, **1971**.

[11] G. Gripenberg, S.-O. Londen and O. Staffans, *Volterra Integral and Functional Equations*, Encyclopedia of Mathematics and its Applications, Cambridge University Press (1990).

[12] L. V. Hau, S. E. Harris, Z. Dutton and C. H. Behroozi, *Light speed reduction to 17 metres per second in an ultracold atomic gas*, Nature, **397** (1999), 594–598.

[13] M. Loéve, *Probability Theory*, van Nostrand, Toronto, **1963**.

[14] W. Ketterle, *Experimental studies of Bose-Einstein condensation*, Physics Today, December 1999, 30–35.

[15] J. Marangos, *Slow light in cool atoms*, Nature, **397** (1999), 559–560.

[16] L. C. G. Rogers and D. Williams, *Diffusions, Markov Processes and Martingales*, vol. 2, J. Wiley & Sons London, **1987**.

[17] B. Saubaméa, M. Leduc and C. Cohen-Tannoudji, *Experimental investigations of non-ergodic effects in subrecoil laser cooling*, Physical Review Letters, **83** (1999), 3796–3799.

[18] C. Townsend, W. Ketterle and S. Stringari, *Bose-Einstein condensation*, Physics World, **10** (1997), 29–34.

MaPhySto[11]
Department of Mathematical Sciences
University of Aarhus
DK-8000 Århus
Denmark
E-mail address: oebn@imf.au.dk

[11] MaPhySto – Centre for Mathematical Physics and Stochastics, funded by a grant from the Danish National Research Foundation.

Progress in Probability, Vol. 52, 19–31

How Random Are Random Walks?

R. Blei

Abstract. Steps of random walks in the real world are not independent, but, rather, depend on hidden variables acting in a complicated manner that is not feasible to deterministic analysis. Proposed is a continuous scale of random walks with interdependent steps, which is calibrated by combinatorial measurements. In the limit, modeling continuous time, this scale of random walks becomes a scale of chaos processes, which is calibrated by tail-probability estimates.

1. A general problem

In the real world, Brownian motion generically refers to random, difficult-to-predict trajectories of particles. A long-standing mathematical model for these motions has been a Wiener process [20]: a process with independent Gaussian increments, which, in a precise sense, is a limit of simple random walks (e.g., [13]). A linear simple random walk – we recall – is a walk on a line, where a step to the right is as likely as a step to the left, and steps are independent. This is an idealized construct. We all agree (I hope...) that *independence-of-steps* is not an intrinsic feature of "random walks" in the real world, but that it conveys that we, observers of "random walks", are baffled by them. No one believes (I trust...) that actual walks have independent steps. Rather, steps depend on "hidden variables", and are interdependent. The problem is: *how can interdependence-of-steps be effectively modeled, gauged, and detected?*

2. Heuristics

Say we are observing haphazard movements of particles; to wit, we are observing physical *Brownian motion.* For example, we could be tracking microscopically small particles suspended in liquid, whose motions are caused by sub-microscopic molecular collisions (cf. [10]). Or, we could be following prices on stock exchanges, whose movements are caused by a multitude of factors hidden from view (cf. [1]). Our goal: model these movements. We consider one such generic particle p, and

2000 *Mathematics Subject Classification.* 60G05, 60K35, 60G50, 49J55.
Key words and phrases. Random walks, interdependence, combinatorial dimension, α-chaos, decision making machine.

suppose it is *free*; that is, we assume that the only forces acting on p are imparted by an ambient environment – all hidden, and all acting in a very complicated manner. We dub the particle *Brownian*, and, to simplify matters, assume its motion is linear. We let its position on the line at time $t = 0$ be the origin, and ask: what can be said about the particle's position $X = X(t)$ at time $t > 0$?

We concede at the outset that we do not know and are unable to determine the particle's extremely complex dynamics. Our perceptions, in effect based on "zero knowledge", are that

i at any instant, p moves to the right or to the left with equal likelihood, and

ii p's trajectories over disjoint time intervals appear unrelated.

We also presume that

iii p's "statistics" over time intervals of equal length are the same.

The last surmise (time-homogeneity) also stems from "zero knowledge": knowing and assuming nothing about forces acting on p, we imagine that Brownian motion is statistically the same in every time interval of the same length. ("Statistics" in **iii** could mean *expected* distance traveled by p, *variance* of its increments, or *probability distributions* of its increments.)

We model these perceptions in a probabilistic framework. To begin, we think of $X(t)$ as a real-valued random variable with finite variance on some probability space. We assume that $\mathbf{E}X(t) = 0$ (assertion **i**); that increments over disjoint time intervals are uncorrelated (assertion **ii**),

$$\mathbf{E}(X(t_1) - X(s_1))(X(t_2) - X(s_2)) = 0, \quad 0 \leq s_1 < t_1 \leq s_2 < t_2 < \infty, \tag{1}$$

and that the variance of $X(t) - X(s)$ is a function of $t - s$ (assertion **iii**),

$$Var(X(t) - X(s)) = v(t - s), \quad 0 \leq s < t \leq \infty, \tag{2}$$

where v is a non-negative function on $[0, \infty)$. From (1) and (2), we conclude

Proposition 2.1. *There exists $c \geq 0$ such that $Var\ X(t) = ct$ for all $t \geq 0$.*

Proof. If ϕ is a real-valued Lebesgue measurable function on $\mathbb{R}$ such that $\phi(x+y) = \phi(x) + \phi(y)$ for all $x \in \mathbb{R}$ and $y \in \mathbb{R}$, then there exists $c \in \mathbb{R}$ so that $\phi(x) = cx$ for all $x \in \mathbb{R}$. Now observe that v is monotone, and that $v(t - s) = v(t) - v(s)$. □

To obtain $Var\ X(t) = ct$ for all $t \geq 0$, we have used a mild, indeed a minimal interpretation of assertions **i**, **ii** and **iii**. Let us now apply a more stringent interpretation, one that in a certain sense is extremal.

3. A first-order approximation: the Wiener process

Assume that path increments over disjoint time intervals are independent random variables. This, in a probabilistic context, is the most extreme interpretation of assertion **ii** (and, insofar that modeling "reality" is our objective, the simplest and the most naive...). We fix time $t > 0$, fix an arbitrary integer $n > 0$, and imagine

$X(t)$ to be "approximately" the result of a simple random walk clocked by discrete time $t/n, 2t/n, \ldots, jt/n, \ldots, (n-1)t/n$: at time $jt/n, j = 0, \ldots, n-1$, we imagine the particle moving a distance s_n to the right or to the left with probability $\frac{1}{2}$, moves are independent, and $s_n = \sqrt{ct/n}$ (cf. Prop. 2.1). Then, $X(t)$ is "approximately"

$$X_n(t) = \sqrt{\frac{ct}{n}} \sum_{j=1}^{n} r_j,$$

where $r_j, j = 1, \ldots, n$, are Rademacher functions defined on $\{-1,1\}^n$. (We view $\{-1,1\}^n$ as a uniform probability space, and $r_j(\omega) = \omega(j), \omega \in \{-1,1\}^n, j = 1, \ldots, n$.) Taking $n \to \infty$, we obtain, by the Central Limit Theorem, that X_n converges in distribution to a Gaussian r.v. with mean 0 and variance ct, and think of this limit as $X(t)$. This motivates

Definition 3.1. *A stochastic process* $W = \{W(t) : t \in [0,1]\}$ *on a probability space* $(\Omega, \mathcal{A}, \mathbb{P})$ *is a* Wiener process *if* W *has the following properties:* $W(0) = 0$ *a.s.* $(\mathbb{P})$*; for* $[s,t] \subset [0,1], W(t) - W(s)$ *is a Gaussian random variable with mean 0 and variance* $t - s$*; for all partitions* $\{0 \leq t_1 < \cdots < t_n \leq 1\}$ *of* $[0,1], W(t_2) - W(t_1), \ldots, W(t_n) - W(t_{n-1})$ *are independent random variables.*

A Wiener process W is a first-order approximation to a perception of Brownian movement: its underlying sample space Ω represents the ensemble of all possible paths of a Brownian particle, and its properties reflect the simplest interpretation of the heuristic assertions **i**, **ii**, and **iii**. The trajectories $W(t), t \in [0,1]$, are almost surely continuous, a property we expect from any reasonable model of physical Brownian motion. (Almost sure sample-path continuity of W follows from a theorem of Kolmogorov – e.g., [2, Th. 12.4] –, or from analysis of stochastic series of W – e.g., [6, Cor. X.10].)

Remark 3.2 (From the viewpoint of physics). The perception that a Brownian particle's position is a Gaussian random variable follows from statistical analysis based in effect on "zero knowledge". This perception is a cornerstone in Wiener's mathematical model of Brownian motion, which Wiener himself viewed as "a first approximation" [19, p. 295].

A Gaussian model of Brownian movement can be derived also in a context of statistical mechanics from "idealized" physical principles. Indeed, this was Albert Einstein's observation in his seminal paper about Brownian movement [11] (the second of his three 1905 papers in *Annalen der Physik*, vol. 17). Einstein's Gaussian model was experimentally affirmed by J. Perrin [18], by verifying that the average distance traveled by Brownian particles in time t was proportional to $\sqrt{t}$. Notably, the latter is predicted by *any* model of Brownian movement that is a homogeneous L^2-bounded process with orthogonal increments (cf. Prop. 2.1).

Let us now agree that a view of Brownian increments as independent random variables is unrealistic. Instead, let us admit that Brownian increments are, somehow, interdependent through "hidden variables" whose precise causal action is – alas! – beyond our comprehension. A question arises: how can this scenario be modeled?

4. Decision Making Machines

We change the paradigm from movements of a generic Brownian particle to a stroll of a drunk. Here is an old tale commonly told to illustrate a simple random walk. A drunk leaves a pub, and walks along a road for one hour. Every $1/N$ hours, he takes a step of length $1/\sqrt{N}$ to the right or to the left with equal probability, *and*, a simple-minded fellow, he takes the steps independently in time. (The step's length $1/\sqrt{N}$ – *normalization* – is a consequence of Prop. 2.1.) With the pub as the origin at time $t = 0$, the drunk's position at time $t = 1/N, 2/N, \ldots, 1$ is a random variable

$$X_t(\omega) = \frac{1}{\sqrt{N}} \sum_{j=1}^{tN} r_j(\omega), \quad \omega \in \{-1,+1\}^N, \tag{3}$$

where ω is any one if the possible 2^N paths which occurs with probability $1/2^N$, and $r_j(\omega)/\sqrt{N} := w(j)/\sqrt{N}$ is the step at time $(j-1)/N$.

Let us embellish the story. In this version, the drunk possesses a "Decision Making Machine" (DMM), and steps to the right or to the left according to it. The machine consists of N labeled switches $1, \ldots, N$ inside a sealed box, and N labeled light bulbs $1, \ldots, N$ on a panel. A randomizing device inside the sealed box turns each of the N switches "on" or "off" independently with probability $\frac{1}{2}$. The randomizing device, activated by a push of a blue button on the panel, produces a state $\omega \in \{-1,+1\}^N$ of the DMM: $\omega(j) = +1$ means that switch j is "on", and $\omega(j) = -1$ means that it is "off"; the DMM's state remains fixed until the next push of the button. The light bulbs are wired to switches so that bulb j is lit if and only if switch j is "on". Armed with this DMM, our drunk starts from the pub, pushes the blue button and walks: at time $(j-1)/N, j = 1, \ldots, N$, he takes a step to the right if bulb j is lit, and to the left if it is not. His position X_t at time $t = j/N, j = 1, \ldots, N$, is given by (3), but in this tale, $\omega \in \{-1,+1\}^N$ is a state of the DMM, produced by a push of a button and fixed throughout the walk.

We now retell the story with a more complex "Decision Making Machine". This model has N switches labeled $1, \ldots, N$, and $N_k := \binom{N}{1} + \binom{N}{2} + \cdots + \binom{N}{k}$ light bulbs labeled $1, \ldots, N_k$, where k is a given integer. We fix a one-to-one correspondence between the light bulbs and all subsets of switches of cardinality $\leq k$,

$$S_j \subset [N], 0 < |S_j| \leq k, j = 1, \ldots, N_k,$$

and define

$$\chi_j = \prod_{i \in S_j} r_i, \quad j = 1, \ldots, N_k.$$

Switches are wired to the bulbs so that bulb j is lit if and only if $\chi_j = 1$. (Such circuits are always feasible; e.g., see [12, §1.6].) Equipped with this DMM, the drunk pushes the blue button, and leaves the pub. He takes a step of length $1/\sqrt{N_k}$ every $1/N_k$ hours (normalization); at time $(j-1)/N_k, j = 1, \ldots, N_k$, to the right if bulb j is lit, and to the left if it is not. The drunk's position at time t, no longer a result of a simple random walk, is given by

$$X_t(\omega) = \frac{1}{\sqrt{N_k}} \sum_{j=1}^{tN_k} \chi_j(\omega), \quad \omega \in \{-1,+1\}^N, \quad t = \frac{1}{N_k}, \frac{2}{N_k}, \ldots, 1. \tag{4}$$

Again, $\omega \in \{-1,+1\}^N$ denotes the DMM's state, produced by a push of a button and fixed throughout the walk. In particular, after one hour, the drunk's position is

$$X_1(\omega) = \frac{1}{\sqrt{N_k}} \sum_{m=1}^{N_k} \sum\nolimits_{1 \le l_1 < \cdots < l_m \le N} r_{l_1}(\omega) \cdots r_{l_m}(\omega), \quad \omega \in \{-1,+1\}^N.$$

Here is a more general tale. The DMM has N switches labeled $1, \ldots, N$, and n light bulbs labeled $1, \ldots, n$, where $0 < N \le n$ are arbitrary integers. We take n distinct non-empty sets of switches, designate them $S_1, \ldots, S_n$, and wire bulbs to switches so that bulb j is lit if and only if

$$\chi_j(\omega) = \prod_{i \in S_j} r_i(\omega) = 1, \quad j = 1, \ldots, n. \tag{5}$$

The clock is calibrated by $1/n$ hours. At time $j/n, j = 0, \ldots, n-1$, the drunk takes a step of length $1/\sqrt{n}$ to the right if bulb j is lit, and to the left if not. Then, his position at time $t = 1/n, 2/n, \ldots, 1$ is

$$X_t(\omega) = \frac{1}{\sqrt{n}} \sum_{j=1}^{tn} \chi_j(\omega), \quad \omega = \{-1,+1\}^N. \tag{6}$$

I refer to (6) as an *F-walk*, where $F = \{S_1, \ldots, S_n\}$.

We note two extremal scenarios:

1. $S_j = \{j\}, j = 1, \ldots, N$. This is the standard simple random walk, indeed the simplest possible.
2. F consists of all non-empty subsets of $\{1, \ldots, N\}$, i.e., $k = N$ in (4). This is the most complex F-walk possible. At the end of each but one of these F-walks, the drunk will be found $1/\sqrt{n}$ units away from the pub.

General F-walks, falling anywhere between these two cases, manifest a varying degree of "randomness complexity". This of course needs to be made precise, and the next question is: how can the "complexity" of F-walks be gauged? This question is addressed in the next two sections.

5. Scales of combinatorial measurements

Let $0 < k \le N$ be integers. For $A \subset \{1, \ldots, N\}$, denote

$$\Gamma_k(A) = \{S : S \subset A, 0 < |S| \le k\}.$$

We write $\Gamma_k(N)$ for $\Gamma_k(\{1,\ldots,N\})$. In this terminology,

$$|\Gamma_k(N)| = N_k := \binom{N}{1} + \binom{N}{2} + \cdots + \binom{N}{k}, \tag{7}$$

and the walks in (4) are $\Gamma_k(N)$-walks. Integers $k > 0$ in this context provide a natural calibration of "complexity" of random walks. The question is: can we extend further this calibration, and precisely measure the "complexity" of *arbitrary* F-walks, $F \subset \Gamma_k(N)$?

Let $F \subset \Gamma_k(N)$ be arbitrary. Assume that elements in F are distinct and that every $i \in \{1,\ldots,N\}$ appears in some $S \in F$. For integers $s > 0$, define

$$\theta_F(s) = \max\left\{|F \cap \Gamma_k(A)| : A \subset \{1,\ldots,N\}, 0 < |A| \leq s\right\}.$$

If $F = \Gamma_m(N), m = 1,\ldots,k$, then for $s = m,\ldots,N$,

$$\theta_F(s) = \binom{s}{1} + \binom{s}{2} + \cdots + \binom{s}{m} \qquad \text{(cf. (7))}. \tag{8}$$

In general, for $F \subset \Gamma_k(N)$ and $s = k,\ldots,N$,

$$s \leq \theta_F(s) \leq s^k. \tag{9}$$

For $a \geq 1$, define

$$d_F(a) = \sup\left\{\theta_F(s)/s^a : k \leq s \leq N\right\}.$$

For $N_j \uparrow \infty$, and $F_j \subset \Gamma_k(N_j), j \in \mathbb{N}$, we define

$$d_{\{F_j\}}(a) = \sup_j d_{F_j}(a),$$

and

$$\dim\{F_j\} = \inf\left\{a : d_{\{F_j\}}(a) < \infty\right\}.$$

I refer to $\dim\{F_j\}$ as the combinatorial dimension of $\{F_j\}$. If $\dim\{F_j\} = \alpha$, and $d_{\{F_j\}}(\alpha) < \infty$, then $\dim\{F_j\} = \alpha$ *exactly*; otherwise, if $d_{\{F_j\}}(\alpha) = \infty$, then $\dim\{F_j\} = \alpha$ *asymptotically*. By (9),

$$1 \leq \dim_{\{F_j\}} \leq k.$$

By (8), if $F_j = \Gamma_m(N_j), m = 1,\ldots,k$, then

$$\dim\{\Gamma_m(N_j)\} = m \text{ exactly}.$$

Constructions of $\{F_j\}$ with non-integer combinatorial dimension are less obvious.

Here is an example whose combinatorial dimension is $\frac{3}{2}$. Let $k = 3, N_j = j^2$, and enumerate $\{1,\ldots,j^2\}$ by $\{1,\ldots,j\} \times \{1,\ldots,j\}$ (any enumeration will do):

$$\{1,\ldots,j^2\} = \{n_{xy} : x = 1,\ldots,j,\ y = 1,\ldots,j\}.$$

For $j \in \mathbb{N}$, define

$$F_j = \left\{\{n_{xy}, n_{yz}, n_{xz}\} : (x,y,z) \in \{1,\ldots,j\}^3\right\} \subset \Gamma_3(j^2). \tag{10}$$

Lemma 5.1. *For all* $j = 2, \dots,$

$$\left(\frac{1}{8}\right)s^{\frac{3}{2}} \le \theta_{F_j}(s) \le s^{\frac{3}{2}}, \quad s = 4, \dots, j^2.$$

Proof. To verify the inequality on the left side, fix $s \in \{4, \dots, j^2\}$, and let l_s be the integer such that $l_s \le \sqrt{s} < l_s + 1$. Let

$$A = \left\{n_{xy} : (x,y) \in \{1, \dots, l_s\}^2\right\}.$$

Then, $|A| \le s$, and

$$|F_j \cap \Gamma_3(A)| = (l_s)^3 \ge (\sqrt{s} - 1)^3 \ge \left(\frac{1}{8}\right)s^{\frac{3}{2}}.$$

To verify the inequality on the right side, fix $s \in \{4, \dots, j^2\}$ and $A = \{n_{x_1y_1}, \dots, n_{x_sy_s}\} \in \{1, \dots, j^2\}$. For convenience, we identify A with the subset $\{(x_1, y_1), \dots, (x_s, y_s)\} \subset \{1, \dots, j\} \times \{1, \dots, j\}$, which we denote also by A. Observe that

$$|F_j \cap \Gamma_3(A)| = \sum_{x,y,z} \mathbf{1}_A(x,y)\mathbf{1}_A(y,z)\mathbf{1}_A(x,z).$$

(Summation is over $\{1, \dots, j\}^3$, and $\mathbf{1}_A$ is the usual indicator function.) By a three-fold application of Cauchy-Schwarz, we deduce

$$|F_j \cap \Gamma_3(A)| \le |A|^{\frac{3}{2}} = s^{\frac{3}{2}}.$$

□

The lemma implies that if $\{F_j\}$ is defined by (10), then $\dim\{F_j\} = \frac{3}{2}$ *exactly*.

The preceding "$\frac{3}{2}$-dimensional" construction can be naturally generalized to produce $\{F_j^{(q)}\}$ for every rational $q = \frac{n}{k} \ge 1$, such that

$$F_j^{(q)} \subset \Gamma_k(j^k), \quad j \in \mathbb{N},$$

and

$$\dim\{F_j^{(q)}\} = q.$$

These constructions are based on designs that first appeared in [3] (see also [4] and [6, Ch. XIII]). These examples, however, do not resolve completely the existence question: for $k \in \mathbb{N}$ and $\alpha \in [1,k]$, does there exist $\{F_j\}$ such that $F_j \subset \Gamma_k(j)$ for $j = k, \dots,$ and $\dim\{F_j\} = \alpha$? The answer is affirmative, but designs hitherto known for arbitrary $\alpha \in [1,k]$ can be produced only randomly. These random designs are based on

Lemma 5.2 ([6, 7, 9]). *For all integers* $k > 1$ *and arbitrary* $\alpha \in (1,k)$*, there exists* $C_{\alpha,k} > 0$ *with the following property: for every* $N \in \mathbb{N}$*, there exist* $F \subset \Gamma_k(N)$ *such that*

$$|F| \ge \left(\frac{1}{2}\right)N^{\alpha}, \tag{11}$$

and

$$\theta_F(s) \le C_{\alpha,k}s^{\alpha}, \quad s = 1, \dots, N. \tag{12}$$

Idea of proof. Define a system of Bernoulli variables $X_{\mathbf{u}}, \mathbf{u} \in \Gamma_k(N)$, such that

$$\mathbb{P}(X_{\mathbf{u}} = 1) = |\Gamma_k(N)|^{\alpha-k}, \quad \mathbb{P}(X_{\mathbf{u}} = 0) = 1 - |\Gamma_k(N)|^{\alpha-k}, \quad \mathbf{u} \in \Gamma_k(N).$$

Consider the random set $F = \Big\{\mathbf{u} \in \Gamma_k(N) : X_{\mathbf{u}} = 1\Big\}$, and produce $C_{\alpha,k} > 0$ (depending on α and k), such that F satisfies (11) and (12) with positive probability. □

6. A scale of limit theorems

Next we consider relations between the combinatorial complexity of $F \subset \Gamma_k(N)$ and the statistical complexity of the corresponding F-walk.

Let $Y \in L^2(\Omega, \mathbb{P})$. For $v > 0$, consider

$$b(Y;v) := \sup_p \|Y\|_{L^p}/p^{v/2},$$

$$c(Y;v) := \liminf_{y\to\infty} \Big((-y^{-2/v}) \log \mathbb{P}(|Y| > y)\Big), \tag{13}$$

and the Orlicz-norm $\|Y\|_{\phi_v}$, where the Orlicz function ϕ_v is given by

$$\phi_v(x) = \exp x^{2/v} - 1, \quad x > 0.$$

These three gauges – of L^p-norm growths, of tail-probabilities, and of Orlicz-norms – are equivalent in the sense that for all $Y \in L^2(\Omega, \mathbb{P})$ and $v > 0$,

$$b(Y;v) < \infty \Longleftrightarrow c(Y;v) > 0 \Longleftrightarrow \|Y\|_{\phi_v} < \infty.$$

We define the index (e.g., [6, Ch. X §9])

$$\delta(Y) := \inf\Big\{v : b(Y;v) < \infty\Big\} = \inf\Big\{v : c(Y;v) > 0\Big\} = \inf\Big\{v : \|Y\|_{\phi_v} < \infty\Big\}. \tag{14}$$

If $c(Y;v) > 0$, then Y is a *sub-v-variable.* If $\delta(Y) = \alpha$, then Y is an *α-variable*; Y is an *exact* α-variable if $c(Y;\alpha) > 0$ ($\delta(Y) = \alpha$ *exactly*), and an *asymptotic* α-variable if $c(Y;\alpha) = 0$ ($\delta(Y) = \alpha$ *asymptotically*). If $\mathbf{E}|Y|^2 = 1, \mathbf{E}Y = 0$, and $\delta(Y) = \alpha$, then Y is a *standard α-variable.* Stated in long hand, the assertion that Y is an α-variable means: for every $v > \alpha$,

$$\begin{gathered}\text{there exists } K_v > 0 \text{ such that for sufficiently large } y > 0,\\ \mathbb{P}\Big(|Y| > y\Big) \le \exp\Big(-K_v y^{2/v}\Big),\end{gathered} \tag{15}$$

and if $0 < v < \alpha$, then (15) fails; or, another way put, Y is a sub-v-variable for all $v > \alpha$, but for no $v < \alpha$.

Here are some canonical examples: if $\|Y\|_{L^\infty} < \infty$, then $\delta(y) = 0$; if $\|Y\|_{L^p} = \infty$ for some $p > 2$, then $\delta(Y) = \infty$; if Y is a standard Gaussian, then $\delta(Y) = 1$ *exactly*; if $Y = X_1 \cdots X_n$, where the X_i's are standard independent Gaussian variables, then $\delta(Y) = \frac{n}{2}$ *exactly.*

The link between the statistical measurements defined above and the combinatorial measurements defined in the previous section rests on a general theorem originally established in a harmonic-analytic framework [4]. We state a version of it adapted to the present setting. For integers $0 < k \le N$, and $F \subset \Gamma_k(N)$, we define

$$\eta_F(v) := \sup\Big\{ b\Big(\sum_{\mathbf{u}\in F} a_{\mathbf{u}}\chi_{\mathbf{u}}; v\Big) : \sum_{\mathbf{u}\in F} |a_{\mathbf{u}}|^2 \le 1\Big\}, \quad v > 0. \tag{16}$$

(In (16), the $\chi_{\mathbf{u}}$'s are the random variables in (5), and L^p-norms are evaluated over $\{-1,1\}^N$ with respect to the uniform probability measure on it.)

Theorem 6.1 ([4],[6, Cor. XIII.28]). *If $F \subset \Gamma_k(N)$, then*

$$16^{-k} d_F(2v) \le \eta_F(v) \le 4^k d_F(2v), \quad v \ge \frac{1}{2}. \tag{17}$$

By iterating the inequality on the right side of (17), we deduce

Corollary 6.2. *If $F \subset \Gamma_k(N)$, and $Y = \sum_{\mathbf{u}\in F} a_{\mathbf{u}}\chi_{\mathbf{u}}$, then for $m \in \mathbb{N}$, and $p \ge 2m$,*

$$\|Y\|_{L^p} \le 4^k d_F(2v)\Big(\frac{p}{m}\Big)^v \|Y\|_{L^{2m}}, \quad v \ge \frac{1}{2}.$$

We use both Th. 6.1 and Cor. 6.2 to establish

Theorem 6.3. *Suppose $N_j \uparrow \infty, F_j \subset \Gamma_k(N_j)(j \in \mathbb{N})$, and that for $\alpha \in [1,k]$,*

$$|F_j| \ge \kappa(N_j)^\alpha \quad \textit{for all } j \in \mathbb{N}, \tag{18}$$

and

$$d_{\{F_j\}}(\alpha) < \infty. \tag{19}$$

(I.e., $\dim\{F_j\} = \alpha$ exactly.) Then there exists a subsequence $N_{j_l} \uparrow \infty$ and a standard exact α-variable Y such that

$$(1/|F_{j_l}|^{\frac{1}{2}}) \sum_{\mathbf{u}\in F_{j_l}} \chi_{\mathbf{u}} \xrightarrow[l\to\infty]{} Y \quad \textit{in distribution.}$$

Proof (sketch). Denote $Y_j = (1/|F_j|^{\frac{1}{2}}) \sum_{\mathbf{u}\in F_j} \chi_{\mathbf{u}}$. Then, $\|Y_j\|_{L^2} = 1$ for all $j \in \mathbb{N}$, and, by (19), the sequence (Y_j) is uniformly integrable. Therefore, there exists a subsequence (Y_{j_l}) that converges in distribution to some $Y, \|Y\|_{L^2} = 1$ and $\mathbf{E}Y = 0$.

The right side of (17) implies

$$\|Y_j\|_{L^p} \le 4^k d_{\{F_j\}}(\alpha) p^{\alpha/2}, \quad p > 2, j \in \mathbb{N},$$

and therefore,

$$b(Y;\alpha) \le 4^k d_{\{F_j\}}(\alpha), \tag{20}$$

i.e., Y is a sub-α-variable.

By an interpolation argument, for each $j \in \mathbb{N}$ and $q \in (1,2)$,

$$\Big\| \prod_{m=1}^{N_j} (1 + r_m) \Big\|_{L^q} \le 4^{N_j/p}, \quad \frac{1}{q} + \frac{1}{p} = 1.$$

Therefore, for all $p > 2$,

$$|F_j|^{\frac{1}{2}} = \mathbf{E}Y_j \prod_{m=1}^{N_j} (1 + r_m) \leq \|Y_j\|_{L^p} 4^{N_j/p}. \tag{21}$$

By taking $p = N_j$ in (21), and applying (18), we obtain

$$\|Y_j\|_{L^{N_j}} \geq \frac{1}{4}|F_j|^{\frac{1}{2}} \geq \frac{1}{4}\kappa(N_j)^{\alpha/2}.$$

Fix an arbitrary $j_0 \in \mathbb{N}$, and apply Cor. 6.2 with $p = N_j$ and $m := N_{j_0}$ to Y_j for $j > j_0$:

$$4^k d_{\{F_j\}}(\alpha)\|Y_j\|_{L^{N_{j_0}}} \geq \frac{1}{4}\kappa(N_{j_0})^{\alpha/2}.$$

This implies $b(Y; v) = \infty$ for all $v < \alpha$, and then, together with (20), that Y is an exact α-variable. □

7. Continuous-time models: α-chaos

Just as the Central Limit Theorem led to the Wiener process – a continuous-time model for the simple random walk – Th. 6.3 analogously leads to continuous-time models for random walks of varying complexity.

The framework is this. A stochastic process $X = \Big\{X(t) : t \in [0,1]\Big\}$ is L^2-*bounded with orthogonal increments* if $\mathbf{E}|X(t)|^2 := \varphi_X(t) < \infty$ and $\mathbf{E}X(t) = 0$ for all $t \in [0,1]$, and (1) holds. Observe that φ_X is an increasing function. Such processes X are canonical integrators in the following sense: if $f \in L^2([0,1], d\varphi_X)$, then the $\int_{[0,1]} f\, dX$ is obtained as the $L^2(\Omega, \mathbb{P})$-limit of integrals – naturally defined – of step functions on $[0,1]$ converging to f in $L^2([0,1], d\varphi_X)$. Note that

$$\Big\| \int_{[0,1]} f\, dX \Big\|_{L^2(\mathbb{P})} = \|f\|_{L^2(d\varphi_X)}.$$

(This integral is a straightforward generalization of the usual Wiener integral; e.g., see [6, Ch. XI].)

For a subspace H of $L^2(\Omega, \mathbb{P})$, we define (cf. (13) and (14))

$$c_{\mathrm{H}}(v) = \inf \Big\{c(Y; v) : Y \in \mathrm{H}, \|Y\|_{L^2} \leq 1\Big\},$$

and

$$\delta_{\mathrm{H}} = \inf \Big\{v : c_{\mathrm{H}}(v) > 0\Big\}.$$

If a process X is a L^2-bounded process and has orthogonal increments, then we consider

$$H(X) = \Big\{ \int_{[0,1]} f\, dX : f \in L^2([0,1], d\varphi_X)\Big\}.$$

We dub X *homogeneous* if X is L^2-bounded with orthogonal increments, and

$$\varphi_X(t) = ct, \quad c = \mathbf{E}|X(1)|^2, \quad t \in [0,1] \quad \text{cf. Proposition 2.1.}$$

Definition 7.1 ([6, Ch. XI], cf. [5]). *An L^2-bounded process X with orthogonal increments is an* α-chaos, *$\alpha \in [1,\infty)$, if $\delta_{H(X)} = \alpha$. If X is an α-chaos, and $c_{H(X)}(\alpha) > 0$, then X is an* exact *α-chaos; otherwise, if $c_{H(X)}(\alpha) = 0$, then X is an* asymptotic *α-chaos.*

If X is homogeneous, and $\delta_{H(X)} = \alpha$ (exact or asymptotic), then X is a homogeneous *α-chaos (exact or asymptotic).*

In the remainder of the article we focus on homogeneous α-chaos processes, which, according to Theorem 6.3, we view as continuous-time models for F-walks whose combinatorial complexity is gauged by *combinatorial dimension* (defined in §5). Almost sure sample-path continuity of homogeneous α-chaos follows from Kolmogorov's theorem (e.g., [2, Th. 12.4]), or from analysis of stochastic series associated with α-chaos [6, Ch. XI].

Here are some canonical examples. For $n \in \mathbb{N}$, consider the *Wiener homogeneous chaos* of order n (cf. [17, 21]),

$$H_n = \left\{ \int_{[0,1]} \cdots \int_{[0,1]} f(s_1,\ldots,s_n)\, dW(s_1)\cdots dW(s_n) : f \in L^2([0,1]^n, m^n) \right\},$$

where integrals are n-fold iterated Itô-Wiener integrals, and m^n is the n-dimensional Lebesgue measure on $[0,1]^n$. Then, $\delta_{H_n} = n$ exactly. By taking indicator functions as integrands in the Itô-Wiener integrals, and by re-parameterizing, we obtain an exact homogeneous n-chaos. To obtain homogeneous α-chaos for arbitrary $\alpha \in [0,\infty)$, we generalize Kakutani's construction of a Wiener process [14, 15]. We take $F \subset \mathbb{N}^k$ with combinatorial dimension $\alpha \in [1,\infty)$, in the (properly adapted) sense of §5 or in the (precise) sense of [4]. Let $L^2_F(\Omega,\mathbb{P}) := L^2_F$ denote the L^2-closure of the linear span of $\left\{ r_{j_1}\cdots r_{j_k} : (j_1,\ldots,j_k) \in F \right\}$. Then, $\delta_{L^2_F} = \alpha$ (cf. Theorem 6.1). Let

$$U : L^2([0,1], m) \mapsto L^2_F(\Omega,\mathbb{P})$$

be a unitary equivalence (any equivalence will do), and define

$$U\mathbf{1}_{[0,t]} = X(t), \quad t \in [0,1].$$

Then, X is a homogeneous α-chaos, which is exact if and only if $\dim F = \alpha$ is exact.

Remark 7.2 (Variations of F-measures associated with α-chaos). I conclude by proposing a measurement of "stochastic complexity" closely related to measurements in §5 and §6, which I believe is feasible and possibly also useful.

Suppose $X = \left\{ X(t) : t \in [0,1] \right\}$ is a process on a probability space $(\Omega, \mathcal{A}, \mathbb{P})$, and $\mathbf{E}|X(t)| < \infty$ for all $t \in [0,1]$. Define

$$\lambda_X(A,(s,t]) = \mathbf{E}\mathbf{1}_A(X(t) - X(s)), \quad A \in \mathcal{A}, 0 \le s < t \le 1.$$

(The analogous measurement in the case of an F-walk is $\chi_{\mathbf{u}}(\omega)/2^N, \mathbf{u} \in F, \omega \in \{-1,1\}^N$.) It is easy to verify that if X is a homogeneous α-chaos, then λ_X determines a scalar measure separately in each coordinate; that is, λ_X is an F_2-measure on $\mathcal{A} \times B$, where B = usual Borel field in $[0,1]$ (e.g., [5, Ch. XI]). It is also easy to verify that λ_X cannot be extended to a bona fide measure on $\sigma(\mathcal{A} \times B)$ (= the σ-field generated by $\mathcal{A} \times B$); specifically, that the *total variation* of λ_X is infinite,

$$\|\lambda_X\|_{(1)} := \sup \Big\{ \sum_{i,j} |\lambda_X(A_i, B_j)| : \\ \{A_i\} \subset \mathcal{A},\ \{B_j\} \subset B,\ \textstyle\sum_i \mathbf{1}_{A_i} \le 1,\ \sum_j \mathbf{1}_{B_j} \le 1 \Big\} = \infty.$$

(Indeed, it is this very fact that gives the subject of stochastic integration a life of its own.) And a natural issue arises how "far" is λ_X from a measure on $\sigma(\mathcal{A} \times B)$?

A straightforward application of Littlewood's $\frac{4}{3}$-inequality [16] implies that $\|\lambda_X\|_{(\frac{4}{3})} < \infty$, where

$$\|\lambda_X\|_{(p)} := \sup \Big\{ \sum_{i,j} |\lambda_X(A_i, B_j)|^p : \textstyle\sum_i \mathbf{1}_{A_i} \le 1,\ \sum_j \mathbf{1}_{B_j} \le 1 \Big\}.$$

But we can be more precise [8]. For $\gamma \in [1, \infty)$, define

$$\theta_\gamma(x) = x / \{\log(1/x)\}^{\gamma/2}, \quad x \in (0,1),$$

and $\theta_\gamma(0) = 0$. If X is a homogeneous α-chaos, $\mathbf{E}|X(1)|^2 = 1$, then for all $\gamma > \alpha$,

$$\|\lambda_X\|_{(\theta_\gamma)} := \sup \Big\{ \sum_{i,j} \theta_\gamma(|\lambda_X(A_i, B_j)|) : \\ \{A_i\} \subset \mathcal{A},\ \{B_j\} \subset B,\ \textstyle\sum_i \mathbf{1}_{A_i} \le 1,\ \sum_j \mathbf{1}_{B_j} \le 1 \Big\} < \infty. \quad (22)$$

In particular, $\|\lambda_X\|_{(p)} < \infty$ for all $p > 1$. These estimates lead to the index

$$\chi_X := \inf \Big\{ \gamma : \|\lambda_X\|_{(\theta_\gamma)} < \infty \Big\},$$

which I view as a gauge of the "stochastic" complexity of X. Statement (22) implies that for homogeneous X, $\chi_X \le \delta_{H(X)}$. All known homogeneous α-chaos X satisfy $\chi_X = \alpha$.

Question. Does there exist a homogeneous α-chaos X such that $\chi_X < \alpha$?

References

[1] L. Bachelier, *Theory of Speculation*, in P. H. Cootner, editor, The Random Character of Stock Market Prices, The MIT Press, 1969.

[2] P. Billingsley, *Convergence of Probability Measures*, John Wiley & Sons, **1968**.

[3] R. C. Blei, *Fractional Cartesian products of sets*, Ann. Inst. Fourier (Grenoble), **29** (1979), 79–105.

[4] R. C. Blei, *Combinatorial dimension and certain norms in harmonic analysis*, Amer. J. Math., **106** (1984), 847–887.

[5] R. C. Blei, *α-Chaos*, J. Funct. Anal., **81** (1988), 279–296.

[6] R. C. Blei, *Analysis in Integer and Fractional Dimensions*, Cambridge University Press, **2001**.

[7] R. C. Blei and T. W. Körner, *Combinatorial dimension and random sets*, Israel J. Math., **47** (1984), 65–74.

[8] R. C. Blei and J.-P. Kahane, *A computation of the Littlewood exponent of stochastic processes*, Math. Proc. Camb. Phil. Soc., **103** (1988), 367–370.

[9] R. C. Blei, Y. Peres and J. H. Schmerl, *Fractional products of sets*, Random Designs and Algorithms, **6** (1995), 113–119.

[10] R. Brown, *On the existence of active molecules in organic and inorganic bodies*, Phil. Mag. (New Series), **4** (1828), 161–173.

[11] A. Einstein, *Investigations on the Theory of Brownian Movement*, edited by R. Fürth, translated by A. D. Cowper, Dover Publications, Inc., **1956**.

[12] H. Enderton, *A Mathematical Introduction to Logic*, Academic Press, New York and London, **1972**.

[13] M. Kac, *Random walk and the theory of Brownian motion*, Amer. Math. Month., **54** (1947), 369–391.

[14] S. Kakutani, *Brownian motion and duality of locally compact Abelian groups*, Proceedings of the International Congress of Mathematicians, 1950, 456–457.

[15] S. Kakutani, *Spectral analysis of stationary Gaussian processes*, Proceedings of the Fourth Berkeley Symposium, vol. II, University of California Press, 1961.

[16] J. E. Littlewood, *On bounded bilinear forms in an infinite number of variables*, Quart. J. Math. Oxford, **1** (1930), 164–174.

[17] D. Nualart, *The Malliavin Calculus and Related Topics*, Springer-Verlag, **1995**.

[18] J. Perrin, *Mouvement brownien et réalité moléculaire*, Ann. Chimie et Physique, 8e série, **18** (1909), 1–114.

[19] N. Wiener, *The average of an analytic functional and the Brownian movement*, Proc. Nat. Acad. Sci., **7** (1921), 294–298.

[20] N. Wiener, *Differential space*, J. Math and Physics, **2** (1923), 131–174.

[21] N. Wiener, *The homogeneous chaos*, Amer. J. Math., **60** (1938), 897–936.

Department of Mathematics
196 Auditorium Road
University of Connecticut, Unit 3009
Storrs, CT 06269
United States
E-mail address: `blei@math.uconn.edu`

Progress in Probability, Vol. 52, 33–44

Classical Solutions for SPDEs with Dirichlet Boundary Conditions

S. Bonaccorsi and G. Guatteri

Abstract. The aim of the paper is to prove some significative results for a given class of stochastic evolution equations by means of a suitable adaptation of techniques (the stochastic characteristics method and a Itô-type formula for backward diffusions) which are already known in the literature, but not so widely used.

We are able to prove both the existence of a stochastic flow related to the equation in spaces of continuous functions and a representation formula for the solution of the equation.

1. Setting of the problem

In this paper we deal with a class of stochastic evolution equations in a bounded domain D in $\mathbb{R}^n$, with finitely many Wiener processes; the coefficients of the white noise term are first order differential operators having compact support.

Let $(\Omega, \mathcal{F}, \{\mathcal{F}_t\}_{t\geq 0}, \mathbb{P})$ be a complete probability space and let $\{W(t)\}_{t\geq 0}$ be a k-dimensional Wiener process on this space; D denotes a bounded domain in $\mathbb{R}^n$ with smooth boundary ∂D. In this paper the following SPDE of parabolic type will be considered:

$$\begin{cases} du = [Au(t,x) + f(x)u(t,x)]\,dt + \sum_{i=1}^{k} B_i u(t,x)\,dW_i(t) & 0 \leq t \leq T \\ u(0,x) = g(x) \quad x \text{ in } D, \qquad u(t,x) = 0 & x \in \partial D. \end{cases} \tag{1}$$

We assume that A and B_i are linear differential operators of the form

$$Au(t,x) = \frac{1}{2}\sum_{i,j=1}^{n} a_{ij}(t,x)\partial_i\partial_j u(t,x) - \sum_{i=1}^{n} b_i(t,x)\partial_i u(t,x)$$

$$B_i u(t,x) = \sum_{j=1}^{n} d_{ij}(t,x)\partial_j u(t,x) \qquad i = 1,\dots,k$$

2000 *Mathematics Subject Classification.* 60H15, 35A05, 35R60.

Key words and phrases. Stochastic partial differential equations, characteristics method, existence of strong solutions, Feyman-Kac representation, existence of the stochastic flow.

with smooth coefficients in "parabolic" Hölder spaces (see the definition in the appendix)

$$\begin{aligned} &a_{ij}(t,x), b_i(t,x) \in C^{\alpha/2,\alpha}([0,T]\times\bar{D}) \\ &d_{ij}(t,x) \in C^{\alpha/2,3+\alpha}([0,T]\times\bar{D}); \quad d_{ij}(t,x)=0 \quad x\in\partial D. \end{aligned} \tag{2}$$

We shall assume as usual that

$$\sum_{i,j=1}^{n}\left[a_{ij}(t,x)-\sum_{l=1}^{k} d_{li}(t,x)d_{lj}(t,x)\right]\xi_i\xi_j \geq \nu|\xi|^2 \qquad \forall \xi\in\mathbb{R}^n. \tag{3}$$

Let us denote $\sigma_1 = (d_{ij})^*$ the $n\times k$ matrix given by the coefficients of the white noise term and, for every $i=1,\dots,k$, $d_i=(d_{i1},\dots,d_{in})^*$. Let σ_2 be a $n\times n$ matrix such that $\sigma_2\sigma_2^* = (a_{ij})-\sigma_1\sigma_1^*$; we put $\sigma=(\sigma_1,\sigma_2)$: notice that $\sigma\sigma^*=(a_{ij})$. We denote $\nabla\sigma_1\cdot\sigma_1(t,x)$ the n-dimensional vector with i-th component

$$\sum_{j=1}^{k}\sum_{h=1}^{n}\partial_h d_{ij}(t,x)\, d_{hj}(t,x).$$

Finally, we are given two functions $f\in C(\bar{D})$, $g\in C^{2+\alpha}(\bar{D})$.

Remark 1.1. We are interested in a SPDE on a bounded domain, so the coefficients can be extended outside D in several ways. With no loss of generality, we can assume that the coefficients b_i, f and g have a smooth extension to $\mathbb{R}^n$ with compact supports.

In the next section we prove an existence and uniqueness result for the classical solution Eq. (1); the main tool used here is an equivalence result between stochastic partial differential equations and deterministic evolution equations with random coefficients, proved by means of the stochastic characteristics. The method allows us to prove the existence of a stochastic flow for the equation in spaces of Hölder continuous functions. The equivalence result given here is a particular case of a more general result which will be treated in a different paper.

The last part of the paper is devoted to prove a stochastic Feynman-Kac formula for the solution of the stochastic equation; this result is a consequence of a general Itô formula in the spirit of [2], which can be useful in order to treat, in a simple and uniform fashion, other classical topics.

2. Existence of the strong solution

Under the assumptions stated above, existence and uniqueness results over finite time horizon for the Eq. (1) have been proved by means of different techniques. We will follow the stochastic characteristics approach of Tubaro [11]. We consider the random field $\xi(t,x)$ of stochastic characteristics and perform the following transformation:

$$v(t,x)=u(t,\xi(t,x)).$$

Then it can be shown that v verifies a deterministic equation of the form

$$\begin{cases} \frac{\partial v}{\partial t}(t,x) = Lv(t,x) + \tilde{f}(t,x)v(t,x), \\ v(0,x) = g(x) \quad x \in D, \qquad v(t,x) = 0 \quad x \in \partial D, \end{cases} \tag{4}$$

where $L(t)$ is a second order differential operator with random coefficients. We prove an equivalence result between problems (1) and (4); we remark that for a SPDE in the whole space, a similar result has been proved by Rozovskiĭ [10]. It is sufficient then to solve Eq. (4) by means of known results in the theory of abstract evolution equations.

We note that Eq. (1) has been studied in [3] by means of the semigroup approach of Da Prato, which leads to the mild solution of Eq. (1). A different approach is based on variational methods, see [4] or [9].

We are concerned, instead, with a classical solution u to Eq. (1) in the sense of [11]: for any t, $u(t,x)$ is two times Hölder continuously differentiable with respect to the variable x with probability 1, and it satisfies Eq. (1) in the integral (strong) sense.

2.1. The stochastic characteristics method

We consider the random field $\xi(t;s,x)$ of stochastic characteristics introduced by Kunita [6]. We define $\xi(t;s,x)$ the solution of the stochastic differential equation

$$\begin{cases} d\xi(t) = \frac{1}{2}\nabla\sigma_1 \cdot \sigma_1(t,\xi(t))\,dt - \sigma_1^*(t,\xi(t))\,dW(t) \\ \xi(s) = x, \qquad x \in \mathbb{R}^d. \end{cases} \tag{5}$$

Problem (5) has, for every $s \geq 0$, a unique solution (see [5, Theorem 3.3]); moreover, the first order derivative $\partial_l \xi(s;t,x)$ verifies a linear stochastic differential equation that, by the regularity of the d_{ij}'s, has a bounded solution. When the initial time is $s = 0$ we will write $\xi(t,x) = \xi(t;0,x)$.

Taking into account the assumptions on the coefficients of (5), by [7, Theorem 4.6.5] we get that the solution $\xi(t;s,x)$ has a modification that is a C^2-diffeomorphism for all $t \geq s$. Further $\xi(t;s,x)$ belongs to $C^{2+\beta}(\mathbb{R}^n)$ for any $\beta < \alpha$.

Let us denote by $\eta(s;t,x)$ the inverse mapping of ξ, i.e. the unique process such that

$$\eta(s;t,\xi(t;s,x)) = \xi(t;s,\eta(s;t,x)) = x \qquad w.p.1$$

The process $\eta(s;t,x)$ is a solution to the Itô backward differential equation

$$d\eta(s) = \frac{1}{2}\nabla\sigma_1 \cdot \sigma_1(t,\eta(s))\,ds + \sigma_1^*(s,\eta(s))\,dW(s)$$

with final-time condition $\eta(t;t,x) = x$. Moreover we define the process $\eta(t,x) := \eta(0;t,x)$. We consider $\eta(t,x)$ the backward flow with respect to ξ; by [7, Corollary 4.6.6] we get that the family of maps $\eta(t,x)$ verifies the following property $\mathbb{P}$-a.s.:

$$\eta(t,x) \in C^{\beta/2,2+\beta}([0,T] \times \mathbb{R}^n)$$

for any $\beta < \alpha$.

We recall that the coefficients $d_{ji}(t,x)$ have compact support in D. In this case, from Eq. (5) it follows that the evolution of the stochastic characteristics $\xi(t,x)$ shows the following behaviour: for any $x \notin D$, $\xi(t,x) \equiv x$ for all t. Therefore, we get that D is η-invariant in the sense of [1].

Theorem 2.1. *Assume that the domain D in η-invariant. Then Problem* (1) *admits a unique classical solution u (see for instance* [11]*) if and only if Problem* (4) *has a unique classical solution v in Hölder spaces in the sense of* [8, Definition 4.1.1], *with the additional assumption that $v(t)$ as a random process is adapted to the filtration $\{\mathcal{F}_t\}$.*

Proof. If Problem (1) admits a solution that is regular in space, applying the Itô Wentzell formula proved in [11, Proposition 2] we get that $u(t,\xi(t,x))$ solves the family of problems (4) in a classical way, where the operator $L(t)$ is defined by

$$L(t) = \frac{1}{2}\sum_{i,j=1}^{n} \tilde{a}_{ij}(t,x)\partial_i\partial_j + \sum_{j=1}^{n} \tilde{a}_j(t,x)\partial_j \tag{6}$$

$$\begin{aligned} \tilde{a}_{ij}(t,x) = &\ \langle(\sigma_2\sigma_2^*)\nabla\eta_i(t,\xi(t,x)), \nabla\eta_j(t,\xi(t,x))\rangle \\ \tilde{a}_j(t,x) = &\ \frac{1}{2}\mathrm{Tr}[(\sigma_2\sigma_2^*)\nabla^2\eta_j(t,\xi(t,x))] \\ &+\langle b(t,\xi(t,x)) - \frac{1}{2}\nabla\sigma_1\cdot\sigma_1(t,\xi(t,x)), \nabla\eta_j(t,\xi(t,x))\rangle \end{aligned}$$

and

$$\tilde{f}(t,x) = f(\xi(t,x)).$$

Vice-versa if Problem (4) admits a classical solution $v(t,x)$ for a.e. $\omega \in \Omega$, it is possible to apply the Itô formula to $v(t,\eta(t,x))$ and we get that Problem (1) has a solution with the required regularity. □

We remark that the assumptions stated above are not the only way to get the stochastic viability for the domain D. Moreover, even the condition of η-invariance is not necessary, provided one considers a weaker type of solution. We hope to return on this in a subsequent paper.

2.2. Deterministic non-autonomous equation

The above result allows us to solve a parabolic initial value problem in cylindrical domain $[0,T]\times\bar{D}$ for almost every ω in order to find a solution for Eq. (1); in the following, we want to approach Eq. (4) in the framework of evolution operators, following the book of Lunardi [8].

The ellipticity condition (3) can be rewritten as $\langle(\sigma_2\sigma_2^*)\xi_i,\xi_j\rangle \geq \nu|\xi|^2$; it is easy to check that also $\tilde{a}_j$ satisfies, with probability 1, a uniform ellipticity condition with a constant depending on ω, taking into account that $\nabla\eta(t,x)$ has rank n with probability 1 (see for instance [11, Remark 2] or [7]). Moreover, from the assumptions in (2) and the regularity of η we get that the coefficients of $L(t)$, as defined in (6), belong to $C^{\beta/2,\beta}([0,T]\times\bar{D})$. This is sufficient in order to apply the following theorem, that is proved in [8].

Theorem 2.2. *Let the initial condition $g \in C_0^{2+\beta}(\bar{D})$; there exists a unique classical solution to Eq. (4) such that*

$$v \in C^{1+\beta/2,2+\beta}([0,T] \times \bar{D}).$$

Remark 2.3. The main tool used in the proof (compare [8, Theorem 5.1.16]) is a contraction principle; the proof can be arranged in order to show that the solution $v(t,\cdot)$, as a random process, is $\mathcal{F}_t$-adapted. See also [11, Remark after Theorem 2]. In particular it is possible to take a random initial data assuming that it is $\mathcal{F}_0$-measurable.

The second part of the needed result is concerned with estimates on the solution.

Theorem 2.4. *Let the initial condition $g \in C_0^{2+\beta}(\bar{D})$; then the unique classical solution to Eq. (4) satisfies*

$$\|v(t)\|_{C_0^{2+\beta}(\bar{D})} \leq C\|g\|_{C_0^{2+\beta}(\bar{D})}.$$

2.3. Existence of the flow

Let $u(t;s,g)$ be the solution at time t of a certain stochastic differential equation in a Banach space X with initial condition $u(s) = g$ given at time $s \geq 0$. Assume that an existence and uniqueness result holds for the SPDE. We define the mapping $U(t,s)$ by $U(t,s)g = u(t;s,g)$; it verifies the semigroup law:

$$U(t,s)g = U(t,r)U(r,s)g, \qquad s < r < t.$$

We say that $U(t,s)$ defines a random evolution operator if

1. $U(t,s)$ is a linear and bounded mapping from X into $L^0(\Omega, \mathcal{F}_t; X)$;
2. $U(t,s)g$ is $\mathcal{F}_s^t$ measurable for any $g \in X$;
3. $U(s,s) = I$ and $U(t,s)g = U(t,r)U(r,s)g$, $\mathbb{P}$-a.s.

We say that a stochastic differential equation defines a stochastic flow in X if there exists a mapping $\omega \to \phi_{s,t}(\omega)$ from Ω to the space of continuous mapping in X such that

$$\phi_{s,t}(\omega)g = U(t,s)g(\omega), \qquad \mathbb{P} - a.s.$$

The problem of existence of the flow for infinite dimensional stochastic equations is still open in the general case, although several particular cases have been treated. On the contrary, the flow property is often trivially satisfied by a deterministic equation, as it is in the case we are considering. The equivalence result allows us to show existence of the flow for the solutions of Eq. (1).

Let us put

$$u(t,x) = v(t,\eta(t,x)). \tag{7}$$

We have, from the results in the previous section

Theorem 2.5. *Assume that the initial condition g belongs to $C_0^{2+\alpha}(\bar{D})$; then there exists a unique solution $u(t,x)$ for Eq. (1), defined by Eq. (7). Moreover $u(t,x)$ defines a stochastic flow in the space $C_0^{2+\beta}(\bar{D})$ for every $\beta < \alpha$.*

From Theorem 2.4 there exists a constant C depending on ω, $\mathbb{P}$-a.s. finite, such that $\|u(t)\|_{C_0^{2+\beta}(\bar{D})} \leq C\|g\|_{C_0^{2+\beta}(\bar{D})}$. From Remark 2.3 we get, moreover, that Problem (1) can be solved for every initial time $s \leq T$ and that the solutions patch well together; we can conclude by stating the existence of a random evolution operator $U(t,s)$ such that $U(t,s)g = u(t;s,g)$ that possesses, almost surely, a regular version as a mapping from $C_0^{2+\beta}(\bar{D})$ in itself.

We remark that existence of the stochastic flow for mild solutions of Eq. (1) in the Hilbert space $L^2(D)$ was proved in [3] by means of a stochastic Feynman-Kac formula. In the next section we show that this formula still holds for a classical solution, thus giving us a representation for $u(t,x)$.

3. A stochastic Feynman-Kac formula

In this section we prove a representation formula of Feynmac-Kac type for a classical solution of Eq. (1). Several versions of the related result are known: see for instance [3] and the references therein. The main tool for the proof is given in Theorem 3.3 below; we remark that a similar result was proved in [2], where it is shown how to use it in order to give simple proofs of known results, such as Feynman-Kac's and Girsanov's formula.

We are given two non-random functions $b(t,x)$ and $\sigma(t,x)$ on $[0,\infty) \times \mathbb{R}^n$, with values in $\mathbb{R}^n$ and $L(\mathbb{R}^{k+n},\mathbb{R}^n)$, respectively. For simplicity, we enumerate the following assumptions on these coefficients.

H1 Functions $b(t,x)$ and $\sigma(t,x)$ are continuous in $[0,T] \times \mathbb{R}^n$.

H2 We assume that $b(t,x)$ and $\sigma(t,x)$ satisfy the following Lipschitz condition:

$$\|\sigma(t,x)\| + |b(t,x)| \leq K(1+|x|)$$
$$\|\sigma(t,x) - \sigma(t,y)\| + |b(t,x) - b(t,y)| \leq K|x-y|,$$

where K is a fixed constant.

H3 For any $t \in [0,T]$, $b(t,\cdot)$ and $\sigma(t,\cdot)$ belong to $C^{2+\alpha}(\mathbb{R}^n)$.

We assume that $\Gamma = \Omega \times \tilde{\Omega}$ is a probability space and $V = (W,\tilde{W})$ is a Wiener process on Γ, where W and $\tilde{W}$ are k- and n-dimensional Wiener processes on Ω and $\tilde{\Omega}$. We denote $\tilde{E}$ the mathematical expectation with respect to the second component. In this framework we consider the backward stochastic Itô equation

$$\zeta(s;t,x) = x - \int_s^t b(r,\zeta(r;t,x))\,dr - \int_s^t \sigma(r,\zeta(r;t,x))\ d_b V(r) \tag{8}$$

where b and σ fulfill the above assumptions. The key point in the next proof is the following identity: for any $s < r < t$ and $x \in \mathbb{R}^n$ we have

$$\zeta(s;t,x) = \zeta(s;r,\zeta(r;t,x)).$$

We denote by $\mathcal{L}_s$ and $\mathcal{M}_s$ the linear operators

$$\mathcal{L}_s\phi(x) = \frac{1}{2}\mathrm{Tr}[\nabla^2\phi(x)\sigma(s,x)\sigma^*(s,x)] - \langle\nabla\phi(x), b(s,x)\rangle, \qquad \phi \in C_b^2(\mathbb{R}^n),$$

and

$$\langle\mathcal{M}_s\phi(x), h\rangle = \langle\nabla\phi(x), \sigma(s,x)h\rangle, \qquad h \in \mathbb{R}^{k+n}, \phi \in C_b^2(\mathbb{R}^n).$$

The following result gives the Itô formula for the function $u(t;s,x) = g(\zeta(s;t,x))$, where g is a smooth function in $\mathbb{R}^n$. The proof follows the ideas in [2].

Theorem 3.1. *Assume* **H1**, **H2** *and* **H3**, *and let* $g(x)$ *be in* $C_b^{2+\alpha}(\mathbb{R}^n)$. *Then, setting*

$$u(t;s,x) = g(\zeta(s;t,x)),$$

the process $u(t;s,x)$ *is a classical solution (see Section 2) for the equation*

$$u(t;s,x) = g(x) + \int_s^t \mathcal{L}_r u(r;s,x)\,dr + \int_s^t \langle\mathcal{M}_r u(r;s,x), dV(r)\rangle.$$

Proof. We fix $s < t$ and consider a partition π of the interval $[s,t]$:

$$s = s_0 < s_1 < \cdots < s_N = t.$$

We also set $|\pi| = \max(s_{k+1} - s_k)$. Then we may write

$$\begin{aligned} g(\zeta(s;t,x)) - g(\zeta(s;s,x)) &= \sum_{k=0}^{N-1} \Big[g(\zeta(s_0;s_{k+1},x)) - g(\zeta(s_0,s_k,x))\Big] \\ &= \sum_{k=0}^{n} \Big[g(\zeta(s_0;s_k,\zeta(s_k;s_{k+1},x))) - g(\zeta(s_0;s_k,x))\Big]. \end{aligned}$$

Using Taylor's formula up to second order, we may write

$$\begin{aligned} g(\zeta(s;t,x)) - g(\zeta(s;s,x)) &= \sum_{k=0}^{N-1} \langle(\nabla g(\zeta(s_0;s_k,\cdot))(x), \zeta(s_k;s_{k+1},x) - x\rangle \\ &+\frac{1}{2}\sum_{k=0}^{N-1} \langle(\nabla^2 g(\zeta(s_0;s_k,\cdot))(x)\Big(\zeta(s_k;s_{k+1},x) - x\Big), \zeta(s_k;s_{k+1},x) - x\rangle. \end{aligned}$$

By formula (8) we get

$$\zeta(s_k;s_{k+1},x) - x = -b(s_k,x)(s_{k+1} - s_k) - \sigma(s_k,x)(V(s_{k+1}) - V(s_k)) + o(|\pi|),$$

and putting this relation in the above formula, we have

$$\begin{aligned} &\sum_{k=0}^{N-1} \langle(\nabla g(\zeta(s_0;s_k,\cdot))(x), \zeta(s_k,s_{k+1},x) - x\rangle \\ &\quad = -\sum_{k=0}^{N-1} \langle(\nabla g(\zeta(s_0;s_k,\cdot))(x), b(s_k,x)\rangle(s_{k+1} - s_k) + \\ &\quad -\sum_{k=0}^{N-1} \langle(\nabla g(\zeta(s_0;s_k,\cdot))(x), \sigma(s_k,x)(V(s_{k+1}) - V(s_k))\rangle \end{aligned}$$

and also

$$\frac{1}{2}\sum_{k=0}^{N-1}\langle(\nabla^2 g(\zeta(s_0;s_k,\cdot))(x)(\zeta(s_k;s_{k+1},x)-x),(\zeta(s_k;s_{k+1},x)-x)\rangle$$

$$=\frac{1}{2}\sum_{k=0}^{N-1}\mathrm{Tr}\Big[(\nabla^2 g(\zeta(s_0;s_k,\cdot))(x)\sigma(s_k,x)\sigma^*(s_k,x)\Big](s_{k+1}-s_k).$$

The conclusion follows by letting $|\pi|$ go to 0. □

3.1. The zero-order term

In order to extend the results proved in the previous theorem, we introduce a continuous function f defined on $\mathbb{R}^n$. In order to simplify the exposition, we do not assume f depending on time (but compare with [2]). We introduce the process

$$\mu^{t,x}(s) := \mu(s,t,x) = \exp\left(\int_s^t f(\zeta(r;t,x))\,dr\right).$$

Under our assumptions, $\mu^{t,x}(s)$ is a solution to the linear stochastic differential equation

$$\begin{cases} d\mu^{t,x}(s) = f(\zeta(s;t,x))\mu^{t,x}(s)\,ds, & 0\le s\le t\\ \mu^{t,x}(t)=1. \end{cases}$$

Now we are interested in the forward Itô differential of this function. To this end, we note the identity

$$\mu(s,t,x)=\mu(r,t,x)\mu(s,r,\zeta(r;t,x)), \qquad s\le r\le t.$$

Proposition 3.2. *Let f be a real valued, continuous function defined in $\mathbb{R}^n$. Then $\mu(s,t,x)$ verifies the equation*

$$\mu(s,t,x)=1+\frac{1}{2}\int_s^t \mathrm{Tr}\Big[\mu(s,r,x)\sigma(x)\sigma^*(x)\Big]\,dr+\int_s^t\langle\nabla\mu(s,r,x),b(x)\rangle\,dr$$
$$+\int_s^t f(x)\mu(s,r,x)\,dr+\int_s^t\langle\nabla\mu(s,r,x),\sigma(x)dV(r)\rangle. \quad (9)$$

Proof. Let π be a partition of $[s,t]$ given by $s=s_0<s_1<\cdots<s_n=t$ and set $|\pi|=\max(s_{k+1}-s_k)$. We write

$$\mu(s,t,x)-\mu(s,s,x)$$
$$=\sum_{k=0}^{n-1}\mu(s,s_{k+1},x)-\mu(s,s_k,x)$$
$$=\sum_{k=0}^{n-1}\mu(s,s_k,\zeta(s_k,s_{k+1},x))\mu(s_k,s_{k+1},x)-\mu(s,s_k,x)$$

$$
\begin{aligned}
&= \sum_{k=0}^{n-1} \mu(s, s_k, \zeta(s_k, s_{k+1}, x)) - \mu(s, s_k, x) \\
&\quad + \sum_{k=0}^{n-1} \Big(\mu(s, s_k, \zeta(s_k, s_{k+1}, x)) - \mu(s, s_k, x)\Big)\Big(\mu(s_k, s_{k+1}, x) - 1\Big) \\
&\quad + \sum_{k=0}^{n-1} \mu(s, s_k, x)\Big(\mu(s_k, s_{k+1}, x) - 1\Big).
\end{aligned}
$$

We argue as in the proof of Theorem 3.1:

$$
\begin{aligned}
\mu(s,t,x) - \mu(s,s,x) &= \sum_{k=0}^{n-1} \mu(s, s_{k+1}, x) - \mu(s, s_k, x) \\
&= -\sum_{k=0}^{n-1} \langle \nabla\mu(s, s_k, \cdot)(x), b(s_k, x)\rangle (s_{k+1} - s_k) \\
&\quad + \sum_{k=0}^{n-1} \langle \nabla^2\mu(s, s_k, \cdot)(x)\sigma(s_k, x), \sigma(s_k, x)\rangle (s_{k+1} - s_k) \\
&\quad - \sum_{k=0}^{n-1} \langle \nabla\mu(s, s_k, \cdot)(x), \sigma(s_k, x)(V(s_{k+1}) - V(s_k))\rangle \\
&\quad + \sum_{k=0}^{n-1} \mu(s, s_k, x) f(\zeta(s_k, s_{k+1}, x))(s_{k+1} - s_k).
\end{aligned}
$$

Taking into account the assumptions on f, ζ and μ we have the thesis as $|\pi| \to 0$. □

3.2. Main theorem

Theorem 3.3. *Define the process*

$$
z(t,x) = g(\zeta(0;t,x)) \exp\left(\int_0^t f(\zeta(s,t,x))\, ds\right).
$$

Then $z(t,x)$ is a classical solution to the problem

$$
\begin{cases} dz(t,x) = \Big[\mathcal{L}_t z(t,x) + f(x) z(t,x)\Big]\, dt + \langle \mathcal{M}_t z(t,x), dV(t)\rangle \\ z(0,x) = g(x). \end{cases} \tag{10}
$$

Proof. We apply Itô's formula to the product of random processes

$$
z(t,x) = g(\zeta(0;t,x))\mu(0,t,x)
$$

and, setting $u(t,x) = g(\zeta(0;t,x))$, we get

$$
dz(t,x) = du(t,x)\mu(0,t,x) + u(t,x) d\mu(0,t,x) + du(t,x) d\mu(0,t,x).
$$

As usual, we can forget about terms of order higher than dt (this procedure can be made formally correct, if necessary). By Theorems 3.1 and 3.2 we have

$$\begin{aligned} dz(t,x) =& \Big[\mathcal{L}_t u(t,x)\,dt - \langle \mathcal{M}_t u(t,x), dV(t)\rangle\Big]\mu(0,t,x) \\ &+ \Big[(\mathcal{L}\mu(0,t,x) + f(x)\mu(0,t,x))\,dt - \langle \mathcal{M}_t\mu(0,t,x), dV(t)\rangle\Big]u(t,x) \\ &+ \langle \mathcal{M}_t\mu(s,t,x), \mathcal{M}_t u(t,x)\rangle dt. \end{aligned}$$

The key point of this proof is the following identity.

$$\begin{aligned} \frac{1}{2}\mathrm{Tr}[\nabla^2\mu(0,t,x)\sigma(t,x)\sigma^*(t,x)]u(t,x) + \frac{1}{2}\mathrm{Tr}[\nabla^2 u(t,x)\sigma(t,x)\sigma^*(t,x)]\mu(0,t,x) \\ + \langle \sigma^*(t,x)\nabla\mu(0,t,x), \sigma^*(t,x)\nabla u(t,x)\rangle = \frac{1}{2}\mathrm{Tr}[\nabla^2 z(t,x)\sigma(t,x)\sigma^*(t,x)]. \end{aligned}$$

If we substitute what we have in the above differential, we get Equation (10) with initial data

$$z(0,x) = g(\zeta(0;0,x))\mu(0,0,x) = g(x).$$

□

Let D be a bounded domain in $\mathbb{R}^n$ with smooth boundary ∂D. We define the random variable

$$\tau(t,x) = \sup\Big\{r \le t :\ \zeta(r,t,x) \in \partial D\Big\}.$$

We have the following identity

$$\{\,\tau(t,x) < s\,\} = \{\zeta(r,t,x) \in D\ \forall r \in [s,t]\}$$

which implies that $\{\,\tau(t,x) < s\,\} \in \mathcal{F}_s^t$.

Given the domain D, we consider the problem

$$\begin{cases} d\phi(t,x) = \Big[\mathcal{L}_t\phi(t,x) + f(x)\phi(t,x)\Big]\,dt + \langle \mathcal{M}_t\phi(t,x), dV(t)\rangle \\ \phi(0,x) = g(x), \quad x \in D; \qquad \phi(t,x) = 0, \quad x \in \partial D. \end{cases} \tag{11}$$

Corollary 3.4. *Assume that $g(x) \in C_0^{2+\alpha}(D)$. Let $\phi(t,x)$ be defined by*

$$\phi(t,x) = g(\zeta(0,t,x))\exp\left(\int_0^t f(\zeta(s,t,x))\,ds\right)\mathbf{1}_{\{\,\tau(t,x)<0\,\}}.$$

Then $\phi(t,x)$ is a solution to Equation (11).

Proof. From the result in Theorem 3.3 we have that with probability 1 the equation

$$z(t;s,x) = g(x) + \int_s^t \mathcal{L}_r z(r;s,x)\,dr + \int_s^t \langle \mathcal{M}_r z(r;s,x), dV(r)\rangle$$

holds for every $s \ge 0$. In particular, it remains valid if we substitute s by the random variable $0 \vee \tau(t,x)$; then the result follows from the identity

$$z(t; 0\vee\tau(t,x), x) = \phi(t,x) \qquad w.p.1.$$

□

3.3. A representation formula for the solution of Eq. (1)

In this last section we show the Feynman-Kac representation for the classical solutions of Eq. (1). We suppose that the coefficients of Eq. (1) verify (2) and the following regularity assumption

$$a_{ij}(t,x), b_i(t,x) \in C^{\alpha/2,2+\alpha}([0,T] \times \bar{D}).$$

Corollary 3.4 applies and the process $\phi(t,s)$ is a classical solution of

$$\phi(t,x) = g(x) + \int_0^t A\phi(s,x) + f(x)\phi(s,x)\,ds + \int_0^t \langle \nabla\phi(s,x), \sigma_1(s,x)dW(s)\rangle + \int_0^t \langle \nabla\phi(s,x), \sigma_2(s,x)d\tilde{W}(s)\rangle.$$

Taking conditional expectation $\tilde{\mathbb{E}}$ we get that

$$u(t,x) = \tilde{\mathbb{E}}\left(g(\zeta(0,t,x))\exp\left(\int_0^t f(\zeta(s,t,x))\,ds\right)\mathbf{1}_{\{\tau(t,x)<0\}}\right)$$

corresponds with the unique classical solution of Eq. (1).

Appendix A. Some function spaces

Let D a bounded domain in $\mathbb{R}^n$ with smooth boundary; we denote $C^\alpha(\bar{D})$, $0 < \alpha < 1$, the space of Hölder continuous functions of order α and $C^{m+\alpha}(\bar{D})$, $m \in \mathbb{N}$, $0 < \alpha < 1$, the space of m-times continuously differentiable functions such that the m-derivative $\partial^m f$ belongs to $C^\alpha(\bar{D})$. If Y is any of the spaces defined above, by Y_0 we denote the subset of all functions $f \in Y$ that vanish on ∂D.

The space of little-Hölder continuous functions is defined by

$$h^\alpha(\bar{D}) = \{f \in C^\alpha(\bar{D}) \mid \lim_{\delta\to 0} \sup_{x,y\in\bar{D},|x-y|<\delta} \frac{|f(x)-f(y)|}{|x-y|^\alpha} = 0\}.$$

One can check that $h^\alpha(\bar{D})$ is the closure of $C^\theta(\bar{D})$ in $C^\alpha(\bar{D})$ for any $\theta > \alpha$.

Let us recall now the definition of the parabolic Hölder spaces used in this paper. For $\alpha, \beta \in]0,1[$ we set

$$C^{\alpha,\beta}([a,b] \times \bar{D}) = \{\, f \in C([a,b] \times \bar{D}) \mid f(\cdot,x) \in C^\alpha([a,b]),\ f(t,\cdot) \in C^\beta(\bar{D}) \,\}$$

and

$$C^{1+\alpha,2+\beta}([a,b] \times \bar{D}) = \{\, f \in C([a,b] \times \bar{D}) \mid f, \partial_t f, \partial^2_{ij} f \in C^{\alpha,\beta}([a,b] \times \bar{D}) \,\}.$$

We finish with the following definition. Let $\alpha, \beta \in]0,1[$. Then we set

$$C^\beta([a,b]; C^\alpha(\bar{D})) = \{\, f \in C([a,b]; C^\alpha(\bar{D})) \mid [f]_{C^\beta([a,b];C^\alpha(\bar{D}))} = \sup_{t\neq s} \frac{\|f(t)-f(s)\|_{C^\alpha(\bar{D})}}{|t-s|^\beta} < \infty \,\}.$$

References

[1] J.-P. Aubin and G. Da Prato, *Stochastic viability and invariance*, Ann. Scuola Norm. Sup. Pisa Cl. Sci. (4) **17** (1990), 595–613.

[2] G. Da Prato and L. Tubaro, *Some remarks about backward Itô formula and applications*, Stochastic Anal. Appl. **16** (1998), 993–1003.

[3] F. Flandoli and K.-U. Schaumlöffel, *Stochastic parabolic equations in bounded domains: random evolution operator and Lyapunov exponents*, Stochastics Rep. **29** (1990), 461–485.

[4] N. V. Krylov and B. L. Rozovskiĭ, *Stochastic evolution equations*, Current problems in mathematics, Vol. 14 (Russian), Akad. Nauk SSSR, **1979**, 71–147, 256.

[5] H. Kunita, *Stochastic differential equations and stochastic flows of diffeomorphisms*, École d'été de probabilités de Saint-Flour, XII – 1982, Lecture Notes Math. **1097** (1984), 143–303.

[6] H. Kunita, *First order stochastic partial differential equations*, Stochastic analysis, North Holland, 1984, 249–269.

[7] H. Kunita, *Stochastic flows and stochastic differential equations*, Cambridge University Press, 1990.

[8] A. Lunardi, *Analytic semigroups and optimal regularity in parabolic problems*, Birkhäuser Verlag, **1995**.

[9] E. Pardoux, *Équations du filtrage non linéaire, de la prédiction et du lissage*, Stochastics **6** (1981/82), 193–231.

[10] B. L. Rozovskiĭ, *Stochastic evolution systems*, Linear theory and applications to linear filtering, Mathematics and its applications (Soviet series) 35, Kluwer Academic Publishers Group, 1990.

[11] L. Tubaro, *Some result on stochastic partial differential equations by the stochastic characteristics method*, Stochastic Analysis and Applications **6** (1988), 217–230.

Dipartimento di Matematica
Università di Trento
38050 Povo
Italia
E-mail address: bonaccor@science.unitn.it

E-mail address: guatteri@science.unitn.it

Progress in Probability, Vol. 52, 45–53

Credit Risk: The Structural Approach Revisited

E. Buffet

Abstract. In Merton's seminal article on credit risk (1974), corporate bonds are viewed as derivatives based on the value of the firm; the latter is implicitly assumed to be a tradable asset. An unavoidable consequence of this approach is that the growth rate of the firm plays no rôle in the pricing of corporate bonds, as it is wiped out by the change to the risk-neutral measure.

We revisit the above structural approach to credit risk, taking the shares of the firm as the basic tradable asset. A difficulty arises from the fact that the value of the assets of the firm is not directly related to the value of its shares; this is resolved by constructing a model in which both quantities derive from the profits of the firm. A simple formula for the profits is derived by making certain assumptions on the structure of the firm, and this yields explicit formulae for the share price and the value of the assets of the firm.

This construction allows us to identify the relevant risk-neutral measure in an unambiguous manner, and thus to price the defaultable bond as a functional of the profits. This leads to an Asian option type of problem.

1. Introduction

In Merton's classic article [8], a **corporate bond** with face value b and maturity T is treated as a derivative with payoff

$$H = b_\wedge V_T = V_T - (V_T - b)^+, \tag{1}$$

where the underlying security V_t is the value of the **assets of the firm**. Assuming for V_t the standard stochastic differential equation

$$dV_t = \alpha V_t dt + \sigma V_t dB_t, \tag{2}$$

under the objective probability measure or, denoting by r the rate of interest

$$dV_t = rV_t dt + \sigma V_t dW_t,$$

2000 *Mathematics Subject Classification.* 60H30, 91B28.

Key words and phrases. Defaultable bond, asian option, exponential martingale, Kazamaki's criterion.

It is a pleasure to thank John Appleby, Connor Griffin, Gabriella Iovino, Brian O'Kelly, Anatoly Patrick and Paul Quigley (all members of the DCU 'Default Club' of 1998–99) for stimulating discussions.

under the risk-neutral measure, one obtains readily a formula of Black-Scholes type for the value of defaultable bonds; the parameter α which characterises the growth rate of the firm does not enter the formula. Various refinements of Merton's model have been proposed [1, 7], but all share the same spirit.

A problem with this approach is that the assets of the firm are rarely tradable; thus the method of pricing by replication, which is implicit in the above is questionable. For the same reason, it is debatable if the probability measure employed for the calculation (namely the one under which $e^{-rt}V_t$ is a martingale) has any economic significance.

We advocate below an alternative approach based on a very simple idea: there must be an indisputable tradable asset underlying the problem, leading to a clear choice of risk-neutral measure. This asset will be the **shares of the firm**. Then, we have to face the difficulty that the value of the assets of the firm is not directly related to that of its shares; the solution lies in the fact that both assets have a common origin, namely the **profits of the firm**. So, we formulate a simple model of the profits; this results in a formula for the value of the shares, allowing us to identify the relevant risk-neutral measure. It then follows that a **defaultable bond** can be priced as an **asian-type** derivative based on the stock of the firm.

In order to carry out this programme, we have made specific assumptions on the structure of the firm; these may not fit all sectors of the economy, and in this sense our model does not have the universality of Merton's theory. However, many variations of the model can be incorporated without affecting the broad scheme presented here.

2. A simple model of the profit rate

Consider for definiteness a firm manufacturing a single product. The main variables are

n_t	:	Number of units sold per unit time
P_t	:	Unit selling price
C_t	:	Unit manufacturing cost
θ_t	:	Overheads (fixed costs) per unit time
X_t	:	Profit per unit time

These are linked by the relation

$$X_t = n_t(P_t - C_t) - \theta_t.$$

We assume the following stochastic differential equations:

$$\begin{cases} dC_t = & \alpha C_t dt + C_t \, \underline{\sigma} \cdot d\underline{B}_t, \\ \frac{dn_t}{n_t} = & -\lambda \frac{dP_t}{P_t} + \beta dt + \underline{\eta} \cdot d\underline{B}_t. \end{cases} \tag{3}$$

where $\lambda > 1$ is the **elasticity of demand**, β is the growth rate in that sector of the economy, α is a measure of inflation and $\underline{B}_t$ is a two-dimensional standard Brownian motion.

We assume further that the firm selects the selling price P_t so as to maximise the profit rate calculated with the noiseless version of (3). This leads to $P_t = \frac{\lambda}{\lambda-1}C_t$ and, after a routine calculation to

$$X_t = \kappa e^{\sigma B_t + (\nu - \frac{\sigma^2}{2})t} - \theta_t, \tag{4}$$

where $B_t = \frac{1}{\sigma}(\underline{\eta} - (\lambda - 1)\underline{\sigma}) \cdot \underline{B}_t$ is a standard scalar Brownian motion derived from $\underline{B}_t$, and κ, σ, ν are related to the original parameters as follows:

$$\kappa = \frac{n_0 c_0}{\lambda - 1}, \quad \sigma = |\underline{\eta} - (\lambda - 1)\underline{\sigma}|, \quad \nu = \beta - \alpha(\lambda - 1) + \underline{\sigma} \cdot (\underline{\eta} - \lambda\underline{\sigma}). \tag{5}$$

Note how, for given values of the original volatilities $|\underline{\eta}|$ and $|\underline{\sigma}|$, the profit growth rate ν and its volatility σ are monotone functions of the correlation $\underline{\sigma} \cdot \underline{\eta}$ (increasing in the case of ν, decreasing in the case of σ).

3. The share price and the risk-neutral measure

We take the time-honoured view [6], [2] that the share price is the expected value of the discounted future dividend payments. So, using Δ as a **discount rate** and denoting by $(\mathcal{F}_t)_{t \geq 0}$ the natural filtration of $\underline{B}_t, t \geq 0$ we have for the share value

$$S_t = \mathbb{E}[\int_t^\infty \gamma(s) X_s e^{\Delta(t-s)} ds | \mathcal{F}_t],$$

where $\gamma(s)$ is the **dividend rate**. Using (4) and assuming for simplicity that γ and θ are constant we obtain (assuming $\Delta > \nu$)

$$S_t = \frac{\gamma}{\Delta - \nu} X_t + \frac{\gamma\theta\nu}{\Delta(\Delta - \nu)}. \tag{6}$$

Remark 3.1. Note how close to the Black-Scholes model our formula for share prices is. In fact, combining (4) and (6) we see that

$$dS_t = (S_t + \frac{\gamma\theta}{\Delta})(\sigma dB_t + \nu dt), \tag{7}$$

so that $S_t + \frac{\gamma\theta}{\Delta}$ follows exactly the standard Black-Scholes model with volatility σ and growth rate ν.

Remark 3.2. We leave the choice of the discount rate open; the special case $\Delta = r$ leads to considerable simplifications, see below.

Because the stock pays dividends, the naked stock is not itself a tradable asset: when buying the stock, you also buy the right to collect the dividends [3]. As in [3], form a portfolio by reinvesting the dividends in the stock. During dt, the number of shares increases by $\gamma \frac{X_t}{S_t} dt$, and so the value of the portfolio at time t is

$$\hat{S}_t = S_t e^{\gamma \int_0^t \frac{X_u}{S_u} du}. \tag{8}$$

This is our **basic tradable asset**; its discounted value

$$Y_t = e^{-rt} \hat{S}_t, \tag{9}$$

obeys the following stochastic differential equation (just combine (7), (8) and (9)):

$$dY_t = \sigma(Y_t + \frac{\gamma\theta}{\Delta}\frac{Y_t}{S_t})(dB_t + \frac{\Delta - r}{\sigma}\frac{S_t}{S_t + \frac{\gamma\theta}{\Delta}}\ dt). \tag{10}$$

We can solve (7) and express the drift term in (10) as

$$\frac{\Delta - r}{\sigma}\frac{S_t}{S_t + \frac{\gamma\theta}{\Delta}}\ dt = \frac{\Delta - r}{\sigma}(1 - \frac{\gamma\theta}{\Delta S_0 + \gamma\theta}e^{-\sigma B_t - (\nu - \frac{\sigma^2}{2})t})dt.$$

Introduce the shorthand

$$\bar{\theta} = \frac{\gamma\theta}{\Delta S_0 + \gamma\theta},$$

and define the **exponential supermartingale**

$$M_t = \exp\Big\{\frac{r - \Delta}{\sigma}\int_0^t (1 - \bar{\theta}e^{-\sigma B_s - (\nu - \frac{\sigma^2}{2})s})dB_s - \frac{(r-\Delta)^2}{2\sigma^2}\int_0^t (1 - \bar{\theta}e^{-\sigma B_s - (\nu - \frac{\sigma^2}{2})s})^2 ds\Big\}. \tag{11}$$

A risk-neutral measure $\mathbb{P}^*$ exists whenever M_t is actually a martingale [4], and in that case

$$M_T = \frac{d\mathbb{P}^*}{d\mathbb{P}}. \tag{12}$$

Proposition 3.3. *M_t is a $\mathbb{P}$-martingale whenever $\Delta \geq r, \nu \geq \sigma^2$.*

Proof. Let

$$Z_t = exp\{\frac{r - \Delta}{2\sigma}\int_0^t (1 - \bar{\theta}e^{-\sigma B_s - (\nu - \frac{\sigma^2}{2})s})dB_s\}.$$

It is known that a sufficient condition for M_t to be a martingale is that Z_t be a submartingale [5], [10]. This in turn will hold provided that $\mathbb{E}[Z_t] < \infty$, because the exponent of Z_t is a martingale (and not just a local martingale). Using Itô's lemma to calculate the stochastic integral in the exponent of Z_t, we obtain

$$Z_t = exp\{\frac{r - \Delta}{2\sigma}[B_t + \frac{\bar{\theta}}{\sigma}(e^{-\sigma B_t - (\nu - \frac{\sigma^2}{2})t} - 1) + \frac{\bar{\theta}}{\sigma}(\nu - \sigma^2)\int_0^t e^{-\sigma B_s - (\nu - \frac{\sigma^2}{2})s}ds]\}.$$

Since $\bar{\theta} \geq 0, \sigma > 0$ we have when $\Delta \geq r, \nu \geq \sigma^2$

$$Z_t \leq e^{\frac{r-\Delta}{2\sigma}(B_t - \frac{\bar{\theta}}{\sigma})},$$

and in particular $\mathbb{E}[Z_t] < \infty$. □

Remark 3.4. The regime $\Delta \geq r$ is the economically relevant one in that investors will not countenance growth rates that are below the riskless rate r.

Remark 3.5. The condition $\nu \geq \sigma^2$ (which is sufficient but not necessary for the existence of $\mathbb{P}^*$) places restrictions on the original growth rates α, β, see (3), (5).

Remark 3.6. In the absence of overheads ($\theta = 0$), (11) reduces to $e^{\frac{r-\Delta}{\sigma}B_t - \frac{(r-\Delta)^2}{2\sigma^2}t}$ and M_t is a martingale for all values of the other parameters.

Remark 3.7. Equation (10) can be written as

$$dY_t = \sigma(Y_t + \frac{\gamma\theta}{\Delta}\frac{Y_t}{S_t})dW_t,$$

where W_t, which is standard Brownian motion under $\mathbb{P}^*$, is related to B_t by

$$\begin{aligned} dW_t &= dB_t + \frac{\Delta - r}{\sigma}\frac{S_t}{S_t + \frac{\gamma\theta}{\Delta}}dt \\ &= dB_t + \frac{\Delta - r}{\sigma}(1 - \bar{\theta}e^{-\sigma B_t - (\nu - \frac{\sigma^2}{2})t})dt. \end{aligned} \tag{13}$$

Remark 3.8. In contrast to (2), the change of measure to $\mathbb{P}^*$ does not, in our model, erase all trace of the growth rate ν. A simple way of seeing this is by considering the special case $\theta = 0$; then (13) reduces to

$$dW_t = dB_t + \frac{\Delta - r}{\sigma}dt,$$

so that (4) can be written as

$$X_t = \kappa e^{\sigma W_t + (\nu + r - \Delta - \frac{\sigma^2}{2})t}. \tag{14}$$

So the growth rate ν of the profit rate under the objective measure $\mathbb{P}$ still figures in the risk-neutral formulation (14).

In order to price corporate bonds, it is essential to express the profit rate X_t in terms of the new standard Brownian motion W_t, or in other words, to invert explicitly the change of variable (13). This is not completely trivial in view of the dependence of the drift on B_t.

Proposition 3.9. *The inversion of formula (13) is*

$$dB_t = dW_t + \left(\frac{r - \Delta}{\sigma}\right)dt + \frac{1}{\sigma}\frac{\bar{\theta}(\Delta - r)e^{-\sigma W_t - (\nu + r - \Delta - \frac{\sigma^2}{2})t}dt}{1 + \bar{\theta}(\Delta - r)\int_0^t e^{-\sigma W_s - (\nu + r - \Delta - \frac{\sigma^2}{2})s}ds} \tag{15}$$

or equivalently when θ_t is constant formula (4) can be written in terms of W_t as

$$X_t = \kappa e^{\sigma W_t + (\nu + r - \Delta - \frac{\sigma^2}{2})t}(1 + \bar{\theta}(\Delta - r)\int_0^t e^{-\sigma W_s - (\nu + r - \Delta - \frac{\sigma^2}{2})s}ds) - \theta. \tag{16}$$

Proof. From (4) and (13):

$$\begin{aligned} dX_t &= \kappa e^{\sigma B_t + (\nu - \frac{\sigma^2}{2})t}(\sigma dB_t + \nu dt) \\ &= (X_t + \theta)(\sigma dW_t + (\nu + r - \Delta)dt + \frac{(\Delta - r)\bar{\theta}}{e^{\sigma B_t + (\nu - \frac{\sigma^2}{2})t}}dt). \end{aligned} \tag{17}$$

Introduce the shorthand

$$\bar{X}_t = X_t + \theta, \quad \bar{\nu} = \nu + r - \Delta, \tag{18}$$

and rewrite accordingly (17) in the form:

$$d\bar{X}_t = \bar{X}_t(\sigma dW_t + \bar{\nu}dt) + \kappa\bar{\theta}(\Delta - r)dt. \tag{19}$$

Next we solve the stochastic differential equation (19) by the method of variation of the parameter, i.e. we look for a solution of the form

$$\bar{X}_t = e^{\sigma W_t + (\bar{\nu} - \frac{\sigma^2}{2})t} U_t.$$

This yields after a straightforward calculation

$$U_t = \kappa\bar{\theta}(\Delta - r) \int_0^t e^{-\sigma W_s - (\bar{\nu} - \frac{\sigma^2}{2})s} ds,$$

and finally

$$\bar{X}_t = \kappa e^{\sigma W_t + (\bar{\nu} - \frac{\sigma^2}{2})t} (1 + \bar{\theta}(\Delta - r) \int_0^t e^{-\sigma W_s - (\bar{\nu} - \frac{\sigma^2}{2})s} ds).$$

This is equivalent to (16). Formula (15) follows by comparing (16) to (4). □

4. The value of the assets of the firm and the price of corporate bonds

We take the view that the value of the firm derives from the **reinvested profits**

$$V_t = V_0 e^{-\delta t} + (1-\gamma) \int_0^t X_s e^{\delta(s-t)} ds. \tag{20}$$

The part of the profits which is not distributed to the shareholders is reinvested in the firm (say by upgrading the production equipment); the effects of such an injection of funds wear off at the rate δ, which can be thought as a **depreciation rate**. We can now price a corporate bond with payoff (1) as

$$P(t,T) = e^{r(t-T)} \mathbb{E}^*[b_\wedge V_T | \mathcal{F}_t], \tag{21}$$

where $\mathbb{E}^*$ is the expectation under the risk-neutral measure $\mathbb{P}^*$ defined in (12).

Proposition 4.1. *The value at time t of a defaultable bond with maturity T and face value b can be expressed as*

$$P(t,T) = be^{r(t-T)} - (1-\gamma)e^{(r+\delta)(t-T)} f(t, A_t, \bar{X}_t), \tag{22}$$

where $\bar{X}_t$ is as in (18) and

$$A_t = \frac{1}{1-\gamma}(be^{\delta(T-t)} - V_t) + \frac{\theta}{\delta}(e^{\delta(T-t)} - 1), \tag{23}$$

$$\begin{aligned} f(t,x,y) \quad &= \quad \mathbb{E}^*[(x - y\int_t^T e^{\sigma(W_s - W_t) + (\bar{\nu}+\delta-\frac{\sigma^2}{2})(s-t)} ds \\ &\quad -\rho \int_t^T du \int_u^T ds\; e^{\sigma(W_s - W_u) + (\bar{\nu} - \frac{\sigma^2}{2})(s-u) + \delta(s-t)})^+], \end{aligned} \tag{24}$$

$$\rho = \kappa\bar{\theta}(\Delta - r). \tag{25}$$

Proof. The strategy is to express (21) in terms of $\mathcal{F}_t$-measurable and $\mathcal{F}_t$-independent quantities. For this purpose, we write the value of the firm at maturity as

$$\begin{aligned} V_T &= V_t e^{\delta(t-T)} + (1-\gamma)\int_t^T X_s e^{\delta(s-T)} ds \\ &= V_t e^{\delta(t-T)} - \tfrac{\theta(1-\gamma)}{\delta}(1-e^{\delta(t-T)}) + (1-\gamma)\int_t^T \bar{X}_s e^{\delta(s-T)} ds \\ &= b - (1-\gamma)e^{\delta(t-T)}(A_t - \int_t^T \bar{X}_s e^{\delta(s-t)} ds). \end{aligned}$$

Hence the value of the bond is

$$P(t,T) = be^{r(t-T)} - (1-\gamma)e^{(r+\delta)(t-T)}\mathbb{E}^*[(A_t - \int_t^T ds \bar{X}_s e^{\delta(s-t)})^+|\mathcal{F}_t]. \quad (26)$$

Next we use (16), (18) and the shorthand

$$X_t^o = e^{\sigma W_t + (\bar{\nu} - \frac{\sigma^2}{2})t},$$

to write the integral in (26) as

$$\int_t^T \bar{X}_s e^{\delta(s-t)} ds = \kappa \int_t^T X_s^o e^{\delta(s-t)} ds + \rho \int_t^T ds X_s^o e^{\delta(s-t)} \int_0^s \frac{du}{X_u^o} \ .$$

After permuting the iterated integrals, the above expression can be rearranged to read

$$\bar{X}_t \int_t^T \frac{X_s^o}{X_t^o} e^{\delta(s-t)} ds + \rho \int_t^T du \int_u^T ds \frac{X_s^o}{X_u^o} e^{\delta(s-t)}.$$

The result follows. □

Remark 4.2. Although the Brownian functional (24) is relatively simple, it does not lend itself to an explicit evaluation. In fact, even the special $\theta = 0$ (no overheads) is equivalent to the problem of pricing and Asian put option. The best computational approach is by reduction to a partial differential equation along the same lines as in [9].

Proposition 4.3. *The function $f(t,x,y)$ defined in (24) obeys the partial differential equation*

$$\begin{aligned} \partial_1 f(t,x,y) &= -\tfrac{1}{2}\sigma^2 y^2 \partial_3^2 f(t,x,y) - (\rho + \bar{\nu} y)\partial_3 f(t,x,y) \\ &\quad + (\delta x + y)\partial_2 f(t,x,y) - \delta f(t,x,y), \end{aligned}$$

with terminal condition

$$f(T,x,y) = x^+.$$

Proof. The discounted bond price $\tilde{P}(t,T) = \mathbb{E}^*[b_\wedge V_T|\mathcal{F}_t]$, is a $\mathbb{P}^*$-martingale. On the other hand from (22)

$$d\tilde{P}(t,T) = -(1-\gamma)e^{\delta(t-T)}\{\delta f dt + df\} \quad (27)$$

$$= -(1-\gamma)e^{\delta(t-T)}\{\delta f dt + \partial_1 f dt + \partial_2 f dA_t + \partial_3 f d\bar{X}_t + \frac{1}{2}\sigma^2 \bar{X}_t^2 \partial_3^2 f dt\}. \quad (28)$$

The two stochastic differentials in (27) are, using (19), (25), (23)

$$\begin{aligned} d\bar{X}_t &= \sigma \bar{X}_t dW_t + (\bar{\nu}\bar{X}_t + \rho)dt, \\ dA_t &= -\delta(\frac{be^{\delta(T-t)}}{1-\gamma} + \frac{\theta}{\delta}e^{\delta(T-t)})dt - \frac{dV_t}{1-\gamma} \\ &= -\delta(A_t + \frac{V_t}{1-\gamma} + \frac{\theta}{\delta})dt - \frac{dV_t}{1-\gamma}. \end{aligned} \tag{29}$$

On the other hand, using (20):

$$\begin{aligned} dV_t &= d(e^{-\delta t}[V_0 + (1-\gamma)\int_0^t X_s e^{\delta s} ds]) \\ &= -\delta V_t dt + (1-\gamma)X_t dt. \end{aligned}$$

Combining (28) and (29) we have $dA_t = -(\delta A_t + \bar{X}_t)dt$, so that (27) reads

$$\begin{aligned} d\tilde{P}(t,T) \quad = \quad & -(1-\gamma)e^{\delta(t-T)}\{\sigma \bar{X}_t \partial_3 f dW_t + [\delta f + \partial_1 f - (\delta A_t + \bar{X}_t)\partial_2 f \\ & +(\bar{\nu}\bar{X}_t + \rho)\partial_3 f + \tfrac{1}{2}\sigma^2 \bar{X}_t^2 \partial_3^2 f]dt\}. \end{aligned}$$

Setting the drift term in the above equal to zero produces the stated equation. □

Remark 4.4. When $\rho = 0$ (because either $\theta = 0$ or $\Delta = r$), the valuation of defaultable bonds can be reduced to a two-variable problem by replacing (22), (24) with

$$\begin{aligned} P(t,T) \quad &= \quad be^{r(t-T)} - (1-\gamma)e^{(r+\delta))(t-T)}\bar{X}_t h(t, \tfrac{A_t}{\bar{X}_t}), \\ h(t,u) \quad &= \quad \mathbb{E}^*[(u - \textstyle\int_t^T \frac{X_s^o}{X_t^o} e^{\delta(s-t)} ds)^+]. \end{aligned}$$

The equation obeyed by h is similar to that obtained in [9]:-

$$\partial_1 h = -\frac{1}{2}\sigma^2 u^2 \partial_2^2 h + (1 + (\bar{\nu} + \delta)u)\partial_2 h - (r + \delta + \bar{\nu})h.$$

References

[1] F. Black and J. C. Cox, *Valuing Corporate Securities: Some Effects of Bond Indenture Provisions,* J. Finance, **31** (1976), 351–367.

[2] O. J. Blanchard and S. Fisher, *Lectures on Macroeconomics, (Chapter 10),* Cambridge (MA): MIT Press, 1989.

[3] M. Baxter and A. Rennie, *Financial Calculus (Section 4.2),* Cambridge: Cambridge University Press, 1996.

[4] R. J. Elliot and P. E. Kopp, *Mathematics of Financial Markets (Chapter 7),* New York: Springer, 1999.

[5] N. Kazamaki, *On a Problem of Girsanov,* Tôhoku Math. Journ., **29** (1977), 597–600.

[6] R. E. Lucas, Jr., *Asset Prices in an Exchange Economy,* Econometrica, **46** (1978), 1429–1445.

[7] F. A. Longstaff and E. S. Schwartz, *A Simple Approach to Valuing Risky Fixed and Floating Debt,* J. Finance, **50** (1995), 789–819.

[8] R. C. Merton, *On the Pricing of Corporate Debt: The Risk Structure of Interest Rates,* J. Finance, **29** (1974), 449–470.

[9] L. C. G. Rogers and Z. Shi, *The Value of an Asian Option,* J. Appl. Prob., **32** (1995), 1077–1088.

[10] D. Revuz and M. Yor, *Continuous Martingales and Brownian Motion (Section VIII.1),* Berlin: Springer, 1991.

School of Mathematical Sciences
Dublin City University
Dublin 9
Ireland

Dublin Institute for Advanced Studies
10, Burlington Road
Dublin 4
Ireland
E-mail address: emmanuel.buffet@dcu.ie

Progress in Probability, Vol. 52, 55–71

Classical Solutions for Kolmogorov Equations in Hilbert Spaces

S. Cerrai

Abstract. We are concerned with backward Kolmogorov equations (both in the evolutionary and in the stationary version) associated with stochastic reaction-diffusion systems with additive noise in bounded domains of $\mathbb{R}^d$, with $d \leq 3$. Our aim is proving that the transition semigroup corresponding to the stochastic system provides the unique classical solutions to the Kolmogorov problems.

1. Introduction

Let H denote the Hilbert space $L^2(\mathcal{O};\mathbb{R}^n)$, where $\mathcal{O}$ is a bounded domain of $\mathbb{R}^d$, with $d \leq 3$. We are here concerned with the study of the parabolic and the elliptic problem associated with the following class of differential operators defined in H

$$\mathcal{L}(x, D) = \frac{1}{2}\mathrm{Tr}\,[BB^\star D^2] + \langle Ax + F(x), D\rangle_H\,, \quad x \in D(A).$$

Here B is a bounded linear operator from H into itself, A is the realization in H of a second order uniformly elliptic operator $\mathcal{A} = (\mathcal{A}_1, \dots, \mathcal{A}_n)$, endowed with suitable boundary conditions $\mathcal{G} = (\mathcal{G}_1, \dots, \mathcal{G}_n)$, and F is the Nemytskii operator (composition type operator) corresponding to a function $f : \overline{\mathcal{O}} \times \mathbb{R}^n \to \mathbb{R}^n$ which satisfies the Caratheodory condition and is twice differentiable with respect to $\sigma \in \mathbb{R}^n$, with bounded derivatives, uniformly for $\xi \in \overline{\mathcal{O}}$.

Our aim is studying the existence of *regular* solutions for the parabolic and the elliptic problem associated with the operator $\mathcal{L}(x, D)$

$$u_t(t, x) = \mathcal{L}(x, D)u(t, x), \quad x \in D(A), \quad t > 0, \qquad u(0, x) = \varphi(x) \tag{1}$$

and

$$\lambda\psi(x) - \mathcal{L}(x, D)\psi(x) = \varphi(x), \qquad x \in D(A), \quad \lambda > 0. \tag{2}$$

2000 *Mathematics Subject Classification.* 47D07, 60H15, 35K10, 35J15.
Key words and phrases. Kolmogorov equations in Hilbert spaces, Schauder estimates in infinite dimension, stochastic reaction-diffusion systems.

Namely, we show that for any initial datum $\varphi \in C_b(H)$ Problem (1) admits a unique *classical* solution $u(t,x)$ which is given by

$$u(t,x) = P_t\varphi(x) = \mathbf{E}\,\varphi(y(t;x)), \qquad (t,x) \in [0,+\infty) \times H,$$

where $y(t;x)$ is the unique solution of the stochastic reaction-diffusion system

$$dy(t) = (Ay(t) + F(y(t)))\,dt + B\,dw(t), \qquad y(0) = x.$$

This means that $u : [0,+\infty) \times H \to \mathbb{R}$ is continuous, the function $u(t,\cdot)$ is in $C_b^2(H)$, for any $t > 0$, the operator $D^2u(t,x)BB^\star$ is *trace-class*, for any $x \in H$, and the function

$$(0,+\infty) \times H \to \mathbb{R}, \qquad (t,x) \mapsto \mathrm{Tr}\,\left[D^2u(t,x)BB^\star\right]$$

is continuous. Moreover for any $x \in D(A)$ the function $u(\cdot,x)$ is differentiable in $(0,+\infty)$ and u fulfills Equation (1).

Under the same hypotheses, we show that if $\varphi \in C_b^1(H)$ then the elliptic problem (2) has a unique *strict* solution ψ which is given by

$$\psi(x) = \int_0^{+\infty} e^{-\lambda t} P_t\varphi(x)\,dt, \qquad x \in H.$$

This means that $\psi \in C_b^2(H)$, the operator $D^2\psi(x)BB^\star$ is *trace-class*, for any $x \in H$, and the mapping $x \mapsto \mathrm{Tr}\,[D^2\psi(x)BB^\star]$ is bounded and continuous. Moreover ψ fulfills Equation (2). We also show that if the datum φ is only continuous, then the function ψ is a *strong* solution, that is there exist two sequences $\{\varphi_n\}$ and $\{\psi_n\}$, converging respectively to φ and ψ in $C_b(H)$, such that ψ_n is the unique strict solution of Problem (2) with datum φ_n.

Notice that the most difficult step in the proof of the results mentioned above consists in proving the *trace-class* property for the operator $D^2(P_t\varphi)(x)BB^\star$, for any $x \in H$ and $t > 0$. Such a property follows from the explicit formula which we have for the first and the second derivatives of $P_t\varphi$ (Bismut-Elworthy formula). The proof is quite delicate and is based on the existence of a suitable complete orthonormal system in H which enjoys some infinite sum properties which will be described in details later on.

In the present paper we extend a previous result (see [2]) proved in the case H is $L^2(I,\mathbb{R}^n)$, where I is a bounded interval of $\mathbb{R}$, and B is an invertible operator with bounded inverse. Here we are dealing with the Hilbert space $H = L^2(\mathcal{O};\mathbb{R}^n)$, where $\mathcal{O}$ is a bounded domain of $\mathbb{R}^d$, with $d \leq 3$. Thus, as the space dimension d is bigger than 1, if A is the realization of a second order uniformly elliptic operator, we cannot assume the operator B to be invertible with a bounded inverse and then the proof of the trace-class property of the operator $D^2(P_t\varphi)(x)BB^\star$ is much more delicate.

As known, the operator $\mathcal{L}(x,D)$ is the diffusion operator associated with the stochastic reaction-diffusion system

$$dy(t) = (Ay(t) + F(y(t)))\,dt + B\,dw(t), \qquad y(0) = x. \tag{3}$$

Since the nonlinear term F is Lipschitz continuous, it is possible to show that, under suitable assumptions on A and B, for any initial datum $x \in H$, system (3) admits a unique *mild* solution $y(\cdot;x) \in L^2(\Omega; C([0,+\infty); H))$, where $(\Omega, \mathcal{F}, \mathcal{F}_t, \mathbf{P})$ is the stochastic basis where the noise $w(t)$ is defined. In [1] and [3, Chapter 4] we have shown that for any fixed $t \geq 0$ the mapping

$$H \to L^2(\Omega, H), \qquad x \mapsto y(t;x)$$

is k-times differentiable, where $k < 1 + 4/d$.

Moreover, we have shown that the following Bismut-Elworthy type formula for the second derivative of $P_t\varphi$ holds

$$\begin{aligned}
&\left\langle D^2(P_t\varphi)(x)h, k\right\rangle_H \\
&= \frac{1}{t}\mathbf{E}\,\langle D\varphi(y(t;x)), D_x y(t;x)h\rangle_H \int_0^t \left\langle B^{-1}D_x y(s;x)k, dw(s)\right\rangle_H \\
&+\frac{1}{t}\mathbf{E}\,\varphi(y(t;x)) \int_0^t \left\langle B^{-1}D_x^2 y(s;x)\,(h,k)\,, dw(s)\right\rangle_H,
\end{aligned} \tag{4}$$

where $D_x y(t;x)h$ and $D_x^2 y(t;x)(h,k)$ are respectively the first and the second mean-square derivatives of $y(t;x)$ along the directions h and k. Thus, in order to prove that the operator $D^2(P_t\varphi)(x)BB^\star$ is nuclear for any $\varphi \in B_b(H)$, in addition to the difficulties arising from infinite sums, in the present paper (where we do not assume B to be invertible) we have also to give a meaning to the stochastic integrals where the unbounded operator B^{-1} appears. To this purpose the crucial step will be proving that there exists a complete orthonormal system $\{e_k\}$ in H such that

$$\sup_{x\in H} \sum_{k=1}^{+\infty} \int_0^t \left|B^{-1}D_x y(s;x)BB^\star e_k\right|_H^2 ds \leq c(t)\,t^{\,\delta}, \qquad \mathbf{P}-\text{a.s.}$$

for some constant $\delta > 0$, and for any bounded operator N

$$\sup_{x\in H} \sum_{k=1}^{+\infty} \left|B^{-1}D_x^2 y(t;x)(Ne_k, BB^\star e_k)\right|_H \leq \|N\|\,c(t),$$

for some continuous increasing function $c(t)$ not depending on N.

Actually, from the two inequalities above and from Formula (4), we can conclude that for any $\varphi \in B_b(H)$ and for any $t > 0$ and $x \in H$ the operator $D^2(P_t\varphi)(x)BB^\star$ is *trace-class* and

$$\sup_{x\in H} \left|\mathrm{Tr}\,\left[D^2(P_t\varphi)(x)BB^\star\right]\right| \leq c\,(t\wedge 1)^{-\alpha} \sup_{x\in H} |\varphi(x)|, \tag{5}$$

where α is a suitable positive constant which we can determine explicitly. Notice that the singularity at $t = 0$ in (5) makes the study of the elliptic problem more delicate. In fact, in this case the solution is given by

$$\psi(x) = \int_0^{+\infty} e^{-\lambda t} P_t\varphi(x)\,dt, \qquad x \in H$$

and then we have to deal with the convergence of some integrals having singularities at $t = 0$.

2. Notations and preliminary facts

Throughout the paper we shall denote by H the separable Hilbert space $L^2(\mathcal{O}, \mathbb{R}^n)$, where $\mathcal{O}$ is a bounded open subset of $\mathbb{R}^d$, with $d \leq 3$. H is endowed with the norm $|\cdot|_H$ and the corresponding scalar product $\langle \cdot, \cdot \rangle_H$. Moreover we denote by $|\cdot|_p$ the norm in $L^p(\mathcal{O}; \mathbb{R}^n)$, for any $p \in [1, +\infty]$, with $p \neq 2$.

We define $B_b(H)$ as the set of all functions $\varphi : H \to \mathbb{R}$ which are Borel and bounded. $B_b(H)$ is a Banach space, endowed with the supremum norm

$$\|\varphi\|_0 = \sup_{x \in H} |\varphi(x)|.$$

$C_b(H)$ is the linear subspace of uniformly continuous functions. For any integer $k \geq 1$ we denote by $C_b^k(H)$ the space of k-times Fréchet differentiable functions which are uniformly continuous, together with their derivatives up to the k-th order. $C_b^k(H)$ is a Banach space, endowed with the norm

$$\|\varphi\|_k = \|\varphi\|_0 + \sum_{h=1}^{k} [\varphi]_h = \|\varphi\|_0 + \sum_{h=1}^{k} \sup_{x \in H} |D^h \varphi(x)|_{\mathcal{L}^h(H)}.$$

Finally, in the sequel we shall denote by $\mathcal{L}(H)$ the Banach space of bounded linear operators in H. $\mathcal{L}_1(H)$ is the subspace of *trace-class* operators, endowed with the norm

$$\|T\|_{\mathcal{L}_1(H)} = \mathrm{Tr}\,\sqrt{TT^\star}$$

and $\mathcal{L}_2(H)$ is the subspace of *Hilbert-Schmidt* operators, endowed with the norm

$$\|T\|_{\mathcal{L}_2(H)} = \sqrt{\mathrm{Tr}\,[TT^\star]}.$$

The nonlinear term F. In the present paper we are dealing with a class of non linear terms F of Nemytskii type. More precisely, we assume that there exists a function $f : \mathcal{O} \times \mathbb{R}^n \to \mathbb{R}^n$, such that

$$F(x)(\xi) = f(\xi, x(\xi)), \qquad \xi \in \mathcal{D}, \qquad x \in H.$$

In the sequel we shall assume that f verifies the following conditions.

Hypothesis 2.1. *The function $f : \mathcal{O} \times \mathbb{R}^n \to \mathbb{R}^n$ is measurable in $\xi \in \mathcal{O}$, for all $\sigma \in \mathbb{R}^n$. Moreover, for almost all $\xi \in \mathcal{O}$, the function $f(\xi, \cdot)$ is of class $C^2(\mathbb{R}^n)$ and it holds*

$$\sup_{\xi \in \mathcal{O}} \sup_{\sigma \in \mathbb{R}^n} |\frac{\partial^j}{\partial \sigma^j} f(\xi, \sigma)| < \infty, \qquad j = 1, 2.$$

It is immediate to check that under these conditions the operator F is Lipschitz continuous in H. Concerning its smoothness, even if we are assuming f to be twice differentiable, F is never Fréchet differentiable, unless $f(\xi,\cdot)$ is linear. However F is Gâteaux differentiable at any $x \in H$ and along any direction $y \in H$ and it holds

$$DF(x)y(\xi) = \frac{\partial}{\partial \sigma} f(\xi, x(\xi))y(\xi), \qquad \xi \in \mathcal{O}.$$

Moreover F is twice Gâteaux differentiable at any $x \in H$, but only along directions y, z such that $yz \in H$. In this case it holds

$$D^2F(x)(y,z)(\xi) = \frac{\partial^2}{\partial \sigma^2} f(\xi, x(\xi))y(\xi)z(\xi), \qquad \xi \in \mathcal{O}.$$

The operator A. We assume that for any $k = 1, \dots, n$ the diffusion operator $\mathcal{A}_k$ is given by

$$\mathcal{A}_k(\xi, D) = \sum_{i,j=1}^{d} a_k^{ij}(\xi) \frac{\partial^2}{\partial \xi_i \partial \xi_j} + \sum_{i=1}^{d} b_k^i(\xi) \frac{\partial}{\partial \xi_i}, \qquad \xi \in \overline{\mathcal{O}}.$$

The coefficients a_k^{ij} and b_k^i are taken uniformly continuous and for any $\xi \in \overline{\mathcal{O}}$ the matrix $[a_k^{ij}(\xi)]$ is symmetric and satisfies the uniform ellipticity condition

$$\inf_{\xi \in \overline{\mathcal{O}}} \sum_{i,j=1}^{d} a_k^{ij}(\xi) h_i h_j \geq \nu |h|^2, \quad h \in \mathbb{R}^d,$$

for some $\nu > 0$. For each k the boundary operator $\mathcal{G}_k$ is defined by

$$\mathcal{G}_k(\xi, D) = \sum_{i=1}^{d} \beta_k^i(\xi) \frac{\partial}{\partial \xi_i} + \gamma_k(\xi) I, \qquad \xi \in \partial \mathcal{O},$$

and β_k^i and γ_k are uniformly continuous functions. In the sequel we shall assume that either $\beta_k^i \equiv 0$ and $\gamma \equiv 1$ on $\partial \mathcal{O}$ (Dirichlet boundary condition), or

$$\inf_{\xi \in \partial \mathcal{O}} \left| \sum_{i=1}^{d} \beta_k^i(\xi) \nu_i(\xi) \right| > 0,$$

where $\nu(\xi) = (\nu_1(\xi), \dots, \nu_d(\xi))$ is the normal vector to the boundary of $\mathcal{O}$ (oblique boundary condition). Here all smoothness assumptions on $\partial \mathcal{O}$ are assumed to hold. For any $p \in [1, +\infty]$, we shall denote by A_p the realization in $L^p(\mathcal{O}; \mathbb{R}^n)$ of the differential operator $\mathcal{A} = (\mathcal{A}_1, \dots, \mathcal{A}_n)$, endowed with the boundary conditions $\mathcal{G} = (\mathcal{G}_1, \dots, \mathcal{G}_n)$. If $p > 1$, the operator A_p generates an analytic semigroup e^{tA_p} in $L^p(\mathcal{O}; \mathbb{R}^n)$. All the semigroups e^{tA_p} are consistent, in the sense that

$$x \in L^p(\mathcal{O}; \mathbb{R}^n) \cap L^q(\mathcal{O}; \mathbb{R}^n) \Longrightarrow e^{tA_p} x = e^{tA_q} x,$$

thus we shall omit the subscript p and we shall write e^{tA}. Notice that by replacing A by $A - \omega$ and F by $F + \omega$, for some $\omega > 0$, we can assume without any loss of generality that A is of negative type. Thus for any $\delta \geq 0$ we can introduce the *fractional power* $(-A)^\delta$ of $-A$. It is possible to show that e^{tA} maps $L^p(\mathcal{O};\mathbb{R}^n)$ into $D((-A)^\delta)$, for any $p>1$ and $t>0$, and

$$\left|(-A)^\delta e^{tA}x\right|_p \leq c_\delta\, t^{-\delta}|x|_p, \qquad x\in L^p(\mathcal{O};\mathbb{R}^n). \tag{6}$$

Moreover, for any $\delta_1\leq\delta\leq\delta_2$ and $x\in D((-A)^{\delta_2})$ it holds

$$\left|(-A)^\delta x\right|_p \leq c\left|(-A)^{\delta_1}x\right|_p^{1-\theta}\left|(-A)^{\delta_2}x\right|_p^{\theta}, \qquad \theta=\frac{\delta-\delta_1}{\delta_2-\delta_1}.$$

Thanks to the characterization of $D((-A)^\delta)$, the Sobolev embedding theorem and the Riesz-Thorin interpolation theorem, this implies that the semigroup e^{tA} maps the space $L^p(\mathcal{O};\mathbb{R}^n)$ into $L^q(\mathcal{O};\mathbb{R}^n)$, for any $t>0$ and for any $1<p\leq q\leq\infty$, and

$$|e^{tA}x|_q \leq c\,t^{-\frac{d(q-p)}{2qp}}|x|_p, \qquad x\in L^p(\mathcal{O};\mathbb{R}^n). \tag{7}$$

For the definitions and the proofs of the results we have recalled, we refer to the monographies [6] and [7].

The operator B**.** We assume that B is in $\mathcal{L}(H)$ and fulfills the following assumptions.

Hypothesis 2.2.

1. *There exists* $\delta>0$ *such that for any* $t\geq 0$

$$\int_0^t s^{-\delta}\operatorname{Tr}\left[e^{sA}BB^\star e^{sA^\star}\right]ds<+\infty. \tag{8}$$

2. *There exists* $\epsilon<1$ *such that*

$$\operatorname{Range}((-A)^{-\epsilon/2})\subset\operatorname{Range}(B^{-1}). \tag{9}$$

Notice that Conditions (8) and (9) are in contrast the one with the other. If we assume that $Ae_k=-\alpha_k e_k$, where $\{e_k\}$ is a complete orthonormal system in H and $\alpha_k\asymp k^{2/d}$ (this is the case for second order uniformly elliptic operators in sufficiently regular domains) and if we take $B=C(-A)^{-\delta}$, for some $\delta>0$ and for some bounded linear operator C which is invertible with bounded inverse, then it is possible to show that they are both fulfilled when $d\leq 3$.

Condition (8) assures that if $w(t)$ is space-time white noise given by

$$w(t)=\sum_{k=1}^{+\infty}e_k w_k(t),$$

where $\{e_k\}$ is a complete orthonormal system in H and $w_k(t)$ are mutually independent Brownian motions defined on the same stochastic basis $(\Omega,\mathcal{F},\mathcal{F}_t,\mathbf{P})$, then there exists a unique mild solution $y(t;x)$ for system (3) which belongs to $L^2(\Omega,C([0,+\infty);H))$ (for a proof see e.g. [4, Theorem 7.6]).

On the other hand, Condition (9) assures that the semigroup P_t has a smoothing effect. Actually, in [1] and in [3, Chapter 4] we have proved that under the above conditions

$$\varphi \in B_b(H) \Longrightarrow P_t\varphi \in C_b^k(H), \qquad t > 0, \tag{10}$$

where $k < 1 + 4/d$, and

$$\|P_t\varphi\|_j \leq c\,(t \wedge 1)^{-\frac{(j-i)(1+\epsilon)}{2}} \|\varphi\|_i, \qquad t > 0, \tag{11}$$

for any $0 \leq i \leq j \leq k$. Whence, since we are assuming $d \leq 3$, the semigroup P_t maps $B_b(H)$ into $C_b^2(H)$. Moreover, we have proved that the following generalization of the Bismut-Elworthy formula holds

$$\langle D(P_t\varphi)(x), h\rangle_H = \frac{1}{t}\mathbf{E}\,\varphi(y(t;x)) \int_0^t \left\langle B^{-1}D_x y(s;x)h, dw(s)\right\rangle_H, \tag{12}$$

where $D_x y(s;x)h$ is the mean-square derivative of $y(s;x)$ along the direction $h \in H$. Notice that, as we are not assuming B to have a bounded inverse, in order to give a meaning to Formula (12) we had preliminarily shown that $D_x y(s;x)h \in D(B^{-1})$, for any $s > 0$, and

$$\mathbf{E}\int_0^t |B^{-1}D_x y(s;x)h|_H^2\, ds < \infty.$$

3. The trace-class property of $D^2(P_t\varphi)(x)BB^\star$

As we have seen in the previous section, in [1] and in [3, Chapter 4] we have proved that for any $\varphi \in B_b(H)$ the function $P_t\varphi$ is in $C_b^2(H)$. Our aim here is to show that the operator $D^2(P_t\varphi)(x)BB^\star$ is in $\mathcal{L}_1(H)$, for any $t > 0$ and $x \in H$, and the mapping

$$(0, +\infty) \times H \to \mathbb{R}, \qquad (t, x) \mapsto \operatorname{Tr}\left[D^2(P_t\varphi)(x)BB^\star\right]$$

is continuous.

In [1, Proposition 3.4] and in [3, Theorem 4.2.4] we have proved that the solution $y(t;x)$ of system (3) is twice mean-square differentiable with respect to the initial datum $x \in H$. This means that, for any fixed $t \geq 0$ the mapping

$$H \to L^2(\Omega; H), \qquad x \mapsto y(t;x),$$

is twice differentiable. Notice that this result does not follow from the classical local inversion theorem (see e.g. [4, Theorem 9.4]), as the functional F is not Fréchet differentiable in H. We have also proved that the first and the second derivatives of $y(t;x)$ are solutions of suitable systems which one formally gets by differentiating the coefficients of system (3) respectively once and twice. Due to this fact, we have shown that there exists an increasing continuous function $c(t)$ such that

$$\sup_{x\in H} |D_x y(t;x)h|_H^2 \leq c(t)\,|h|_H^2, \qquad \sup_{x\in H} \left|D_x^2 y(t;x)(h,k)\right|_H^2 \leq c(t)\,|h|_H^2 |k|_H^2, \tag{13}$$

P-a.s. and

$$\sup_{x\in H}\int_0^t\left(\left|B^{-1}D_xy(s;x)h\right|_H^2+\left|B^{-1}D_x^2y(s;x)(h,k)\right|_H^2\right)ds$$
$$\leq c(t)\,t^{1-\epsilon}|h|_H^2\left(1+|k|_H^2\right)$$

P-a.s., for any $h,k\in H$ and $t\geq 0$.

In [1, Lemma 3.3] and in [3, Lemma 4.2.3] we have also proved that, for any $x,h\in H$ and $t>0$, the derivative $D_xy(t;x)h$ is in $L^p(\mathcal{O};\mathbb{R}^n)$, for each $p\geq 2$, and

$$\sup_{x\in H}\sup_{t\in(0,T]}t^{\frac{d(p-2)}{4p}}\left|D_xy(t;x)h\right|_p\leq\mu_p(T)|h|_H,\qquad \mathbf{P}-\text{a.s.}\tag{14}$$

where $\mu_p(T)$ is a continuous increasing function.

In the sequel, in order to prove that the operator $D^2(P_t\varphi)(x)BB^\star$ is in $\mathcal{L}_1(H)$, for any $t>0$ and $x\in H$, in addition to Hypotheses 2.1 and 2.2 we shall assume the following conditions.

Hypothesis 3.1. *There exists a complete orthonormal basis $\{e_k\}$ in H such that*

$$\sum_{k=1}^{+\infty}\left|e^{tA}BB^\star e_k\right|_H\leq c\,t^{-\beta},\qquad t\in(0,1],\tag{15}$$

for some

$$\beta<1\wedge\left(2-\frac{d+2\epsilon}{4}\right).\tag{16}$$

Moreover, there exists some $\gamma<1-\epsilon$ such that

$$\|e^{tA}BB^\star\|_{\mathcal{L}_2(H)}^2\leq c\,t^{-\gamma},\qquad t\in(0,1].\tag{17}$$

Remark 3.2.

1. It is immediate to check that Condition (17) implies Condition (8).
2. Assume that $Ae_k=-\alpha_ke_k$, where $\{e_k\}$ is a complete orthonormal system in H and

$$\alpha_k\asymp k^{2/d},\qquad k\in\mathbb{N}.\tag{18}$$

If we define $Be_k=\alpha_k^{-\epsilon/2}e_k$, for some $\epsilon<1$ as in Hypothesis 2.2, we have

$$\sum_{k=1}^{+\infty}\left|e^{tA}BB^\star e_k\right|_H=\sum_{k=1}^{+\infty}e^{-t\,\alpha_k}\alpha_k^{-\epsilon}.$$

By easy calculations it is possible to show that for any $\beta>0$ and $k\in\mathbb{N}$

$$e^{-t\,\alpha_k}\alpha_k^{-\epsilon}\leq\left(\frac{\beta}{\epsilon}\right)^\beta\alpha_k^{-(\epsilon+\beta)}t^{-\beta},\qquad t>0.$$

Hence we have

$$\sum_{k=1}^{+\infty}\left|e^{tA}BB^\star e_k\right|_H\leq c_\beta\,t^{-\beta}\sum_{k=1}^{+\infty}\alpha_k^{-(\epsilon+\beta)}$$

and, since we are assuming (18), we can conclude that (15) is fulfilled for any

$$\beta > \frac{d}{2} - \epsilon.$$

Notice that the condition above is compatible with (16). Actually it suffices that there exists some $\epsilon < 1$ such that

$$\frac{d}{2} - \epsilon < 1 \quad \text{and} \quad \frac{d}{2} - \epsilon < 2 - \frac{d+2\epsilon}{4},$$

which is verified for

$$\epsilon > \left(\frac{3}{2}d - 4\right) \vee \left(\frac{d}{2} - 1\right).$$

Under the same hypotheses for A and B, the condition (17) reads

$$\sum_{k=1}^{+\infty} e^{-2\,t\alpha_k} \alpha_k^{-2\epsilon} \leq c_\gamma\, t^{-\gamma}$$

and with the same arguments used above it is possible to show that it is verified for

$$2\,\epsilon + \gamma > \frac{d}{2}.$$

Such a condition is compatible with $\gamma < 1 - \epsilon$, once one takes $\epsilon > d/2 - 1$.

Now, before proving the main result of this section, we give some preliminary results.

Lemma 3.3. *Under Hypotheses 2.1, 2.2 and 3.1, if $\{e_k\}$ is the basis of H introduced in Hypothesis 3.1, there exists a suitable continuous increasing function $c(t)$ such that*

$$\sup_{x \in H} \sum_{k=1}^{+\infty} \int_0^t \left|B^{-1} D_x y(s;x) BB^\star e_k\right|_H^2 ds \leq c(t)\, t^{1-\epsilon-\gamma}, \qquad \mathbf{P} - a.s. \tag{19}$$

Moreover the series converges uniformly with respect to $x \in H$.

Proof. We set $v_k(t) = D_x y(t;x) BB^\star e_k$. As seen in [1] and [3, Chapter 4], the process $v_k(t)$ is the unique mild solution of the problem

$$\frac{d}{dt} v(t) = Av(t) + DF(y(t;x))v(t), \qquad v(0) = BB^\star e_k,$$

that is

$$v_k(t) = e^{tA} BB^\star e_k + \int_0^t e^{(t-s)A} DF(y(s;x)) v_k(s)\, ds. \tag{20}$$

Since $\partial/\partial\sigma f(\xi,\cdot)$ is bounded, uniformly with respect to $\xi \in \mathcal{O}$, we have

$$|DF(y(s;x))v_k(s)|_H \leq c\,|v_k(s)|_H, \tag{21}$$

so that

$$|v_k(t)|_H \leq \left|e^{tA} BB^\star e_k\right|_H + c \int_0^t |v_k(s)|_H\, ds.$$

Thanks to (15), this implies that for any $n \in \mathbb{N}$

$$\sum_{k=1}^{n} |v_k(t)|_H \leq c\,t^{-\beta} + c \int_0^t \sum_{k=1}^{n} |v_k(s)|_H \, ds.$$

It is easy to check that the integral equation

$$\alpha(t) = c\,t^{-\beta} + c \int_0^t \alpha(s)\, ds$$

admits a unique continuous solution $\alpha : (0, +\infty) \to (0, +\infty)$ such that

$$\sup_{t \in (0,1]} t^{\beta} \alpha(t) < \infty.$$

Then by comparison we have

$$\sum_{k=1}^{n} |v_k(t)|_H \leq \alpha(t), \qquad \mathbf{P}-\text{a.s.}$$

By taking the limit as n goes to infinity, we get

$$\sup_{x \in H} \sum_{k=1}^{+\infty} |D_x y(t;x) B B^\star e_k|_H \leq c(t)\, t^{-\beta}, \qquad \mathbf{P}-\text{a.s.}, \tag{22}$$

for some continuous increasing function $c(t)$. Notice that the convergence of the series is uniform with respect to $x \in H$ as the integral equation verified by $\alpha(t)$ is independent of $x \in H$.

Now, recalling (6) and Condition (9), due to (20) we have that $v_k(t) \in D(B^{-1})$, for any $t > 0$, and

$$\begin{aligned}
\left|B^{-1} v_k(t)\right|_H &\leq \left|B^{-1} e^{tA} B B^\star e_k\right|_H + \int_0^t \left|B^{-1} e^{(t-s)A} DF(y(s;x)) v_k(s)\right|_H ds \\
&\leq c\,t^{-\epsilon/2} \left|e^{\frac{t}{2}A} B B^\star e_k\right|_H + c \int_0^t (t-s)^{-\epsilon/2} \left|DF(y(s;x)) v_k(s)\right|_H ds.
\end{aligned}$$

Thanks to (21), this yields

$$\left|B^{-1} v_k(t)\right|_H^2 \leq c\,t^{-\epsilon} \left|e^{\frac{t}{2}A} B B^\star e_k\right|_H^2 + c\,t \int_0^t (t-s)^{-\epsilon} |v_k(s)|_H^2 \, ds$$

and then, by using Condition (17) and (22), for any $n \in \mathbb{N}$ we have

$$\sum_{k=1}^{n} \left|B^{-1} v_k(t)\right|_H^2 \leq c\,t^{-(\epsilon+\gamma)} + c(t) \int_0^t (t-s)^{-\epsilon} s^{-\beta} \, ds.$$

Hence, as $\epsilon + \gamma < 1$ and $\beta, \epsilon < 1$, from the arbitrariness of n we can conclude that

$$\int_0^t \sum_{k=1}^{+\infty} \left|B^{-1} v_k(s)\right|_H^2 ds \leq c(t)\, t^{1-\epsilon-\gamma}, \qquad \mathbf{P}-\text{a.s.}$$

for some continuous increasing function $c(t)$. □

Now we prove an analogous result for the second order derivative.

Lemma 3.4. *Let $N \in \mathcal{L}(H)$ and let $\{e_k\}$ be the basis of H introduced in the Hypothesis 3.1. Then, under Hypotheses 2.1, 2.2 and 3.1, for any $t \geq 0$ we have*

$$\sup_{x\in H} \sum_{k=1}^{+\infty} \left|B^{-1}D_x^2 y(t;x)(Ne_k, BB^\star e_k)\right|_H \leq \|N\|\, c(t), \qquad \mathbf{P}-a.s. \tag{23}$$

for some continuous increasing function $c(t)$ independent of N. Moreover the convergence of the series is uniform with respect to $x \in H$.

Proof. We set $u_k(t) = D_x^2 y(t;x)(Ne_k, BB^\star e_k)$. In [1] and in [3, Chapter 4] we have proved that $u_k(t)$ is the unique solution of the problem

$$\begin{aligned} u(t) = &\int_0^t e^{(t-s)A} DF(y(t;x))u(s)\, ds \\ &+ \int_0^t e^{(t-s)A} D^2F(y(s;x))\left(D_x y(s;x)Ne_k, D_x y(s;x)BB^\star e_k\right) ds. \end{aligned}$$

As $\partial^2/\partial\sigma^2 f(\xi,\cdot)$ is bounded, uniformly with respect $\xi \in \mathcal{O}$, by using (7) and the Hölder inequality, for any $p \in (1,2)$ we have

$$\begin{aligned} &\left|e^{(t-s)A} D^2F(y(s;x))\left(D_x y(s;x)Ne_k, D_x y(s;x)BB^\star e_k\right)\right|_H \\ &\leq c\,(t-s)^{-\frac{d(2-p)}{4p}} \left|D_x y(s;x)Ne_k\right|_{\frac{2p}{2-p}} \left|D_x y(s;x)BB^\star e_k\right|_H . \end{aligned} \tag{24}$$

Thanks to (14) and to the boundedness of $DF(x)$, this implies that

$$\begin{aligned} &|u_k(t)|_H \leq c \int_0^t |u_k(s)|_H\, ds \\ &+c(t)\, \mu_{\frac{2p}{2-p}}(t)\|N\| \int_0^t (t-s)^{-\frac{d(2-p)}{4p}} s^{-\frac{d(p-1)}{2p}} \left|D_x y(s;x)BB^\star e_k\right|_H ds. \end{aligned}$$

Therefore, by using (22), for any $n \in \mathbb{N}$ we have

$$\sum_{k=1}^n |u_k(t)|_H \leq c \int_0^t \sum_{k=1}^n |u_k(s)|_H\, ds + c(t)\|N\| \int_0^t (t-s)^{-\frac{d(2-p)}{4p}} s^{-\left(\beta+\frac{d(p-1)}{2p}\right)} ds.$$

As p can be chosen arbitrarily close to 1 and $\beta < 1$, we can apply the Gronwall Lemma and we can conclude that there exists a continuous increasing function $c(t)$, which is independent of $x \in H$ and N, such that for any $t \geq 0$

$$\sum_{k=1}^n |u_k(t)|_H \leq \|N\|\, c(t), \qquad \mathbf{P}-\text{a.s.}$$

Moreover, as the estimate is independent of $n \in N$, we conclude that

$$\sum_{k=1}^{+\infty} |D_x^2 y(t;x)\,(Ne_k, BB^\star e_k)\,|_H \leq \|N\|\, c(t), \qquad \mathbf{P}-\text{a.s.} \tag{25}$$

and the convergence of the series is uniform with respect to $x \in H$.

Now, due to (6) and (9), it is easy to check that $u_k(t) \in D(B^{-1})$, for any $t \geq 0$, and

$$\left|B^{-1}u_k(t)\right|_H \leq c\int_0^t (t-s)^{-\epsilon/2}\left|e^{\frac{t-s}{2}A}DF(y(s;x))u_k(s)\right|_H ds$$
$$+c\int_0^t (t-s)^{-\epsilon/2}\left|e^{\frac{t-s}{2}A}D^2F(y(s;x))\left(D_xy(s;x)Ne_k, D_xy(s;x)BB^\star e_k\right)\right|_H ds.$$

If we fix p such that

$$\frac{1}{p} < \frac{4+d-2\epsilon}{2d}, \tag{26}$$

we have $\epsilon/2 + d(2-p)/(4p) < 1$ and then, by using (14) and (24), it follows

$$\left|B^{-1}u_k(t)\right|_H \leq c\int_0^t (t-s)^{-\epsilon/2}|u_k(s)|_H\, ds$$
$$+c\,\mu_{\frac{2p}{2-p}}(t)\|N\|\int_0^t (t-s)^{-\left(\frac{\epsilon}{2}+\frac{d(2-p)}{4p}\right)}s^{-\frac{d(p-1)}{2p}}\,|D_xy(s;x)BB^\star e_k|_H\, ds$$

Since we are assuming that β fulfills (16), it is easy to check that there exists some $p \in (1,2)$ such that (26) is verified and

$$\beta + \frac{d(p-1)}{2p} < 1.$$

Therefore, thanks to (22) and (25), for any $n \in \mathbb{N}$ we get

$$\sum_{k=1}^n \left|B^{-1}u_k(t)\right|_H \leq c(t)\|N\|\int_0^t (t-s)^{-\epsilon/2}\, ds$$
$$+c(t)\,\mu_{\frac{2p}{2-p}}(t)\|N\|\int_0^t (t-s)^{-\left(\frac{\epsilon}{2}+\frac{d(2-p)}{4p}\right)}s^{-\left(\frac{d(p-1)}{2p}+\beta\right)}\, ds,$$

which implies (23) due to the arbitrariness of $n \in \mathbb{N}$. □

Theorem 3.5. *Under Hypotheses 2.1, 2.2 and 3.1, the operator $D^2(P_t\varphi)(x)BB^\star$ is in $\mathcal{L}_1(H)$, for any $\varphi \in B_b(H)$ and for any $t > 0$ and $x \in H$. Moreover, the mapping*

$$(0,+\infty)\times H \to \mathbb{R}, \qquad (t,x) \mapsto \mathrm{Tr}\left[D^2(P_t\varphi)(x)BB^\star\right],$$

is continuous and

$$\sup_{x\in H}\left|\mathrm{Tr}\left[D^2(P_t\varphi)(x)BB^\star\right]\right| \leq c\,(t\wedge 1)^{-\left(1+\epsilon+\frac{\gamma}{2}\right)}\|\varphi\|_0, \qquad t>0. \tag{27}$$

Proof. In Dunford-Schwartz [5] it is proved that if T is in $\mathcal{L}(H)$ and there exists a constant K such that for any finite rank operator N

$$|\mathrm{Tr}\,[NT]| \leq K\,\|N\|,$$

then T is in $\mathcal{L}_1(H)$ and

$$|\mathrm{Tr}\,T| \leq K.$$

Thus, if we show that for any finite rank operator N

$$\sup_{x\in H} \left|\mathrm{Tr}\ \left[D^2(P_t\varphi)(x)BB^\star N\right]\right| \leq c(t)\|N\|\, t^{-\left(1+\epsilon+\frac{\gamma}{2}\right)}\|\varphi\|_0, \qquad t>0,$$

we are done. Due to (12), if $\{e_k\}$ is the complete orthonormal system introduced in Hypothesis 3.1, for any $\varphi \in C_b(H)$ we have

$$\langle D(P_t\varphi)(x), BB^\star e_k\rangle_H = \frac{1}{t}\mathbf{E}\,\varphi(y(t;x)) \int_0^t \left\langle B^{-1}D_x y(s;x)BB^\star e_k, dw(s)\right\rangle_H.$$

Thus, if $\varphi \in C_b^1(H)$, we can differentiate each side with respect to $x \in H$ along the direction $N^\star e_k$ and we get

$$\begin{aligned}
&\left\langle D^2(P_t\varphi)(x)BB^\star e_k, N^\star e_k\right\rangle_H \\
&= \frac{1}{t}\mathbf{E}\ \langle D\varphi(y(t;x)), D_x y(t;x)N^\star e_k\rangle_H \int_0^t \left\langle B^{-1}D_x y(s;x)BB^\star e_k, dw(s)\right\rangle_H \\
&+\frac{1}{t}\mathbf{E}\,\varphi(y(t;x)) \int_0^t \left\langle B^{-1}D_x^2 y(s;x)\left(BB^\star e_k, N^\star e_k\right), dw(s)\right\rangle_H.
\end{aligned}$$

This implies that for any $n \in \mathbb{N}$

$$\begin{aligned}
&\sum_{k=1}^n \left\langle D^2(P_t\varphi)(x)BB^\star e_k, N^\star e_k\right\rangle_H \\
&= \frac{1}{t}\mathbf{E}\left\langle D\varphi(y(t;x)), D_x y(t;x)N^\star\left(\sum_{k=1}^n \int_0^t \left\langle B^{-1}D_x y(s;x)BB^\star e_k, dw(s)\right\rangle_H e_k\right)\right\rangle_H \\
&+\frac{1}{t}\mathbf{E}\,\varphi(y(t;x)) \int_0^t \sum_{k=1}^n \left\langle B^{-1}D_x^2 y(s;x)\left(BB^\star e_k, N^\star e_k\right), dw(s)\right\rangle_H.
\end{aligned}$$

By using the first inequality in (13) we have

$$\begin{aligned}
&\left|\sum_{k=1}^n \left\langle D^2(P_t\varphi)(x)BB^\star e_k, N^\star e_k\right\rangle_H\right| \\
&\quad\leq \frac{1}{t}c(t)\,\|\varphi\|_1\|N^\star\|\left(\mathbf{E}\left|\sum_{k=1}^n \int_0^t \left\langle B^{-1}D_x y(s;x)BB^\star e_k, dw(s)\right\rangle_H e_k\right|_H^2\right)^{1/2} \\
&\quad+\frac{1}{t}\|\varphi\|_0\left(\mathbf{E}\left|\int_0^t \sum_{k=1}^n \left\langle B^{-1}D_x^2 y(s;x)\left(BB^\star e_k, N^\star e_k\right), dw(s)\right\rangle_H\right|^2\right)^{1/2} \\
&\quad= J_1(t) + J_2(t).
\end{aligned}$$

It is immediate to check that

$$\mathbf{E}\left|\sum_{k=1}^{n}\int_0^t \left\langle B^{-1}D_xy(s;x)BB^\star e_k, dw(s)\right\rangle_H e_k\right|_H^2$$
$$= \mathbf{E}\sum_{k=1}^{n}\int_0^t \left|B^{-1}D_xy(s;x)BB^\star e_k\right|_H^2\,ds$$

and then, due to (19), we have

$$J_1(t) = c(t)\,\|\varphi\|_1\|N^\star\|\,t^{-\frac{(1+\epsilon+\gamma)}{2}}, \qquad t>0.$$

Moreover, thanks to (23)

$$J_2(t) \leq \frac{1}{t}\|\varphi\|_0\left(\mathbf{E}\int_0^t\left|\sum_{k=1}^{n}B^{-1}D_x^2y(s;x)\,(BB^\star e_k, N^\star e_k)\right|_H^2 ds\right)^{1/2}$$
$$\leq c(t)\,\|\varphi\|_0\|N^\star\|\,t^{-1/2}.$$

Therefore, as $n\in\mathbb{N}$ is arbitrary and N is a finite rank operator, it follows that for any $\varphi\in C_b^1(H)$ the operator $D^2(P_t\varphi)(x)BB^\star$ is in $\mathcal{L}_1(H)$ and

$$\sup_{x\in H}\left|\mathrm{Tr}\left[D^2(P_t\varphi)(x)BB^\star\right]\right| \leq c(t)\,\|\varphi\|_1\,t^{-\frac{(1+\epsilon+\gamma)}{2}}, \qquad t>0.$$

Moreover, due to Lemmata 3.3 and 3.4 the convergence of the series is uniform with respect to $x\in H$ and then the mapping

$$(0,+\infty)\times H\to\mathbb{R}, \qquad (t,x)\mapsto \mathrm{Tr}\left[D^2(P_t\varphi)(x)BB^\star\right]$$

is continuous. Notice that if $t>1$ we have $P_t\varphi = P_1(P_{t-1}\varphi)$ so that, by using (11) with $j=i=1$, we conclude

$$\left|\mathrm{Tr}\left[D^2(P_t\varphi)(x)BB^\star\right]\right| \leq c\,\|\varphi\|_1\,(t\wedge 1)^{-\frac{(1+\epsilon+\gamma)}{2}}, \qquad t>0. \tag{28}$$

Finally, let $\varphi\in B_b(H)$. By the semigroup law and (11), with $j=1$ and $i=0$, we have

$$\left|\mathrm{Tr}\left[D^2(P_t\varphi)(x)BB^\star\right]\right| \leq c\,\|P_{t/2}\varphi\|_1\left(\frac{t}{2}\wedge 1\right)^{-\frac{(1+\epsilon+\gamma)}{2}}$$
$$\leq c(t)\,\|\varphi\|_0\,(t\wedge 1)^{-\left(1+\epsilon+\frac{\gamma}{2}\right)},$$

and this implies (27). □

4. The Kolmogorov equations

We use here the trace-class property of the operator $D^2(P_t\varphi)(x)BB^\star$ proved in the previous section, in order to show that the parabolic and the elliptic problems associated to the differential operator $\mathcal{L}(x,D)$ have *regular* solutions.

4.1. The parabolic problem

First of all we give the definitions of *classical* and *strict* solutions for the parabolic problem

$$\begin{cases} \dfrac{\partial u}{\partial t}(t,x) = \dfrac{1}{2}\mathrm{Tr}\,\left[D^2u(t,x)BB^\star\right] + \langle Ax + F(x), Du(t,x)\rangle_H\,, \; x \in D(A),\; t > 0 \\ u(0,x) = \varphi(x), \qquad x \in H. \end{cases} \tag{29}$$

Definition 4.1. *A continuous function $u : [0, +\infty) \times H \to \mathbb{R}$ is a* classical *solution for Problem* (29) *if*

1. *for any $t > 0$ the function $u(t,\cdot)$ is in $C_b^2(H)$ and for any $x \in H$ the operator $D^2u(t,x)BB^\star$ is in $\mathcal{L}_1(H)$. Moreover the function*
$$(0, +\infty) \times H \to \mathbb{R}, \qquad (t,x) \mapsto \mathrm{Tr}\,\left[D^2u(t,x)BB^\star\right]$$
is continuous;
2. *for any $x \in D(A)$ the function $u(\cdot, x)$ is differentiable in $(0, +\infty)$;*
3. *the function u fulfills Equation (29).*

Definition 4.2. *A continuous function $u : [0, +\infty) \times H \to \mathbb{R}$ is a* strict *solution for Problem* (29) *if it satisfies conditions 1, 2 and 3 of Definition 4.1, with $t > 0$ and $(0, +\infty)$ replaced respectively by $t \geq 0$ and $[0, +\infty)$.*

Now we can give the existence and uniqueness theorem in the parabolic case.

Theorem 4.3. *Assume Hypotheses 2.1, 2.2 and 3.1.*

1. *If $\varphi \in C_b^2(H)$, the operator $D^2\varphi(x)BB^\star$ is in $\mathcal{L}_1(H)$ for any $x \in H$ and the mapping $x \mapsto \mathrm{Tr}\,[D^2\varphi(x)BB^\star]$ is bounded and continuous, then the function*
$$u(t,x) = P_t\varphi(x)$$
is the unique strict solution of Problem (29).
2. *If $\varphi \in C_b(H)$, then u is the unique classical solution.*

Proof. Due to the Theorem 3.5, we can proceed as in [2] (proofs of Proposition A-1 and Theorem 3.6) and as in [3] (proofs of Proposition 5.2.8 and Theorem 5.2.9). □

4.2. The elliptic problem

Now, we are concerned with the stationary equation

$$\lambda\psi(x) - \frac{1}{2}\mathrm{Tr}\,\left[D^2\psi(x)BB^\star\right] - \langle Ax + F(x), D\psi(x)\rangle_H = \varphi(x), \qquad x \in D(A), \tag{30}$$

with $\lambda > 0$.

Definition 4.4.

1. *A function $\psi : H \to \mathbb{R}$ is a* strict *solution of Problem* (30) *if $\psi \in C_b^2(H)$, the operator $D^2\psi(x)BB^\star$ is in $\mathcal{L}_1(H)$, for any $x \in H$, and the mapping $x \mapsto \mathrm{Tr}\,[D^2\psi(x)BB^\star]$ is bounded and continuous. Moreover ψ fulfills Equation* (30).

2. *A function* $\psi : H \to \mathbb{R}$ *is a* strong *solution of Problem* (30) *if there exist two sequences* $\{\varphi_n\}$ *and* $\{\psi_n\}$, *converging respectively to* φ *and* ψ *in* $C_b(H)$, *such that* ψ_n *is the unique strict solution of the problem*

$$\lambda\psi(x) - \frac{1}{2}\mathrm{Tr}\,\left[D^2\psi(x)BB^\star\right] - \langle Ax + F(x), D\psi(x)\rangle_H = \varphi_n(x), \quad x \in D(A),$$

for any $n \in \mathbb{N}$.

Theorem 4.5. *Assume Hypotheses* 2.1, 2.2 *and* 3.1 *and fix* $\lambda > 0$.

1. *If* $\varphi \in C_b^1(H)$, *then there exists a unique strict solution for Problem (30) which is given by*

$$\psi(x) = \int_0^{+\infty} e^{-\lambda t} P_t\varphi(x)\,dt, \qquad x \in H. \tag{31}$$

2. *If* $\varphi \in C_b(H)$, *the function* ψ *defined above is the unique strong solution of* (30).

Proof. By proceeding as in [2, Theorem 4.3] and as in [3, Theorem 5.3.3], by using (10) and (11) it is possible to show that if $\varphi \in C_b^1(H)$, then the function ψ defined by (31) belongs to $C_b^2(H)$.

Moreover $D^2\psi(x)BB^\star \in \mathcal{L}_1(H)$, for any $x \in H$. Indeed, due to (28), since we are assuming $\gamma + \epsilon < 1$, for any finite rank operator N we have

$$\left|\mathrm{Tr}\,\left[D^2\psi(x)BB^\star N\right]\right| \leq c\,\|N\|\,\|\varphi\|_1 \int_0^{+\infty} e^{-\lambda t}(t\wedge 1)^{-\frac{1+\gamma+\epsilon}{2}}\,dt \leq c\,\|N\|\,\|\varphi\|_1\,\lambda^{\frac{\gamma+\epsilon-1}{2}}.$$

Then, by using the Dunford-Schwartz lemma recalled in the proof of Theorem 3.5, we conclude that $D^2\psi(x)BB^\star \in \mathcal{L}_1(H)$ and

$$\sup_{x\in N}\left|\mathrm{Tr}\,\left[D^2\psi(x)BB^\star\right]\right| \leq c\,\|\varphi\|_1\,\lambda^{\frac{\gamma+\epsilon-1}{2}}.$$

By proceeding as in the proof of the Theorem 4.3 of [2] and as in proof of the Theorem 5.3.3 of [3], this allows us to conclude that if $\varphi \in C_b^1(H)$ then ψ is the unique strict solution of (30) and if $\varphi \in C_b(H)$ then ψ is the unique strong solution of (30). □

References

[1] S. Cerrai, *Differentiability with respect to initial datum for solutions of SPDE'S with no Fréchet differentiable drift term*, Communications in Applied Analysis, **2** (1998), 249–270.

[2] S. Cerrai, *Kolmogorov equations in Hilbert spaces with non smooth coefficients*, Communications in Applied Analysis, **2** (1998), 271–297.

[3] S. Cerrai, *Second Order PDE's in Finite and Infinite Dimensions. A probabilistic Approach*, Lecture Notes in Mathematics, Springer Verlag, to appear.

[4] G. Da Prato and J. Zabczyk, *Stochastic Equations in Infinite Dimensions*, Cambridge University Press, Cambridge, **1992**.

[5] N. Dunford and J.T. Schwartz, *Linear Operators*, vol. II, Interscience, New York, London, **1963**.

[6] E.B. Davies, *Heat Kernels and Spectral Theory*, Cambridge University Press, Cambridge, **1989**.

[7] A. Lunardi, *Analytic Semigroups and Optimal Regularity in Parabolic Problems*, Birkhäuser Verlag, Basel, **1995**.

[8] S. Peszat and J. Zabczyk, *Strong Feller property and irreducibility for diffusion processes on Hilbert spaces*, Annals of Probability, **23** (1995), 157–172.

Dipartimento di Matematica per le Decisioni
Università di Firenze
Via C. Lombroso 6/17
I-50134 Firenze
Italy

Progress in Probability, Vol. 52, 73–88

Monotone Gradient Systems in L^2 Spaces

G. Da Prato

Abstract. We prove maximal dissipativity of some gradient systems having a convex potential.

1. Introduction and setting of the problem

We are here concerned with the following operator in a separable Hilbert space H.

$$N_0\varphi = \frac{1}{2}\,\mathrm{Tr}\,[D^2\varphi] - \langle x, AD\varphi\rangle - \langle \partial_0 U(x), D\varphi\rangle,\ \ \varphi \in \mathcal{E}_A(H), \tag{1}$$

where $A : D(A) \subset H \to H$ is self-adjoint, strictly negative and such that A^{-1} is of trace class, and $U : H \to (-\infty, +\infty]$ is a convex lower semicontinuous proper function[1]. Moreover $\partial U(x)$ is the subdifferential of U at x :

$$\partial U(x) = \{y \in H :\ U(x+h) - U(x) \geq \langle y, h\rangle,\ \forall\, h \in H\},$$

and $\partial_0 U(x)$ is the element of minimal norm of $\partial U(x)$. Finally $\mathcal{E}_A(H)$ is the linear span of all exponential functions $e^{i\langle h,x\rangle}$, $x \in H$, with $h \in D(A)$.

Our goal is to show that N_0 is essentially self-adjoint in $L^2(H, \nu)$ where ν is the "Gibbs" measure $\nu(dx) = Z^{-1}\mu(dx)$, μ is the gaussian measure with mean 0 and covariance operator $Q = -\frac{1}{2}\,A^{-1}$, and Z is a normalization constant.

The semigroup P_t generated by the closure of N_0 is related to the solution of the differential stochastic inclusion

$$dX \in (AX - \partial U(X))dt + dW(t),\ \ X(0) = x \in H. \tag{2}$$

In fact when Equation (2) has a solution, we have a natural candidate for the semigroup (P_t) given by $P_t\varphi(x) = \mathbb{E}\,[\varphi(X(t,x))]$ for all φ Borel and bounded. For conditions that assure existence and uniqueness of a solution of (2) see [5, 6, 8, 17].

The problem of self-adjointness of the operator N_0 has been extensively studied. We recall the approach by the Dirichlet forms theory, see [2, 18] and references therein, and by the parabolic criterion of Berezanskii, [3, 19].

Another approach was introduced in [11], see also [13]. It consists in proving the density of $(\lambda - N_0)(\mathcal{E}_A(H))$ for any $\lambda > 0$, through a suitable approximation of

2000 *Mathematics Subject Classification.* 47D07, 35K90.
Key words and phrases. Gradient systems, maximal dissipativity operators, stochastic equations.
[1] That is not identically equal to $+\infty$.

U by smooth convex functions. In fact this imply, by the Lumer-Phillips theorem, see [20], that the closure N of N_0 is self-adjoint.

In particular self-adjointness of N_0 when U is convex and regular was proved in [11]. However the regularity of U is a too strong assumption, it does not allow applications to reaction-diffusion equations and to dissipative equations with multivalued operators, see Examples 1.2, 1.3 below.

In the present paper we prove self-adjointness of N_0, for a general lower semicontinuous convex function, see Hypothesis 1.1 below. This result seems not to be covered by previous papers. Moreover we pay much attention to the study of the domain of N. We show that it is always included in $W^{2,2}(H,\nu)$, and in some case we are able to characterize $D(N)$ as a suitable proper subset of $W^{2,2}(H,\nu)$. Finally we show that the semigroup $P_t = e^{tN}$ is strong Feller.

In this paper no attempt is done to apply the infinite dimensional estension of the Fukushima theory considered in [1, 2, 21], to construct a diffusion process $X(t,x)$ on H, up to sets of capacity zero, such that $P_t\varphi(x) = \mathbb{E}[\varphi(X(t,x))]$ for any bounded and Borel function φ. The fact that P_t is strong Feller could be useful to show that $X(t,x)$ is in fact defined in all H, [15].

We shall assume:

Hypothesis 1.1.

(i) $A : D(A) \subset H \to H$ *is self-adjoint, and there exists* $\omega > 0$ *such that* $A \le -\omega$.

(ii) $Q := -\dfrac{1}{2}\,A^{-1}$ *is of trace class.*

(iii) $U : H \to (-\infty, +\infty]$ *is convex, lower semicontinuous and proper.*

(iv) $\mu(\text{dom }(U)) = \mu(\text{dom }(\partial U)) = 1$ ([2]).

(v) *There exists* $\varepsilon > 0$ *such that*

$$\int_H |\partial_0 U(x)|^{2+2\varepsilon} e^{-\lambda U(x)}\mu(dx) < +\infty,\ \forall\, \lambda > 0.$$

Let us define the measure ν. We set $\nu(dx) = \rho(x)\mu(dx)$, where

$$\rho(x) = Z^{-1}e^{-2U(x)},\ x \in H,$$

and

$$Z = \int_H e^{-2U(y)}\mu(dy).$$

Since U is convex there exist $c_1 \in \mathbb{R}$ and $x_0 \in H$ such that:

$$U_\alpha(x) \ge c_1 + \langle x, x_0\rangle,\ x \in H,\ \alpha > 0. \tag{3}$$

Consequently Z is finite. Moreover, by Hypothesis 1.1–(iv), we have $Z > 0$ and

$$\nu(\text{dom }(U)) = \nu(\text{dom }(\partial U)) = 1,$$

since $\nu << \mu$.

---

[2] dom (U) is the set of all $x \in H$ such that $U(x)$ is finite, and dom $(\partial_0 U)$ is the set of all $x \in H$ such that $\partial U(x)$ is not empty.

Example 1.2 (Reaction-diffusion equations). Let $H = L^2(0,1)$, and let A be the self-adjoint operator in H defined by

$$Ax = D_\xi^2 x, \quad x \in D(A) = H^2(0,1) \cap H_0^1(0,1),$$

$m \in \mathbb{N}$, and

$$U(x) = \begin{cases} \dfrac{1}{2m} \displaystyle\int_0^1 |x(\xi)|^{2m} d\xi \text{ if } x \in L^{2m}(0,1), \\ +\infty \text{ otherwise.} \end{cases}$$

Then we have

$$\partial_0 U(x) = x^{2m-1}, \quad \text{dom } (\partial U) = L^{4m-2}(0,1).$$

Let us check Hypothesis 1.1. The spectrum $\sigma(A)$ of A is given by

$$\sigma(A) = \left\{-\pi^2 n^2 : \; n \in \mathbb{N}\right\}.$$

Moreover setting $e_k(\xi) = \sqrt{\frac{2}{\pi}} \sin k\xi$, (e_k) is a complete orthonormal system on $L^2(0,1)$. Thus (i) is fulfilled with $\omega = \pi^2$. (ii) and (iii) are clear. Finally (iv) and (v) (with any ε) follow from the well-known inequality

$$\int_H |x|^{2n}_{L^{2n}(0,1)} \mu(dx) \le C_n,$$

valid, with a suitable constant C_n, for any $n \in \mathbb{N}$.

Example 1.3 (Multi-valued operators). Let again $H = L^2(0,1)$, and let A be as in the previous example. Let $\beta : \mathbb{R} \to (-\infty, +\infty]$ be a lower semicontinuous proper convex function such that its subdifferential $\partial\beta(\xi)$ is not empty for any $\xi \in \mathbb{R}$, and there exists $c > 0$ and $m \in \mathbb{N}$ such that $\partial\beta(\xi) \le c(1 + |\xi|^m)$. Let $U : L^2(0,1) \to (-\infty, +\infty]$ be defined by

$$U(x) = \begin{cases} \displaystyle\int_0^1 \beta(x(\xi)) d\xi \text{ if } \beta(x) \in L^1(0,1), \\ +\infty \text{ otherwise.} \end{cases}$$

Then one can show, see e.g. [4, Proposition 2.8], that U is convex and lower semicontinuous on $L^2(0,1)$. Moreover $y \in \partial U(x)$ if and only if $x(\xi) \in \partial\beta(\xi)$ a.e. on $[0,1]$, and

$$\overline{\text{dom } (U)} = \left\{x \in L^2(0,1) : \; x(\xi) \in \overline{\text{dom } (\beta)}\right\}.$$

Consequently $L^m(0,1) \subset \text{dom } \partial U$, and $\mu(\text{dom } (\partial U)) = 1$. Therefore Hypotheses 1 are fulfilled.

We end this section by reviewing some properties of the Ornstein-Uhlenbeck semigroup, and Yosida approximations of U, needed later.

Let us recall the definition of the Sobolev space $W^{1,2}(H, \mu)$. Since A^{-1} is of trace class, there exists an orthonormal basis (e_k) in H, and a sequence of positive numbers (α_k) such that

$$Ae_k = -\alpha_k e_k, \quad k \in \mathbb{N}.$$

We set $\lambda_k = \frac{1}{2\alpha_k}$, so that $Qe_k = \lambda_k e_k,\quad k \in \mathbb{N}$. For any $k \in \mathbb{N}$ we consider the derivative operator D_k in the direction e_k with domain $\mathcal{E}_A(H)$. It is well known, see e.g. [14], that D_k is closable in $L^2(H,\nu)$. Now we define $W^{1,2}(H,\mu)$ as the linear subspace of all functions $\varphi \in L^2(H,\mu)$ such that $D_k\varphi \in L^2(H,\mu)$ for all $k \in \mathbb{N}$ and

$$\sum_{k=1}^{\infty}\int_H |D_k\varphi|^2 d\mu < +\infty \quad \left(\text{resp. } \sum_{k=1}^{\infty}\int_H \alpha_k|D_k\varphi|^2 d\mu\right) < +\infty.$$

1.1. Ornstein-Uhlenbeck semigroup

Let us consider the semigroup (R_t) on $L^p(H,\mu)$, $p \geq 1$, given by

$$R_t\varphi(x) = \int_H \varphi(e^{tA}x + y)N_{Q_t}(dy),\quad \varphi \in L^p(H,\mu), \tag{4}$$

where N_{Q_t} is the gaussian measure of mean 0 and covariance operator Q_t given by

$$Q_t = \int_0^t e^{2sA}ds = Q(1 - e^{2tA}),\ t \geq 0.$$

$\mu = N_Q$ is an invariant measure for (R_t). We denote by L_p the infinitesimal generator of (R_t) on $L^p(H,\mu)$, and recall that $\mathcal{E}_A(H)$ is a core for L_p. Moreover

$$L_p\varphi = \frac{1}{2}\operatorname{Tr}[D^2\varphi] + \langle x, AD\varphi\rangle,\ \varphi \in \mathcal{E}_A(H).$$

Proposition 1.4. *(R_t) is strong Feller. Moreover the following statements hold.*

(i) *If $f \in L^p(H,\mu)$, and $t > 0$, we have $R_t f \in W^{1,p}(H,\mu)$, and*

$$\int_H |DR_t f|^p d\mu \leq (et)^{-p/2}\int_H |f|^p d\mu.$$

(ii) *If $f \in L^p(H,\mu)$, and $\lambda > 0$. Then $R(\lambda, L)f \in W^{1,p}(H,\mu)$ and*

$$\int_H |DR(\lambda,L)f|^p d\mu \leq (\pi/(e\lambda))^{-p/2}\int_H |f|^p d\mu.$$

The following identity is well known, see e.g. [14].

Proposition 1.5. *Let $\varphi \in D(L_2)$, $\psi \in W^{1,2}(H,\mu)$. Then*

$$\int_H (L\varphi)\psi d\mu = -\frac{1}{2}\int_H \langle D\varphi, D\psi\rangle d\mu. \tag{5}$$

We finish with a characterization of the domain of L_2, see [10]. In its formulation we use the following notation.

$$W_A^{1,2}(H,\mu) = \left\{\varphi \in L^2(H,\mu) : \int_H |(-A)^{1/2}D\varphi|^2 < +\infty\right\}.$$

Proposition 1.6. *We have*

$$D(L_2) = W^{2,2}(H,\mu) \cap W_A^{1,2}(H,\mu). \tag{6}$$

1.2. Yosida aproximations

We here introduce the Yosida approximations of U; they are defined by

$$U_\alpha(x) = \inf\left\{U(y) + \frac{|x-y|^2}{2\alpha},\ \ y \in H\right\},\ x \in H,\ \alpha > 0.$$

U_α is Fréchet differentiable and DU_α is Lipschitz continuous. Moreover the following properties are well known, see e.g. [9],

$$\begin{aligned} &(i)\ \ U_\alpha(x) \le U(x),\ \ x \in H,\ \alpha > 0.\\ &(ii)\ \ \lim_{\alpha\to 0} DU_\alpha(x) = \partial_0 U(x),\ x \in \text{ dom }(\partial U),\ \alpha > 0.\\ &(iii)\ |DU_\alpha(x)| \le |\partial_0 U(x)|,\ x \in \text{ dom }(\partial U),\ \alpha > 0. \end{aligned} \tag{7}$$

Lemma 1.7. *We have*

$$e^{-\lambda U}, e^{-\lambda U_\alpha} \in L^1(H,\mu),\ \forall\,\lambda > 0,\ \forall\,\alpha > 0.$$

Moreover

$$\lim_{\alpha\to 0} e^{-\lambda U_\alpha} = e^{-\lambda U} \text{ in } L^1(H,\mu),\ \forall\,\lambda > 0,$$

$$\lim_{\alpha\to 0}\int_H |\partial_0 U(x) - DU_\alpha(x)|^{2+2\varepsilon}\nu(dx) = 0.$$

Proof. Taking into account (3) we have

$$e^{-\lambda U_\alpha(x)} \le e^{\lambda(|c_1|+|x_0|\,|x|)},\ \ x \in H,\ \alpha \ge 0.$$

Now the conclusion follows from the dominated convergence theorem since

$$\int_H e^{\lambda(|c_1|+|x_0|\,|x|)}\mu(dx) < +\infty.$$

□

2. Symmetry of N_0

We start by introducing the Sobolev spaces in $W^{1,2}(H,\nu)$. We are going to show that D_k is closable in $L^2(H,\nu)$. For this we need a lemma and an integration by parts formula.

Lemma 2.1. *Assume that Hypothesis 1.1 holds. Then* $\rho \in W^{1,2}(H,\mu)$, $\log\rho \in W^{1,2}(H,\mu)$, *and*

$$D\log\rho(x) = -2\rho(x)\partial_0 U(x),\ \mu \text{ a.e,} \tag{8}$$

where $\rho(x) = Z^{-1}e^{-2U(x)},\ x \in H.$

Proof. We first notice that $\rho \in L^p(H,\mu)$ for any $p \geq 1$, since

$$\int_H \rho^p d\mu = Z^{-p} \int_H e^{-2pU} d\mu < +\infty,$$

by Lemma 1.7. Let us prove now that $\rho \in W^{1,2}(H,\mu)$. Set

$$\rho_\alpha(x) = \frac{1}{Z_\alpha}\, e^{-2U_\alpha(x)},\ x \in H,\ \alpha > 0,$$

where

$$Z_\alpha = \int_H e^{-2U_\alpha(y)}\mu(dy),\ \alpha > 0,$$

where U_α is defined by (7). Then by Lemma 1.7 we have that $Z_\alpha \to Z$, and $\rho_\alpha \to \rho$ in $L^p(H,\mu)$, as $\alpha \to 0$. We claim that $(D\rho_\alpha)$ is a Cauchy sequence in $L^2(H,\mu;H)$. We have in fact $D\rho_\alpha(x) = -2\rho_\alpha(x)DU_\alpha(x)$, and moreover, for $\alpha,\beta > 0$,

$$\begin{aligned}
&\int_H |D\rho_\alpha - D\rho_\beta|^2 d\mu = 4\int_H |\rho_\alpha DU_\alpha - \rho_\beta DU_\beta|^2 d\mu \\
&\leq 8\int_H |\rho_\alpha - \rho_\beta|^2 |DU_\alpha|^2 d\mu + 8\int_H |\rho_\beta|^2 |DU_\alpha - DU_\beta|^2 d\mu \\
&\leq 8\left(\int_H |DU_\alpha|^{2+2\varepsilon} d\mu\right)^{1/(1+\varepsilon)} \left(\int_H |\rho_\alpha - \rho_\beta|^{2+2/\varepsilon} d\mu\right)^{\varepsilon/(1+\varepsilon)} \\
&+8\left(\int_H |DU_\alpha - DU_\beta|^{2+2\varepsilon} d\mu\right)^{1/(1+\varepsilon)} \left(\int_H |\rho_\beta|^{2+2/\varepsilon} d\mu\right)^{\varepsilon/(1+\varepsilon)}.
\end{aligned}$$

Therefore $(D\rho_\alpha)$ is a Cauchy sequence in $L^2(H,\mu;H)$, and consequently $\rho \in W^{1,2}(H,\mu)$. Finally, since

$$D\log\rho_\alpha(x) = -2\rho_\alpha D_\alpha U(x),\ x \in H,$$

identity (8) follows. □

Lemma 2.2. *Assume that Hypothesis 1.1 holds, let $\varphi,\psi \in \mathcal{E}_A(H)$, and $k \in \mathbb{N}$. Then we have*

$$\int_H (D_k\varphi)\psi d\nu = -\int_H \varphi(D_k\psi)d\nu + \int_H \left(\frac{x_k}{\lambda_k} - D_k\log\rho\right)\varphi\psi d\nu. \tag{9}$$

Proof. Since $\rho \in W^{1,2}(H,\mu;H)$, we have $\psi\rho \in W^{1,2}(H,\mu;H)$, and

$$\int_H (D_k\varphi)\psi d\nu = \int_H (D_k\varphi)\psi\rho d\mu.$$

Now, using the well-known integration by parts formula for gaussian measures see e.g. [14],

$$\int_H (D_k\varphi)\psi\rho d\mu = -\int_H \varphi D_k(\psi\rho)d\mu + \frac{1}{\lambda_k}\int_H x_k\varphi\psi\rho d\mu,$$

we find

$$\int_H (D_k\varphi)\psi d\nu = -\int_H \varphi D_k(\psi\rho)d\mu + \frac{1}{\lambda_k}\int_H x_k\varphi\psi\rho d\mu,$$
$$= -\int_H \varphi D_k(\psi)d\nu - \int_H \varphi D_k(\log\rho)d\nu + \frac{1}{\lambda_k}\int_H x_k\varphi\psi d\nu.$$

□

Proposition 2.3. *For any $k \in \mathbb{N}$ the operator D_k is closable.*

We shall still denote by D_k its closure.

Proof. Let (φ_n) be a sequence in $\mathcal{E}_A(H)$, and $g \in L^2(H,\nu)$ such that

$$\varphi_n \to 0, \quad D_k\varphi_n \to g, \text{ in } L^2(H,\nu).$$

We have to show that $g = 0$. Let $\psi \in \mathcal{E}_A(H)$. Then by Lemma 2.2 we have

$$\int_H (D_k\varphi_n)\psi d\nu = -\int_H \varphi_n(D_k\psi)d\nu + \int_H \left(\frac{x_k}{\lambda_k} - D_k\log\rho\right)\varphi_n\psi d\nu. \tag{10}$$

But, since $\rho \in W^{1,2}(H,\mu)$ and $\int_H |x_k|^2 d\nu < +\infty$, we have[3]

$$\left|\int_H \left(\frac{x_k}{\lambda_k} + 2D_kU\right)\varphi_n\psi d\nu\right| \le \frac{\|\psi\|_0}{\lambda_k}\left|\int_H |x_k|\,|\varphi_n| d\nu\right|$$
$$+2\|\psi\|_0\left|\int_H |D_kU|\,|\varphi_n| d\nu\right| \to 0, \text{ as } n \to \infty.$$

Finally, letting n tend to ∞ in (10) gives $g = 0$, since $\mathcal{E}_A(H)$ is dense in $L^2(H,\nu)$.

□

Now we can define $W^{1,2}(H,\nu)$ (resp. $W^{1,2}_A(H,\nu)$) as the linear subspace of all functions $\varphi \in L^2(H,\nu)$ such that $D_k\varphi \in L^2(H,\nu)$ for all $k \in \mathbb{N}$ and

$$\sum_{k=1}^\infty \int_H |D_k\varphi|^2 d\nu < +\infty \quad \left(\text{resp. } \sum_{k=1}^\infty \int_H \alpha_k |D_k\varphi|^2 d\nu\right) < +\infty.$$

Endowed with the inner products

$$\langle\varphi,\psi\rangle_{W^{1,2}(H,\nu)} = \langle\varphi,\psi\rangle_{L^2(H,\nu)} + \sum_{k=1}^\infty \int_H (D_k\varphi)(D_k\psi)d\nu,$$

$$\langle\varphi,\psi\rangle_{W^{1,2}_A(H,\nu)} = \langle\varphi,\psi\rangle_{L^2(H,\nu)} + \sum_{k=1}^\infty \int_H \alpha_k(D_k\varphi)(D_k\psi)d\nu,$$

$W^{1,2}(H,\nu)$ and $W^{1,2}_A(H,\nu)$ are Hilbert spaces. In an analogous way we can define $W^{k,2}(H,\nu)$, $k \ge 2$.

We prove finally that N_0 is symmetric.

[3] $\|\psi\|_0 = \sup_{x\in H} |\psi(x)|$.

Proposition 2.4. *For any $\varphi, \psi \in \mathcal{E}_A(H)$ we have*

$$\int_H (N_0\varphi)\psi d\nu = -\frac{1}{2}\int_H \langle D\varphi, D\psi\rangle d\nu. \tag{11}$$

In particular N_0 is symmetric on $L^2(H,\nu)$.

Proof. We have

$$\int_H (N_0\varphi)\psi d\nu = \int_H (L_2\varphi)\psi\rho d\mu - \int_H \langle \partial_0 U, D\varphi\rangle\psi d\nu. \tag{12}$$

But by Lemma 2.1 $\psi\rho \in W^{1,2}(H,\mu)$, and

$$D(\psi\rho) = \rho D\psi - 2\rho\psi\partial_0 U.$$

Thus by Proposition 1.5 we have

$$\begin{aligned}\int_H (L_2\varphi)\psi\rho d\mu &= -\frac{1}{2}\int_H \langle D\varphi, D(\psi\rho)\rangle d\mu\\ &= -\frac{1}{2}\int_H \langle D\varphi, D\psi\rangle d\nu + \int_H \langle \partial_0 U, D\varphi\rangle\psi d\nu.\end{aligned}$$

Comparing with (12) gives the conclusion. □

3. Self-adjointness of N

Since N_0 is symmetric it is closable, let denote by N its closure. We are going to show that the image of $\lambda - N_0$ is dense in $L^2(H,\nu)$ for any $\lambda > 0$. By the quoted result of Lumer and Phillips, [20], this implies that N is self-adjoint.

For this aim we need some regular approximations of U. Since U_α is not of class C^2 in general, we introduce an additional approximation $U_{\alpha,\varepsilon}$, by setting

$$U_{\alpha,\varepsilon}(x) = R_\varepsilon U(x),\ x \in H,\ \varepsilon > 0,$$

where (R_ε) is the Ornstein-Uhlenbeck operator defined by (4). It is easy to see that $U_{\alpha,\varepsilon}$ is convex and of C^∞ class (since (R_t) is strong Feller, see [14]), and consequently that $DU_{\alpha,\varepsilon}$ is Lipschitz continuous and monotone. Therefore we have, for $x, y \in H$,

$$\langle D^2 U_{\alpha,\varepsilon}(x)y, y\rangle \geq 0,\ \ x, y \in H,\ \alpha,\ \varepsilon > 0. \tag{13}$$

Lemma 3.1. *For any $\varphi \in \mathcal{E}_A(H)$ we have*

$$\lim_{\alpha,\varepsilon\to 0}\langle DU_{\alpha,\varepsilon}, D\varphi\rangle = \langle \partial_0 U, D\varphi\rangle,\ \text{in } L^2(H,\nu). \tag{14}$$

Proof. By (7)-(ii) we have

$$\lim_{\alpha,\varepsilon\to 0} DU_{\alpha,\varepsilon}(x) = \partial_0 U(x),\ \ x \in \text{dom}\,(\partial U).$$

Then (14) follows from the uniform integrability of the sequence $(DU_{\varepsilon,\alpha})$. We have in fact, taking into account that (R_t) is a contraction semigroup in $L^{2+2\varepsilon}(H,\mu)$, and recalling (7)-(iii),

$$\begin{aligned}\int_H |DU_{\varepsilon,\alpha}|^{2+\varepsilon}d\nu &= Z^{-1}\int_H |R_\varepsilon DU_\alpha|^{2+\varepsilon}e^{-2U}d\mu\\ &\le Z^{-1}\left(\int_H |R_\varepsilon DU_\alpha|^{2+2\varepsilon}d\mu\right)^{(2+\varepsilon)/(2+2\varepsilon)}\left(\int_H e^{-(4+4/\varepsilon)U}d\mu\right)^{\varepsilon/(2+2\varepsilon)}\\ &\le Z^{-1}\left(\int_H |\partial_0 U|^{2+2\varepsilon}d\mu\right)^{(2+\varepsilon)/(2+2\varepsilon)}\left(\int_H e^{-(4+4/\varepsilon)U}d\mu\right)^{\varepsilon/(2+2\varepsilon)}<\infty\end{aligned}$$

by Hypothesis 1.1–(v). The proof is complete. □

A basic tool to prove self-adjointness of N is the following approximating equation,

$$\lambda\varphi_{\alpha,\varepsilon} - L_4\varphi_{\alpha,\varepsilon} - \langle DU_{\alpha,\varepsilon}, D\varphi_{\alpha,\varepsilon}\rangle = f,\ \lambda>0. \tag{15}$$

Lemma 3.2. *Let $f\in C_b^1(H)$ ([4]), $\lambda,\alpha,\varepsilon>0$. Then Equation (15) has a unique mild solution $\varphi_{\alpha,\varepsilon}\in C_b^1(H)\cap D(L_4)$. Moreover the following estimate holds*

$$\|D\varphi_{\alpha,\varepsilon}\|_0 \le \frac{1}{\lambda}\|Df\|_0,\ \lambda,\alpha,\varepsilon>0. \tag{16}$$

The reason to look for a solution of (15) in $D(L_4)$ is that, in the proof of Theorem 3.3 below, we will need the inclusion $L^4(H,\mu)\subset L^2(H,\nu)$. We have in fact

$$\int_H |\varphi|^2 d\nu \le Z^{-1}\left(\int_H |\varphi|^4 d\mu\right)^{1/2}\left(\int_H e^{-4U}d\mu\right)^{1/2}.$$

We could replace 4 with $2+\varepsilon$, with any $\varepsilon>0$.

Proof. Since $DU_{\alpha,\varepsilon}$ is Lipschitz continuous, the stochastic differential equation

$$dX = (AX - DU_{\alpha,\varepsilon}(X))dt + dW,\quad X(0)=x, \tag{17}$$

has a unique solution $X_{\alpha,\varepsilon}(\cdot,x)$. Set

$$\varphi_{\alpha,\varepsilon}(x) = \int_0^{+\infty}\mathbb{E}\left[f(X_{\alpha,\varepsilon}(t,x))\right]dt,\ x\in H.$$

We claim that $\varphi_{\alpha,\varepsilon}$ belongs to $D(L_4)\cap C_b^1(H)$ and is a solution of (15). In fact $\varphi_{\alpha,\varepsilon}\in C_b^1(H)$ since

$$\langle D\varphi_{\alpha,\varepsilon}(x),h\rangle = \int_0^{+\infty}\mathbb{E}\left[\langle Df(X_{\alpha,\varepsilon}(t,x)),\eta^h_{\alpha,\varepsilon}(t,x)\rangle\right]dt,\ x,h\in H,$$

where $\eta^h_{\alpha,\varepsilon}(\cdot,x)$ is the solution to the equation

$$\frac{d\eta^h_{\alpha,\varepsilon}}{dt} = A\eta^h_{\alpha,\varepsilon} - D^2U_{\alpha,\varepsilon}(X)\eta^h_{\alpha,\varepsilon},\ \eta^h_{\alpha,\varepsilon}(0,x)=h. \tag{18}$$

---

[4]By $C_b^1(H)$ we mean the space of all mappings from H into $\mathbb{R}$ that are uniformly continuous together with their Fréchet derivative. We set $\|D\varphi\|_0 = \sup_{x\in H}|D\varphi(x)|$ for all $\varphi\in C_b^1(H)$.

Moreover, since $f - D\varphi_{\alpha,\varepsilon}$ is Lipschitz continuous (and consequently belongs to $L^4(H,\mu)$), it is not difficult to show that $\varphi_{\alpha,\varepsilon} \in D(L_4)$ and (15) holds.

Finally, recalling (13), we have $|\eta^h_{\alpha,\varepsilon}| \le |h|$, $h \in H$, and (16) follows. □

Now we are ready to prove,

Theorem 3.3. *Assume that Hypothesis 1.1 holds. Then N_0 is essentially self-adjoint in $L^2(H,\nu)$. Moreover the following identity holds,*

$$\int_H (N\varphi)\psi d\nu = -\frac{1}{2}\int_H \langle D\varphi, D\psi\rangle d\nu, \quad \varphi,\psi \in D(N). \tag{19}$$

Proof. We are going to prove that for $\lambda > 0$, the closure of $(\lambda - N_0)(\mathcal{E}_A(H))$ contains $C^1_b(H)$. Since $C^1_b(H)$ is dense in $L^2(H,\nu)$, this will prove the theorem by [20].

Let in fact $f \in C^1_b(H)$, $\lambda, \alpha, \varepsilon > 0$, and let $\varphi_{\alpha,\varepsilon}$ be the solution to (15). We know that $\varphi_{\alpha,\varepsilon}$ belongs to $D(L_4)$. Since $\mathcal{E}_A(H)$ is a core for L_4 there exists a sequence $(\varphi^{(n)}_{\alpha,\varepsilon}) \subset \mathcal{E}_A(H)$ such that

$$\varphi^{(n)}_{\alpha,\varepsilon} \to \varphi_{\alpha,\varepsilon}, \quad L_4\varphi^{(n)}_{\alpha,\varepsilon} \to L_4\varphi_{\alpha,\varepsilon}, \text{ in } L^4(H,\mu), \tag{20}$$

as n tends to ∞. Moreover we know from Proposition 1.4 that $D(L_4) \subset W^{1,4}(H,\mu)$ with continuous inclusion. Consequently we have

$$D\varphi^{(n)}_{\alpha,\varepsilon} \to D\varphi_{\alpha,\varepsilon}, \text{ in } L^4(H,\mu;H), \tag{21}$$

as n tends to ∞.

We claim that

$$\lim_{n\to\infty} \langle DU_{\alpha,\varepsilon}, D\varphi^{(n)}_{\alpha,\varepsilon}\rangle = \langle DU_{\alpha,\varepsilon}, D\varphi_{\alpha,\varepsilon}\rangle, \text{ in } L^2(H,\nu). \tag{22}$$

In fact, since $DU_{\alpha,\varepsilon}$ is Lipschitz continuous, there exists a constant $C_{\alpha,\varepsilon} > 0$ such that $|DU_{\alpha,\varepsilon}(x)| \le C_{\alpha,\varepsilon}(1+|x|)$, $x \in H$. It follows

$$\begin{aligned}
&\int_H |\langle DU_{\alpha,\varepsilon}, D\varphi^{(n)}_{\alpha,\varepsilon} - D\varphi_{\alpha,\varepsilon}\rangle|^2 d\nu \\
&\le Z_{-1}C^2_{\alpha,\varepsilon}\int_H (1+|x|)^2 |D\varphi^{(n)}_{\alpha,\varepsilon} - D\varphi_{\alpha,\varepsilon}|^2 e^{-2U} d\mu \\
&\le Z_{-1}C^2_{\alpha,\varepsilon}\left(\int_H (1+|x|)^4 e^{-4U} d\mu\right)^{1/2}\left(\int_H |D\varphi^{(n)}_{\alpha,\varepsilon} - D\varphi_{\alpha,\varepsilon}|^4 d\mu\right)^{1/2},
\end{aligned}$$

and so (22) follows.

Set now

$$f^{(n)}_{\alpha,\varepsilon} = \lambda\varphi^{(n)}_{\alpha,\varepsilon} - L_4\varphi^{(n)}_{\alpha,\varepsilon} + \langle DU_{\alpha,\varepsilon}, D\varphi^{(n)}_{\alpha,\varepsilon}\rangle.$$

Then, since $L^4(H,\mu) \subset L^2(H,\nu)$, with continuous embedding, we have, taking into account (22), that $f^{(n)}_{\alpha,\varepsilon} \to f$ in $L^2(H,\nu)$. But we have

$$\lambda\varphi^{(n)}_{\alpha,\varepsilon} - N_0\varphi^{(n)}_{\alpha,\varepsilon} = f^{(n)}_{\alpha,\varepsilon} + \langle(\partial_0 U - DU_{\alpha,\varepsilon}), D\varphi^{(n)}_{\alpha,\varepsilon}\rangle,$$

and so

$$\lim_{n\to\infty}(\lambda - N_0)(\varphi_{\alpha,\varepsilon}^{(n)}) = f + \langle \partial_0 U - DU_{\alpha,\varepsilon}, D\varphi_{\alpha,\varepsilon}\rangle,$$

in $L^2(H,\nu)$. This implies

$$f + \langle \partial_0 U - DU_{\alpha,\varepsilon}, D\varphi_{\alpha,\varepsilon}\rangle \in \overline{(\lambda - N_0)(\mathcal{E}_A(H))}. \tag{23}$$

We claim that

$$\lim_{\alpha,\varepsilon\to 0}\langle \partial_0 U - DU_{\alpha,\varepsilon}, D\varphi_{\alpha,\varepsilon}\rangle = 0, \text{ in } L^2(H,\nu). \tag{24}$$

We have in fact by (16),

$$\int_H |\langle \partial_0 U - DU_{\alpha,\varepsilon}, D\varphi_{\alpha,\varepsilon}\rangle|^2 d\nu \le \frac{1}{\lambda^2}\int_H |\partial_0 U - DU_{\alpha,\varepsilon}|^2 d\nu.$$

Now (24) follows by Lemma 3.1. Therefore, letting α,ε tend to 0, in (23), we have

$$f \in \overline{(\lambda - N_0)(\mathcal{E}_A(H))}.$$

Therefore $\overline{(\lambda - N_0)(\mathcal{E}_A(H))}$ contains $C_b^1(H)$ and, consequently coincides with $L^2(H,\nu)$.

The proof is complete. □

We prove now a generalization of Theorem 3.3 in $L^p(H,\mu)$.

Theorem 3.4. *Assume that Hypothesis 1.1 holds and let $p \ge 1$. Then N_0 is m-dissipative on $L^p(H,\nu)$.*

Proof. We first prove that N_0 is dissipative on $L^p(H,\nu)$. Le us consider the transition semigroup

$$P_t^{\alpha,\varepsilon} f(x) = \mathbb{E}\left[f(X_{\alpha,\varepsilon}(t,x)\right], \; f \in C_b(H),$$

where $X_{\alpha,\varepsilon}(\cdot,x)$ is the solution of (15). Let us consider the probability measure $\nu_{\alpha,\varepsilon}$, defined by

$$\nu_{\alpha,\varepsilon}(dx) = Z_{\alpha,\varepsilon}^{-1} e^{-2U_{\alpha,\varepsilon}(x)}, \; x \in H,$$

where

$$Z_{\alpha,\varepsilon} = \int_H e^{-2U_{\alpha,\varepsilon}(x)}\mu(dy).$$

It is easy to see that $\nu_{\alpha,\varepsilon}$ is invariant for the semigroup $(P_t^{\alpha,\varepsilon})$, that, consequently, can be uniquely extended to a strongly continuous semigroup of contractions on $L^p(H,\nu_{\alpha,\varepsilon})$. We denote by $N_{\alpha,\varepsilon}$ its infinitesimal generator.

Now let $\varphi \in \mathcal{E}_A(H)$, $\lambda > 0$, and set $f = \lambda\varphi - N_0\varphi$. Then we have

$$\lambda_\varphi - N_{\alpha,\varepsilon}\varphi = f + \langle DU_{\alpha,\varepsilon} - \partial_0 U, D\varphi\rangle.$$

Since $N_{\alpha,\varepsilon}$ is dissipative on $L^p(H,\nu_{\alpha,\varepsilon})$, we have

$$\int_H |\varphi|^p d\nu_{\alpha,\varepsilon} \le \frac{1}{\lambda^p}\int_H |f + \langle DU_{\alpha,\varepsilon} - \partial_0 U, D\varphi\rangle|^p d\nu_{\alpha,\varepsilon}.$$

Letting $\alpha,\varepsilon \to 0$ gives

$$\int_H |\varphi|^p d\nu \le \frac{1}{\lambda^p} \int_H |f|^p d\nu,$$

so that N_0 is dissipative on $L^p(H,\nu)$.

Finally m-dissipativity is proved arguing as in Theorem 3.3. □

Remark 3.5. If DU is uniformly continuous and bounded, we can consider N_0 as an operator on $C_b(H)$ and conclude that it is closable and its closure is m-dissipative in $C_b(H)$.

Remark 3.6. Identity (19) has important consequences. First, setting $\psi = 1$ we see that ν is an invariant measure for the semigroup $P_t = e^{tN}$, $t \ge 0$. Moreover it implies immediately that

$$D((-N)^{1/2}) = W^{1,2}(H,\nu).$$

Finally it implies that ν is ergodic and strongly mixing, see [12], that is for any $f \in L^2(H,\nu)$, we have

$$\lim_{t\to+\infty} P_t f = \int_H f d\nu, \text{ in } L^2(H,\nu).$$

4. The domain of N

We want here to give some information about the domain of N. It will be interesting to give a characterization of $D(N)$, generalizing formula (6) for $D(L_2)$. As we shall see this is possible under suitable additional assumptions.

We start with an identity.

Lemma 4.1. *Let $\varphi \in \mathcal{E}_A(H)$, set*

$$f_{\alpha,\varepsilon} = L_4\varphi - \langle DU_{\alpha,\varepsilon}, D\varphi\rangle, \quad \alpha,\varepsilon > 0.$$

Then the following identity holds:

$$\begin{aligned}\frac{1}{2}\int_H \mathrm{Tr}\,[(D^2\varphi)^2]d\nu + \int_H |(-A)^{1/2}D\varphi|^2 d\nu + \int_H \langle D^2U_{\alpha,\varepsilon}\cdot D\varphi, D\varphi\rangle d\nu \\ = 2\int_H f_{\alpha,\varepsilon}^2 d\nu + 2\int_H \langle DU_{\alpha,\varepsilon} - \partial_0 U, D\varphi\rangle f_{\alpha,\varepsilon} d\nu \\ + \int_H \langle D^2\varphi\cdot D\varphi, DU_{\alpha,\varepsilon} - \partial_0 U\rangle d\nu. \quad (25)\end{aligned}$$

Proof. Let $h \in \mathbb{N}$. Then we have

$$L_4 D_h\varphi - \alpha_h D_h\varphi - \langle DU_{\alpha,\varepsilon}, DD_h\varphi\rangle - \langle DD_hU_{\alpha,\varepsilon}, D\varphi\rangle = D_h f_{\alpha,\varepsilon}$$

and so

$$ND_h\varphi - \alpha_h D_h\varphi + \langle(\partial_0 U - DU_{\alpha,\varepsilon}), DD_h\varphi\rangle - \langle DD_hU_{\alpha,\varepsilon}, D\varphi\rangle = D_h f_{\alpha,\varepsilon}. \quad (26)$$

Multiplying both sides by $D_h\varphi$, integrating with respect to ν, and taking into account identity (19), gives

$$-\frac{1}{2}\int_H |DD_h\varphi|^2 d\nu - \int_H \alpha_h |D_h\varphi|^2 d\nu - \int_H \langle DD_h U_{\alpha,\varepsilon}, D\varphi\rangle D_h\varphi d\nu$$
$$= \int_H D_h f_{\alpha,\varepsilon} D_h\varphi d\nu - \int_H \langle(\partial_0 U - DU_{\alpha,\varepsilon}), DD_h\varphi\rangle D_h\varphi d\nu.$$

Using again (19) we have

$$\int_H Df_{\alpha,\varepsilon} D\varphi d\nu = -2\int_H N_0\varphi f_{\alpha,\varepsilon} d\nu$$
$$= -2\int_H f_{\alpha,\varepsilon}^2 d\nu - 2\int_H \langle \partial_0 U - DU_{\alpha,\varepsilon}, D\varphi\rangle f_{\alpha,\varepsilon} d\nu \quad (27)$$

Now, summing up on h in Equation (26), and taking into account Equation (27), the conclusion follows. □

Proposition 4.2. *Assume that Hypothesis 1.1 holds. Then we have*

$$D(N) \subset W^{2,2}(H,\nu) \cap W_A^{1,2}(H,\nu). \quad (28)$$

Proof. Let $\varphi \in \mathcal{E}_A(H)$, $\alpha, \varepsilon > 0$, and set

$$f = N_0\varphi, \quad f_{\alpha,\varepsilon} = L_4\varphi - \langle DU_{\alpha,\varepsilon}, D\varphi\rangle.$$

Then, by (25) we find, recalling (13),

$$\frac{1}{2}\int_H \text{Tr}\,[(D^2\varphi)^2] d\nu + \int_H |(-A)^{1/2} D\varphi|^2 d\nu$$
$$\le 2\int_H f_{\alpha,\varepsilon}^2 d\nu + 2\int_H \langle DU_{\alpha,\varepsilon} - \partial_0 U, D\varphi\rangle f_{\alpha,\varepsilon} d\nu$$
$$+ \int_H \langle D^2\varphi \cdot D\varphi, DU_{\alpha,\varepsilon} - \partial_0 U\rangle d\nu.$$

Now, letting α, ε tend to 0, it follows

$$\frac{1}{2}\int_H \text{Tr}\,[(D^2\varphi)^2] d\nu + \int_H |(-A)^{1/2} D\varphi|^2 d\nu \le 2\int_H f^2 d\nu. \quad (29)$$

Finally let $\varphi \in D(N)$. Since, by Theorem 3.3, $\mathcal{E}_A(H)$ is a core for N, there exists a sequence $(\varphi_n) \subset \mathcal{E}_A(H)$ such that

$$\varphi_n \to \varphi, \quad N\varphi_n \to N\varphi, \text{ in } L^2(H,\nu).$$

Set $f_n = \lambda\varphi_n - N\varphi_n$. Now let $m, n \in \mathbb{N}$. Replacing in (29) φ with $\varphi_n - \varphi_m$, and f with $f_n - f_m$ we have

$$\frac{1}{2}\int_H \text{Tr}\,[(D^2(\varphi_n - \varphi_m))^2] d\nu + \int_H |(-A)^{1/2} D(\varphi_n - \varphi_m)|^2 d\nu \le 2\int_H (f_n - f_m)^2 d\nu.$$

Therefore the sequence (φ_n) is Cauchy both in $W^{2,2}(H,\nu)$ and in $W_A^{1,2}(H,\nu)$ and the conclusion follows. □

A precise characterization of $D(N)$ holds under additional assumptions.

Proposition 4.3. *Assume, besides Hypothesis 1.1, that $U \in C^2(H)$. Then we have*

$$D(N) = \left\{ \varphi \in W^{2,2}(H,\nu) \cap W_A^{1,2}(H,\nu) : \int_H \langle D^2U \cdot D\varphi, D\varphi d\nu \rangle < +\infty \right\}. \quad (30)$$

Proof. Letting α, ε tend to 0 in (25) gives

$$\frac{1}{2} \int_H \mathrm{Tr}\, [(D^2\varphi)^2] d\nu + \int_H |(-A)^{1/2} D\varphi|^2 d\nu + \int_H \langle D^2 U D\varphi, D\varphi \rangle d\nu = 2 \int_H (N\varphi)^2 d\nu.$$

Now the conclusion follows arguig as in the proof of Proposition 4.2. □

Remark 4.4. A characterization of $D(N)$ can also be obtained if there exists in some sense the limit $\lim_{\alpha,\varepsilon\to 0} D^2 U_{\alpha,\varepsilon}$. Consider for instance Example 1.2. In this case we have

$$\langle D^2 U(x) y, y \rangle = (2m-1) \int_0^1 |x(\xi)|^{2m-2} y(\xi) z(\xi) d\xi,$$

for any $y, z \in L^\infty(0,1)$, $x \in L^{2m-2}(0,1)$. Then we find

$$D(N) = \Bigg\{ \varphi \in W^{2,2}(H,\nu) \cap W_A^{1,2}(H,\nu) : \int_H \int_0^1 |x(\xi)|^{2m-2} |D\varphi(x)(\xi)|^2 \nu(dx) d\xi < +\infty \Bigg\}.$$

5. Strong Feller property

We recall that (P_t) is said to be *strong Feller* if for any f bounded and Borel and for any $t > 0$ one has $P_t f \in C_b(H)$.

Proposition 5.1. *Assume that Hypothesis 1.1 holds. Then the semigroup $P_t = e^{tN}$, $t \geq 0$, is strong Feller.*

Proof. We first prove that

$$|P_t f(x) - P_t f(y)| \leq t^{-1/2} \|f\|_0 |x-y|, \; f \in C_b^1(H), \; x, y \in H, \; t > 0. \quad (31)$$

Let $f \in C_b^1(H)$, $\lambda, \alpha, \varepsilon > 0$, and let $\varphi_{\alpha,\varepsilon}$ be the solution to (15). It follows

$$\varphi_{\alpha,\varepsilon} = R(\lambda, N) \left[f + \langle DU_{\alpha,\varepsilon} - \partial_0 U, D\varphi_{\alpha,\varepsilon} \rangle \right],$$

and so by (24),

$$\lim_{\alpha,\varepsilon\to 0} \varphi_{\alpha,\varepsilon} = R(\lambda, N) f, \text{ in } L^2(H,\nu).$$

Let now $X_{\alpha,\varepsilon}$ be the solution to the stochastic equation (17). Then by a formula due to Bismut and Elworthy, see [7, 16], we have for $h \in H$,

$$\langle D\mathbb{E}(f(X_{\alpha,\varepsilon}(t,x)),h\rangle = \frac{1}{t}\,\mathbb{E}\left[f(X_{\alpha,\varepsilon}(t,x)\int_0^t \langle \eta^h_{\alpha,\varepsilon}(s,x), dW(s)\rangle\right],$$

where $\eta^h_{\alpha,\varepsilon}(\cdot,x)$ is the solution to (18). It follows

$$|D\mathbb{E}(f(X_{\alpha,\varepsilon}(t,x))| \le t^{-1/2}\,\|f\|_0,$$

which yields

$$|D\varphi_{\alpha,\varepsilon}(t,x)| \le \sqrt{\frac{\pi}{\lambda}}\,\|f\|_0,$$

and consequently

$$|R(\lambda,N)f(x) - R(\lambda,N)f(y)| \le \sqrt{\frac{\pi}{\lambda}}\,\|f\|_0,\ x,y \in H$$

Now (31) follows by taking the inverse Laplace transform.

Now let f be a real bounded Borel mapping in H, and let (f_n) be a sequence in $C^1_b(H)$ pointwise convergent to f and such that $\|f_n\|_0 \le \|f_n\|_0$. Then by (31) it follows

$$|P_tf_n(x) - P_tf_n(y)| \le t^{-1/2}\|f\|_0|x-y|, n \in \mathbb{N}\ x,y \in H,\ t>0.$$

Letting n tend to ∞ we see that P_tf is Lipschitz continuous. □

References

[1] S. Albeverio and R. Hoegh-Krohn, *Dirichlet forms and diffusion processes on rigged Hilbert spaces*, Z. Wahrsch. verw. Geb. **40** (1977), 1–57.

[2] S. Albeverio, V. Kondratiev and M. Röckner, *Dirichlet operators via stochastic analysis*, J. Funct. Analysis, **128** (1995), 102–138.

[3] Yu. M. Berezanskii, *Self-adjoint operators in spaces of functions of infinitely many variables*, Transl. Math. Monography, Vol. 63, Amer. Mat. Soc. Providence RI, 1986.

[4] V. Barbu, *Nonlinear semigroups and differential equations in Banach spaces*, Noordhoff, **1976**.

[5] V. Barbu and A. Rascanu, *Parabolic variational inequalities with singular inputs*, Differential Integral equations, **10** (1997), 67–87.

[6] A. Bensoussan and A. Rascanu, *Stochastic variational equations in infinite dimensional spaces*, Numer. Funct. Anal. Optimiz., **18** (1997), 19–54.

[7] J. M. Bismut, *Large Deviations and the Malliavin Calculus,* Birkhäuser, **1984**.

[8] E. Cepa, *Multivalued stochastic differential equations*, C.R. Acad. Sci. Paris, Ser 1, Math., **319** (1994), 1075–1078.

[9] G. Da Prato, *Applications croissantes et équations d'évolutions dans les espaces de Banach,* Academic Press, **1976**.

[10] G. Da Prato, *Regularity results for Kolmogorov equations on $L^2(H,\mu)$ spaces and applications*, Ukrainian Mathematical Journal, m. 49 (1997), 448–457.

[11] G. Da Prato, *The Ornstein-Uhlenbeck generator perturbed by the gradient of a potential*, Bollettino UMI, (8) I-B, 501–519, 1998.

[12] G. Da Prato and B. Goldys, *Invariant measures of non symmetric dissipative stochastic systems*, Preprint Scuola Normale Superiore di Pisa, 1999.

[13] G. Da Prato and L. Tubaro, *A new method to prove self-adjointness of some infinite dimensional Dirichlet operator,* Probab. Theory Relat. Fields, **118** (2000), 131–145.

[14] G. Da Prato and J. Zabczyk, *Ergodicity for Infinite Dimensional Systems,* London Mathematical Society Lecture Notes, n.229, Cambridge University Press, 1996.

[15] J. Dohmann, *Strong Feller properties and path regularity of Markov processes*, Diploma Thesis, University of Bielefeld, **2001**.

[16] K. D. Elworthy, *Stochastic flows on Riemannian manifolds.* In M.A. Pinsky and V. Wihstutz, editors, Diffusion processes and related problems in analysis, Vol. II, 33–72, Birkhäuser, 1992.

[17] U. G. Haussman and E. Pardoux, *Stochastic variational inequalities of parabolic type*, Appl. Math. Optimiz., **20** (1989), 163–192.

[18] V. Liskevich and M. Röckner, *Strong uniqueness for a class of infinite dimensional Dirichlet operators and application to stochastic quantization*, Ann. Scuola Norm. Sup. Pisa Cl. Sci. (4), **27** (1999), 69–91.

[19] H. Long and I. Simao, *Kolmogorov equations in Hilbert spaces with application to essential self-adjointness of symmetric diffusion operators*, Osaka Journal of Mathematics, (to appear).

[20] G. Lumer and R. S. Phillips, *Dissipative operators in a Banach space*, Pacific J. Math, **11** (1961), 679–698.

[21] Z. M. Ma and M. Röckner, *Introduction to the Theory of (Non Symmetric) Dirichlet Forms*, Springer-Verlag, **1992**.

Scuola Normale Superiore di Pisa
Piazza dei Cavalieri 7
56126, Pisa
Italy

Progress in Probability, Vol. 52, 89–110

Catalytic and Mutually Catalytic Super-Brownian Motions

D. A. Dawson and K. Fleischmann

Abstract. Catalytic branching processes describe the evolution of two types of material (populations) called catalyst and reactant. The catalyst evolves autonomously, but catalyzes the reactant. The individuals of both populations share the features of motion, growth and death. In mutually catalytic models however there is an additional feedback from the reactant to the catalyst destroying completely the basic independence assumption of branching theory. Recent results for continuum models of this type are surveyed.

1. Introduction: discrete models

We are dealing with large stochastic systems, we call *populations*, which are placed in some space, have features of motion, growth, and death. Traditionally in branching theory, the basic assumption is that disjoint parts develop independently. This independence assumption allows the use of a lot of mathematical tools, which has made possible the development of a huge mathematical theory. Probabilists have been busy with this for a long time, nevertheless research is still ongoing in this field, and there remain many interesting problems.

1.1. Critical binary branching simple random walk

Let us mention first one of the simplest models of this kind, the *critical binary branching simple random walk* in $\mathbb{Z}^2$. Initially, start with a chaotic population of particles, placed in $\mathbb{Z}^2$, we imagine as (possibly multiple) green points, say. We impose two kinds of dynamics on this population, namely motion and branching. So first of all, the particles (points) move chaotically. This means, each particle, *independently* of the others, may jump with a fixed rate to one of its neighboring sites, chosen at random (that is, with equal probability). But additionally, also

---

2000 *Mathematics Subject Classification.* 60K35; 60G57, 60J80.
Key words and phrases. Mutually catalytic branching, catalytic super-random walk, catalyst, reactant, superprocess, measure-valued branching, absolute continuity, collision measure, collision local time, self-similarity, martingale problem, segregation of types, coexistence of types, self-duality, finite time extinction, ultimate extinction, biodiversity, cyclic reaction.
This work has been supported by an NSERC grant, a Max Planck award, and the DFG.
Revised version of the WIAS-Preprint No. 546 of January 17, 2000; ISSN 0946-8633.

with a certain constant rate, called the *branching rate*, say ϱ, each particle may split in a critical binary fashion. This means, with an exponential clock of rate ϱ, it doubles, or dies, both with the same chance. As mentioned, the particles evolve *independently*, in particular, this holds for newly born particles (created by the splitting). The only dependence assumption is, that the (two) offspring start from the same position where they had been created.

As time goes on, at some places the particles completely disappear, by the killing aspect of the branching, and in other regions of space, the population grows by the doubling of particles. This way *clusters* are formed (see Figure 1), which move, grow, but may also disappear.

Models of this type have been studied in great detail in one or another setting, in particular in terms of the related diffusion approximation, the continuous super-Brownian motion, and a lot is known about them.

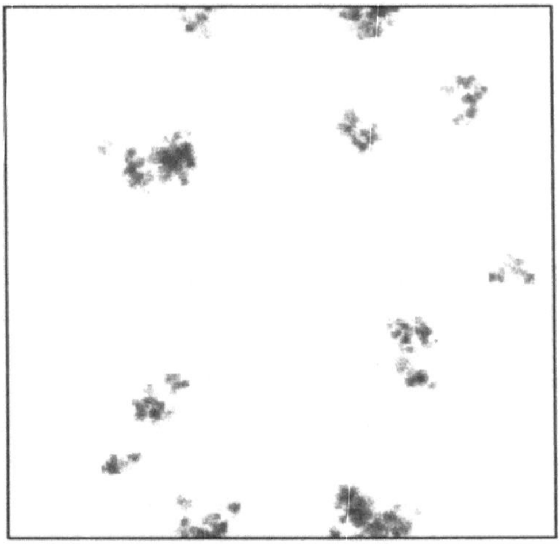

FIGURE 1. Simulation picture of the critical binary branching simple random walk (produced by Achim Klenke, Erlangen).

1.2. Imposing a random medium

The basic independence assumption, of course, is not so exciting, and one would like to get rid of it. How to do this? A first step in this direction is to additionally impose a *random medium*, thus dependencies occur by the randomness of the medium. In the previous model, the evolution mechanisms had been the same everywhere in time and space. We now imagine that they might change in time and space, even in a random way. But for simplification, we keep the rule, that given the medium, the population evolves according to the basic independence assumptions of branching theory, more precisely, we retain the *conditional independence*.

For instance, in the previous model of critical binary branching simple random walk, we proceed as follows. We allow the *branching rate ϱ to fluctuate* in time and space. So we imagine, branching is controlled by some *catalyst.* In some part of space, there might be a catalyst, which allows branching of the population, we now call the *reactant,* and in other parts there might be no catalyst, and the reactant will only move chaotically there. More precisely, the branching rate of the reactant is assumed to be given by the amount $\varrho_t(b)$ of catalyst at time t at site $b \in \mathbb{Z}^2$, and, in particular, $\varrho_t(b)$ might also be zero, suppressing in this case the branching of the reactant.

So first we need a model for the catalyst ϱ. Why not take the previous model in the constant medium as the catalyst?

1.3. Critical binary branching simple random walk reactant with a critical binary branching simple random walk catalyst

In this scenario the imagined green points are now our catalyst, to which we add an initially independent chaotic population of reactant particles thought of as red points situated in $\mathbb{Z}^2$. Then both, catalyst and reactant, are assumed to perform critical binary branching simple random walks on $\mathbb{Z}^2$. More precisely, the catalyst is assumed to evolve autonomously, with branching rate which is constant in time and space, say a constant $\gamma > 0$, whereas the reactant is assumed to be controlled by the catalyst, that is, its branching rate is given by the varying number $\varrho_t(b)$ of catalytic particles at the present time t at the site $b \in \mathbb{Z}^2$. The resulting model is a simple example of a *catalytic branching random walk,* denoted by X^{ϱ}, say.

What long-term effect will the catalyst have on the reactant population? Recall that at some places the catalyst ϱ will be absent, then the reactant X^{ϱ} will only move chaotically. So outside the green clumps, we have only dispersion of the red reactant, similarly as in the heat flow. In other regions we have a large concentration of catalyst, giving a large branching rate for the reactant, hence speeding up the critical branching of reactant. Thus, inside the catalytic clumps, the huge branching rate mainly causes killing of the reactant, so holes of the reactant are created. But at the boundary layer of the green clumps, where the catalyst highly fluctuates, there also the reactant highly fluctuates, creating a kind of *hot spots* of reactant.

1.4. Mutually catalytic critical binary branching simple random walk

The real challenge is to handle branching models with still more interaction. For instance, concerning the previous model the natural desire would be, that both substances catalyze each other. Let us recall that the green particles had been identified with the local branching rate of the red ones. But why should not conversely the red particles serve as the local branching rate of the green ones, instead of the constant in time and space branching rate γ so far? Then we would have two types of matter with a true interaction. In this case, we speak of a *mutually catalytic branching random walk.*

Consequently, initially we start with two chaotic substances, imagined as blue and red points, say. They move chaotically, but, in addition, catalyze each other. So the blues are the local branching rate for the reds, and vice versa.

What behavior do we expect in the long run? Again some clumps of materials should occur, which could lead to a local extinction of the other substance. So we could have some effect of separation of matter. But if a substance disappears locally, the interaction is gone locally, and the remaining population is only smeared out in such an area. On the other hand, at the *interface* of both materials a lot of activities are expected, presumably leading again to hot spots of both types, which are expected to play a more important role than in the former one-sided catalytic branching model.

1.5. Outline

In the remainder of this paper, models of the types described above will be made precise. With this aim, *continuum models* will be used as much as possible. This, in particular, requires replacement of the lattice space by Euclidean space, replacement of the simple symmetric random walk by a Brownian motion of particles, and passing to a high density/small particle mass limit, simultaneously speeding up the branching rates, etc. (*diffusion approximation*).

After modelling, some of the features already indicated will be expressed in terms of theorems, and some comments concerning the methods involved are in order.

For basic facts on superprocesses we refer to [4]. Recent surveys on catalytic processes are given in [11, 21, 23]. A few simulation pictures on the catalytic branching model of Subsection 1.3 are given in [19], and on both, the catalytic and mutually catalytic branching model, can be found via the home page of Achim Klenke:

`http://www.mi.uni-erlangen.de/~klenke`.

2. Catalytic branching

In this section we will start with the modelling of the super-Brownian catalyst, followed by the super-Brownian reactant with a super-Brownian catalyst. For the latter model, results on the long-term behavior and on the local structure are surveyed and compared with those concerning the ordinary super-Brownian motion.

2.1. Super-Brownian catalyst ϱ

Consider a *continuous super-Brownian motion (SBM)* $\{\varrho_t : t \geq 0\}$ in $\mathbb{R}^d$ starting with the Lebesgue measure ℓ, that is, with a uniformly smeared out mass. This ϱ is a certain continuous measure-valued diffusion process. Let us first recall the intuitive description: Each infinitesimally small part of the population, we still call a "particle", moves chaotically, that is Brownian, but additionally, with an everywhere constant rate, say $\gamma > 0$, we have a critical binary splitting, also in a diffusion limit sense.

The latter effect alone can be made precise by the following diffusion ζ in $\mathbb{R}_+$:

$$d\zeta_t \;=\; \sqrt{\gamma\zeta_t}\, dW_t, \quad \zeta_0 = 1, \tag{1}$$

with W a one-dimensional standard Wiener process, and $\gamma > 0$ a constant, the so-called *branching rate.* This is the famous *Feller's branching diffusion* without drift, arising as a high density small mass speeded up limit of critical binary Galton-Watson processes. Loosely speaking, this ζ can be seen as a "zero-dimensional super-Brownian motion".

Adding to (1) a chaotic motion component in space, leads to the formal stochastic equation

$$d\varrho_t(b) \;=\; \frac{1}{2}\Delta\varrho_t(b)dt + \sqrt{\gamma\varrho_t(b)}\, dW_t(b), \quad \varrho_0(b) = 1, \qquad b \in \mathbb{R}^d, \tag{2}$$

with $dW_t(b)$ a d-dimensional standard white noise. Roughly speaking, we keep Feller's branching diffusions, "independently" at each site $b \in \mathbb{R}^d$, but allow the particles additionally to move chaotically. Actually this equation makes sense rigorously only in dimension $d = 1$. Nevertheless, interpreting $\varrho_t(b)$ as generalized densities of measures $\varrho_t(db)$, and applying formally Itô's formula, from (2) one can derive the following representation of *log-Laplace transition functionals:*

$$-\log E\left\{\exp\langle\varrho_t, -\varphi\rangle \,\middle|\, \varrho_r\right\} \;=\; \left\langle\varrho_r, v(r,\cdot)\right\rangle, \qquad 0 \le r \le t. \tag{3}$$

Here given the populations state ϱ_r at time r, a suitable test function $\varphi : \mathbb{R}^d \to \mathbb{R}_+$ is integrated with the desired measure $\varrho_t(db)$ at time t. On the right-hand side of formula line (3), the so-called *log-Laplace function* $v(r,a)$, as a function of the time variable $r \ge 0$ and space variable $a \in \mathbb{R}^d$, is the unique solution to the following relatively simple *diffusion-reaction equation:*

$$\begin{gathered} -\frac{\partial}{\partial r}v(r,a) \;=\; \frac{1}{2}\Delta v(r,a) - \frac{1}{2}\gamma v^2(r,a), \qquad 0 < r < t, \quad a \in \mathbb{R}^d, \\ \text{with terminal condition} \quad v(t-,\cdot) = \varphi \ge 0. \end{gathered} \tag{4}$$

Here the Laplacian Δ acts on the space variable a, and reflects the chaotic motion of particles, whereas the square term is related to the critical binary splitting, independently at each space point. Recall, the constant $\gamma > 0$ is called the branching rate.

This procedure actually indicates one possibility of a rigorous approach to continuous SBM as the (unique in law) continuous measure-valued Markov process ϱ with log-Laplace transition functional (3) where the log-Laplace function v solves uniquely the diffusion-reaction equation (4). This continuous SBM ϱ is nowadays counted as one of the basic models in probability theory.

2.2. Super-Brownian reactant X^ϱ

Recall that this continuous SBM ϱ will serve as our rigorous description of the catalyst. What about a related reactant which we will denote by X^ϱ? We want it to also be a super-Brownian motion, but with the constant rate γ replaced by

$\varrho_r(db)$. So first of all, in the diffusion-reaction equation (4) we replace formally the constant γ by the randomly fluctuating rate $\varrho_r(\mathrm{d}a)$:

$$-\frac{\partial}{\partial r}v(r,a) = \frac{1}{2}\Delta v(r,a) - \frac{1}{2}\varrho_r(\mathrm{d}a)v^2(r,a), \qquad 0<r<t, \quad a\in\mathbb{R}^d, \\ \text{with terminal condition} \quad v(t-,\cdot)=\varphi\geq 0. \tag{5}$$

Consequently, we have a diffusion-reaction equation with a random coefficient in the reaction term. The point is, that these reaction term coefficients are now even measures $\varrho_r(da)$. Actually, we should write more carefully $\frac{\varrho_r(\mathrm{d}a)}{\mathrm{d}a}(a)$ instead of $\varrho_r(da)$. Of course, such derivatives would be relatively harmless, if these measures were absolutely continuous. But the super-Brownian motion ϱ has in fact singular states, except in dimension $d=1$. Indeed, by the critical binary splitting component within ϱ, a lot of extinction is going on, which creates measures which are "thin" in a sense (if $d\geq 2$). On the other hand, these measures $\varrho_r(da)$ are well known to have carrying Hausdorff dimension $2\wedge d$ in any dimension d of space. Note that these singularities of SBM ϱ fit very well to mimic situations where catalysts are distributed heterogeneously in space as, for instance, enzyme catalysts in glycolysis, see [27].

In any case, by this irregularity of the catalyst ϱ, the diffusion-reaction equation (5) with random coefficients is not a standard equation. Nevertheless, it can be handled, namely as an integral equation. In fact, if one formally integrates this equation, in terms of the semigroup $\{S_t : t\geq 0\}$ of the heat flow, and writes the heat flow S probabilistically as the expectation of functionals of a Brownian path W (independent of ϱ), then the following formal quantity will be involved:

$$\mathrm{d}r\int_{\mathbb{R}^d}\varrho_r(\mathrm{d}b)\delta_b(W_r) \;=:\; L_{[W,\varrho]}(\mathrm{d}r).$$

Intuitively, one measures the time points r where a typical Brownian reactant particle's path W is at the position b of a catalytic particle. Such a quantity (if meaningful) is called the *collision local time* $L_{[W,\varrho]}$ of W and ϱ. Since the $\varrho_r(db)$ are rather singular in higher dimensions, it is not immediately clear that this quantity makes sense rigorously. At the first sight, one might think, $L_{[W,\varrho]}$ exists non-trivially only in dimension one, since two independent Brownian particles, one from the catalyst and one from the reactant, will meet only in dimension 1. But we have to take into account, that ϱ describes a whole cloud of particles, which results after taking a diffusion limit. So ϱ is a more involved object than a countable number of moving and branching Brownian particles. Recall that each ϱ_t has carrying Hausdorff dimension $d\wedge 2$.

Actually, the random measures $L_{[W,\varrho]}(\mathrm{d}r)$ make sense (via a regularization procedure) and are non-trivial if and only if $d\leq 3$ ([1, 18]). In higher dimensions, they degenerate to 0; then a typical Brownian reactant particle W does not meet the thin catalyst ϱ, that is, it goes through it without hitting. More precisely, the occupation density of W at ϱ is 0, if $d>3$.

So from now on we want to restrict to dimensions $d \le 3$. And in these dimensions, a super-Brownian reactant X^ϱ with catalyst ϱ really exists as a continuous measure-valued Markov process whose log-Laplace transition functional can be expressed in terms of a log-Laplace function which is the unique solution (in a mild sense) of the diffusion-reaction equation (5) with random coefficients. For simplification, let's start with a uniform initial state: $X_0^\varrho = \ell$, the Lebesgue measure on $\mathbb{R}^d$. Detailed statements follow in the next theorem taken from ([9]). Let $\left(W, \Pi_{r,a}, r \ge 0, a \in \mathbb{R}^d\right)$ denote a standard Brownian motion in $\mathbb{R}^d$ (independent of ϱ), and p the related heat kernel:

$$\mathrm{p}_s(a) \;:=\; (2\pi s)^{-d/2} \exp\left[-\frac{|a|^2}{2s}\right], \qquad s > 0, \quad a \in \mathbb{R}^d, \tag{6}$$

corresponding to the heat flow semigroup S.

Theorem 2.1 (Non-degenerate existence of X^ϱ). *Assume $d \le 3$. Given the super-Brownian catalyst ϱ with $\varrho_0 = \ell$, the super-Brownian reactant X^ϱ with $X_0^\varrho = \ell$ exists non-trivially as a (time-inhomogeneous) continuous measure-valued Markov process $\{X_t^\varrho : t \ge 0\}$, with log-Laplace transition functional*

$$-\log E\left\{\exp\langle X_t^\varrho, -\varphi\rangle \,\middle|\, X_r^\varrho\right\} \;=\; \left\langle X_r^\varrho, v(r,\cdot)\right\rangle, \qquad 0 \le r \le t, \tag{7}$$

where, for each suitable test function $\varphi \ge 0$, the log-Laplace function v is the unique solution of the integral equation

$$v(r,a) = \Pi_{r,a}\left[\varphi(W_t) - \frac{1}{2}\int_r^t L_{[W,\varrho]}(\mathrm{d}s)\, v^2(s, W_s)\right], \qquad 0 \le r \le t, \quad a \in \mathbb{R}^d. \tag{8}$$

X^ϱ has the heat flow as its expectation:

$$E\left\{X_t^\varrho \,\middle|\, X_r^\varrho\right\} \;=\; S_{t-r} X_r^\varrho, \qquad 0 \le r \le t,$$

and, for test functions φ_1 and φ_2, as well as $0 \le r \le t_1, t_2$, the covariance is given by

$$\begin{aligned} &\mathrm{Cov}\left\{\langle X_{t_1}^\varrho, \varphi_1\rangle, \langle X_{t_2}^\varrho, \varphi_2\rangle \,\middle|\, X_r^\varrho\right\} \\ &\quad = \int_{\mathbb{R}^d} X_r^\varrho(\mathrm{d}a) \int_r^{t_1 \wedge t_2} \mathrm{d}s \int_{\mathbb{R}^d} \varrho_s(\mathrm{d}b)\, \mathrm{p}_{s-r}(b-a) S_{t_1-s}\varphi_1(b)\, S_{t_2-s}\varphi_2(b). \end{aligned}$$

On the other hand, if $d \ge 4$, the solution v to (8) degenerates to the heat solution (since $L_{[W,\varrho]} = 0$), hence X^ϱ from (7) degenerates to the heat flow. Recall that by a formal differentiation (8) reduces to (5). Let us skip further details on the construction of X^ϱ.

This reactant process X^ϱ has interesting features, and it is now relatively well understood.

2.3. Long-term behavior of X^ϱ

After construction of X^ϱ, first the long-term behavior was studied. The following theorem is a combination of results taken from [9, 10, 16, 19, 20]. Recall that for simplicity we assumed uniform initial states $\varrho_0 = \ell = X_0^\varrho$.

Theorem 2.2 (Persistent convergence). *In all dimensions* ($d \le 3$), *given* ϱ, *we have persistent convergence in law:*

$$X_T^\varrho \xrightarrow[T\uparrow\infty]{\mathcal{L}} \text{ some } X_\infty^\varrho \text{ with } EX_\infty^\varrho = \ell.$$

Additionally,

$$X_\infty^\varrho \;=\; \begin{cases} \ell, & \text{if } \; d=1, \\ \zeta\ell \text{ with } \operatorname{Var}\zeta > 0, & \text{if } \; d=2, \\ \text{has countably infinite local biodiversity}, & \text{if } \; d=3. \end{cases}$$

Here *countably infinite local biodiversity* means, that given ϱ, the random measure X_∞^ϱ does not have a deterministic component, and, in the cluster representation of the infinitely divisible random measure X_∞^ϱ, infinitely many clusters contribute to each finite region ([20]).

Consequently, in all dimensions, a limit population X_∞^ϱ exists, and has full expectation ℓ. Moreover, in dimension one the limit population is uniform and deterministic, since the catalytic clumps disappear locally; in $d = 2$ the limit X_∞^ϱ is still uniform but with a random intensity ζ, since the large but rare catalytic clumps come back from time to time, creating some randomness even in the limit; whereas in $d = 3$ (where, more precisely, the catalyst is assumed to be in equilibrium from the beginning), X_∞^ϱ exhibits countably infinite local biodiversity (genetic abundance). We mention that the statements in the theorem remain valid in particular in the *annealed* model (if one mixes the laws of the reactant X^ϱ with respect to the law of the catalyst ϱ), and that in this case the variance $\operatorname{Var}\zeta$ of the random intensity ζ is even *infinite*.

Let us stress that in all three dimensions the reactant X^ϱ behaves quite differently from the ordinary SBM ϱ (catalyst). In fact, as well known,

$$\varrho_T \xrightarrow[T\uparrow\infty]{\mathcal{L}} \text{some } \varrho_\infty \text{ for all } d \ge 1, \tag{9}$$

however

$$\varrho_\infty \;=\; \begin{cases} 0, & \text{if } \; d \le 2, \\ \text{steady state with } \textit{finite} \text{ local biodiversity}, & \text{if } \; d \ge 3. \end{cases} \tag{10}$$

Let us mention that it was a matter of dispute for a while whether in dimension two the mean will be preserved for the reactant X^ϱ at $T = \infty$ (persistence), since here huge catalytic clumps come back again and again to a finite window of observation causing an extreme killing of reactant there.

2.4. Local structure of the reactant X^ϱ off the catalyst

In contrast to the usual SBM ϱ, the reactant X^ϱ always has absolutely continuous states, and moreover in dimension two and three the density functions are almost everywhere extremely smooth and actually satisfy the heat equation ([19]):

Theorem 2.3 (Smooth density field). *For* $d = 2, 3$, *given* ϱ,

$$X_t^\varrho(\mathrm{d}b) \;=\; \xi_t^\varrho(b)\mathrm{d}b, \qquad t > 0,$$

where, with (conditional) probability one, ξ^{ϱ} is Lebesgue almost everywhere a C^{∞}-function (in both variables t and b) and satisfies the heat equation (both off the catalyst ϱ, which lives on a Lebesgue zero set in $\mathbb{R}_+ \times \mathbb{R}^d$).

Of course, these heat solutions ξ^{ϱ} might be rather irregular at the boundaries where the catalyst ϱ acts. Note that this theorem does not tell anything about the boundary layer behavior. In particular, it does not explain the occurrence of hot spots at the interface of reactant and catalyst seen in the mentioned simulations.

2.5. On the hot spots

One possible way to gain information about the hot spots is to study the collision local time $L_{[\varrho,X^{\varrho}]}$ between catalyst and reactant, which measures the times r where ϱ_r and X_r^{ϱ} are at the same sites $a \in \mathbb{R}^d$ (in the sense of occupation densities). $L_{[\varrho,X^{\varrho}]}$ exists as a limiting measure on $\mathbb{R}_+ \times \mathbb{R}^d$ in a non-trivial way in all dimensions. The approximating measures are formed by means of the heat kernels p from (6). This is part of the following theorem, taken from [12].

Theorem 2.4 (Collision local time $L_{[\varrho,X^{\varrho}]}$). *The following statements hold in the annealed model.*

(a) (Non-trivial existence): *In $d \leq 3$, the collision local time*

$$L_{[\varrho,X^{\varrho}]}\big(\mathrm{d}(r,a)\big) \;=\; \lim_{\varepsilon\downarrow 0} \mathrm{d}r\,\varrho_r(\mathrm{d}a)\int_{\mathbb{R}} X_r^{\varrho}(\mathrm{d}b)\mathrm{p}_{\varepsilon}(b-a) \tag{11}$$

between ϱ and X^{ϱ} exists non-trivially.

(b) (Spatial marginal collision densities): *In $d = 2$ and for $T > 0$,*

$$L_{[\varrho,X^{\varrho}]}\big([0,T],\mathrm{d}b\big) \quad \textit{is absolutely continuous a.s.} \tag{12}$$

(c) (Random ergodic limit): *In $d = 2$,*

$$\frac{1}{T}L_{[\varrho,X^{\varrho}]}\big([0,T]\times(\cdot)\big) \underset{T\uparrow\infty}{\overset{\mathcal{L}}{\Longrightarrow}} \eta\ell(\cdot), \tag{13}$$

where with respect to the annealed law, η has expectation one and a variance in $(0,\infty)$.

Consequently, besides existence in all dimensions, for $d = 2$ the spatial marginal measures of $L_{[\varrho,X^{\varrho}]}$ are even absolutely continuous. Finally, in the long run one has a kind of ergodic limit, but with some additional noise: The averaging (in $d = 2$) does not lead to a deterministic quantity, the expectation ℓ. Roughly speaking, in the time average, the hot spots are everywhere present, and they grow in all scales. Often this is called a *diffusive behavior*, see, for instance, [2].

Remarkably, the collision local time $L_{[\varrho,X^{\varrho}]}$ is also absolutely continuous in time. To make this precise, fix a continuous function $f \geq 0$ on $\mathbb{R}$ with compact support and $\int_{\mathbb{R}} \mathrm{d}r\, f(r) = 1$, and, for $t, \varepsilon > 0$, define

$$K_t^{\varepsilon}(\mathrm{d}a) \;:=\; \int_{\mathbb{R}_+} L_{[\varrho,X^{\varrho}]}\big(\mathrm{d}(r,a)\big)\,\frac{1}{\varepsilon}f\big(\varepsilon^{-1}(r-t)\big).$$

The following theorem is also proved in [12].

Theorem 2.5 (Two-dimensional collision measures). *Let $d = 2$.*

(a) **(Existence of collision measures):** *For each $t > 0$ there is a random measure K_t such that for reasonable test functions $\varphi \geq 0$ the following L^2-convergence holds:*

$$\langle K_t^{\varepsilon}, \varphi \rangle \underset{\varepsilon \downarrow 0}{\longrightarrow} \langle K_t, \varphi \rangle .$$

(b) **(Carrying Hausdorff dimension two):** *For each $t > 0$ fixed, with probability one, K_t has carrying Hausdorff dimension two.*

(c) **(Representation of collision local time):** *To the family $K = \{K_t : t > 0\}$ of random measures of* (a)*, there is an adapted version denoted by the same symbol K, such that*

$$L_{[\varrho, X^{\varrho}]}\big(\mathrm{d}(r,a)\big) = \mathrm{d}r K_r(\mathrm{d}a), \ \ a.s.$$

Remark 2.6 (Open problems). The L^2-methods used in [12] to derive the $d = 2$ results of Theorem 2.4 (b) and Theorem 2.5 (a) do *not* apply in three dimensions, and the behavior for $d = 3$ remains open.

We mention that the collision local time $L_{[\varrho, X^{\varrho}]}$ is also important for the formulation of the martingale problem for the reactant X^{ϱ} (see [12]).

2.6. Long-term behavior for finite reactant populations

So far we assumed that X^{ϱ} starts from the Lebesgue measure. But also in the case of finite reactant populations the reactant process X^{ϱ} exists for almost all ϱ (starting from $\varrho_0 = \ell$), and has interesting features different from the ordinary super-Brownian motion ϱ. Recall that for finite initial measures ϱ_0, the total mass process $t \mapsto \|\varrho_t\| := \varrho_t(\mathbb{R}^d)$ is Feller's branching diffusion (1) without drift. Hence, with probability one,

$$\varrho_T \ = \ 0 \quad \text{for some } T < \infty. \tag{14}$$

In contrast to this finite time extinction of the ordinary super-Brownian motion ϱ, from [9] and [20] we have the following results.

Theorem 2.7 (Finite reactant populations). *Assume that $0 < \|X_0^{\varrho}\| < \infty$. Then for almost all ϱ (starting from $\varrho_0 = \ell$), the following statements are true.*

(a) **(Persistent convergence):** *If $d = 1$, then*

$$\|X_T^{\varrho}\| \underset{T \uparrow \infty}{\overset{\mathcal{L}}{\longrightarrow}} \textit{some } \|X_{\infty}^{\varrho}\|, \tag{15}$$

where $\|X_{\infty}^{\varrho}\|$ has full expectation $E\|X_0^{\varrho}\|$ and a variance in $(0,\infty)$.

(b) **(Long-term extinction):** *If $d = 2, 3$, then*

$$\|X_T^{\varrho}\| \underset{T \uparrow \infty}{\longrightarrow} 0 \ \ a.s. \tag{16}$$

(c) **(Finite time survival):** *If $d = 3$, then*

$$X_T^{\varrho} \neq 0 \quad \textit{for all}\, T < \infty, \ \ a.s. \tag{17}$$

In fact, in dimension 1, an intrinsic reactant particle has only a *finite time of interference* with the catalyst ([9]), which explains at least heuristically the persistent convergence in (a). In $d = 2, 3$, however, the total collision local time $L_{[W,\varrho]}(\mathbb{R}_+)$ of an intrinsic reactant particle with path W and the catalyst ϱ is *infinite* almost surely ([20]), which intuitively makes plausible the long-term extinction as in (b). This can actually be turned into a rigorous proof by exploiting the fact that, via an individual time change, Feller's branching diffusion is "*embedded*" in the reactant's total mass process ([13]). Finally, in $d = 3$, the finite time survival of (c) is caused by some *instantaneous propagation of reactant matter* property ([20]), which is based on the fact that lines are polar for the support of three-dimensional ordinary super-Brownian motion.

Remark 2.8 (Open problem). The latter fact raises the interesting question whether or not the complement of two-dimensional ordinary super-Brownian motion is *connected*: a positive answer would extend the finite time survival of (c) also to $d = 2$.

3. Mutually catalytic branching

Before we will turn to a mutually catalytic model, let us have an alternative look at the model we just dealt with.

3.1. Heuristic writing of (ϱ, X^ϱ)

Symbolically, the previous model can be described by the following system of stochastic evolution equations:

$$\begin{aligned} \text{catalyst:} \qquad \mathrm{d}\varrho_t &= \frac{1}{2}\Delta\varrho_t \mathrm{d}t + \sqrt{\gamma\varrho_t}\,\mathrm{d}W_t^1, \\ \text{reactant:} \qquad \mathrm{d}X_t^\varrho &= \frac{1}{2}\Delta X_t^\varrho \mathrm{d}t + \sqrt{\varrho_t X_t^\varrho}\,\mathrm{d}W_t^2, \end{aligned} \tag{18}$$

where the $\mathrm{d}W^i$ are independent standard space-time white noises. In fact, the first of these equations is a repetition of (2). Recall, the Laplacian Δ stands for the chaotic motion, so the catalytic measures are smeared out, whereas the root term reflects the critical binary branching of catalyst, in fact in the diffusion limit sense. The second equation in (18) is the symbolic counterpart concerning the reactant: the constant branching rate γ is simply replaced by the measure-valued path ϱ.

Clearly, these symbolic equations (18) can be turned into martingale problems, which indeed can successfully be handled. In fact, for the reactant case, the collision local time $L_{[\varrho,X^\varrho]}$ of Theorem 2.4 occurs ([12]) instead of the suspicious product $\varrho_t X_t^\varrho$ of generalized densities in (18).

Altogether, one has 2 substances, the first one, the catalyst, is autonomous, whereas the second one, the reactant, is catalyzed by the first one. Sometimes such situation is also called a "one-way interaction". However we now want the two substances to catalyze each other. A formal symmetrization of the previous model (ϱ, X^ϱ) would be a pair $\mathbf{X} = (X^1, X^2)$ of processes satisfying

$$\mathrm{d}X_t^i = \frac{1}{2}\Delta X_t^i \mathrm{d}t + \sqrt{X_t^1 X_t^2}\mathrm{d}W_t^i, \qquad i = 1, 2, \tag{19}$$

(which was proposed by Carl Mueller years ago, as remarked in [15]).

At the first sight, there is a serious mathematical problem: How to make sense of such a model (19)? The main obstacle is, that the basic assumption of branching theory is lost, namely the branching property: In the earlier model (ϱ, X^ϱ) we had an independent evolution of disjoint reactant parts, given the catalyst. And this basic conditional independence assumption was the source for the log-Laplace calculus, say. All this is gone in such type of a model $\mathbf{X} = (X^1, X^2)$. In particular, one loses the very useful connection to diffusion-reaction equations. In other words, a completely new mathematical approach is needed.

3.2. Mutually catalytic simple super-random walk X on $\mathbb{Z}^d$

After some years of struggle, [15] succeeded in constructing such a mutually catalytic model, in fact, first of all in a discrete space setting. To make this precise, we call a function f defined on $\mathbb{Z}^d$ *tempered*, if

$$\|f\|_\lambda := \sum_{b \in \mathbb{Z}^d} |f(b)|\, \mathrm{e}^{-\lambda|b|} < \infty \quad \text{for all } \lambda > 0.$$

Topologize the set $\mathcal{M}_{\text{tem}} = \mathcal{M}_{\text{tem}}(\mathbb{Z}^d)$ of all non-negative tempered functions (considered also as measures) by

$$\mathrm{d}_{\text{tem}}^{\mathcal{M}}(f, g) := \sum_{n=1}^{\infty} 2^{-n}\left(\|f - g\|_{1/n} \wedge 1\right), \qquad f, g \in \mathcal{M}_{\text{tem}}.$$

Theorem 3.1 (Existence of $\mathbf{X}$ on $\mathbb{Z}^d$). *On the lattice space $\mathbb{Z}^d$ of each dimension ($d \geq 1$), the mutually catalytic simple super-random walk $\mathbf{X} = (X^1, X^2)$ exists: To each $\mathbf{x} = (x^1, x^2) \in \mathcal{M}_{\text{tem}}^2$, there is a weakly unique continuous $\mathcal{M}_{\text{tem}}^2$-valued solution $\mathbf{X} = (X^1, X^2)$ to the following system of stochastic equations*

$$\mathrm{d}X_t^i(b) = \frac{1}{2}\Delta X_t^i(b)\mathrm{d}t + \sqrt{\gamma X_t^1(b) X_t^2(b)}\mathrm{d}W_t^i(b), \quad b \in \mathbb{Z}^d,\ i = 1, 2, \tag{20}$$

satisfying $\mathbf{X}_0 = \mathbf{x}$, where $\{W^i(b) : b \in \mathbb{Z}^d,\ i = 1, 2\}$ are independent standard Wiener processes, and $\gamma > 0$ is a constant, the so-called collision rate.

In other words, in this model the motion of "particles" is the simple random walk on $\mathbb{Z}^d$, in (20) expressed by the discrete Laplacian Δ. On the other hand, the noise is modelled by independent Wiener processes. Thus, first of all, independently at each space site b one has Feller's branching diffusions without drift [as in (1)], but "additionally" with a rate controlled by the local concentration of mass of the opposite substance.

A solution to (20) can be constructed by standard methods as in [29], for instance. For uniqueness of solutions however, as stressed in [15], a self-duality property was essentially used, recently established in [26].

So first of all, such mutually catalytic process $\mathbf{X}$ in $\mathbb{Z}^d$ makes sense non-trivially. Moreover, it has interesting features, which can be found in [15]. Some of them are related in a sense to the previous (ϱ, X^{ϱ}) model, others are significantly different. But we want to skip all these things at this stage, since our main interest here concerns related continuum models that we now want to discuss.

3.3. Existence of a one-dimensional continuum version

The existence of a one-dimensional continuum version was established in [15], with uniqueness also based on [26]. In order to make this precise, let $\mathcal{C}_{\text{tem}} = \mathcal{C}_{\text{tem}}(\mathbb{R})$ denote the set of all *tempered* continuous functions f defined on $\mathbb{R}$, that is continuous functions such that

$$|f|_\lambda \;:=\; \sup_{b\in\mathbb{R}} \bigl|f(b)\bigr|\,\mathrm{e}^{-\lambda|b|} \;<\; \infty \quad \text{for all } \lambda > 0. \tag{21}$$

Topologize $\mathcal{C}_{\text{tem}}$ by

$$\mathrm{d}^{\mathcal{C}}_{\text{tem}}(f,g) \;:=\; \sum_{n=1}^{\infty} 2^{-n}\left(|f-g|_{1/n} \wedge 1\right), \qquad f,g \in \mathcal{C}_{\text{tem}}.$$

Theorem 3.2 (Existence of $\mathbf{X}$ on $\mathbb{R}$). *The mutually catalytic super-Brownian motion $\mathbf{X} = \bigl(X^1, X^2\bigr)$ in $\mathbb{R}$ exists: To each $\mathbf{x} = \bigl(x^1, x^2\bigr) \in \bigl(\mathcal{C}^+_{\text{tem}}\bigr)^2$, there is a weakly unique continuous $\bigl(\mathcal{C}^+_{\text{tem}}\bigr)^2$-valued solution $\mathbf{X} = \bigl(X^1, X^2\bigr)$ to the following system of stochastic partial differential equations*

$$\mathrm{d}X^i_t(b) \;=\; \frac{1}{2}\Delta X^i_t(b)\mathrm{d}t + \sqrt{\gamma X^1_t(b) X^2_t(b)}\,\mathrm{d}W^i_t(b), \quad b \in \mathbb{R}, \quad i = 1,2, \tag{22}$$

satisfying $\mathbf{X}_0 = \mathbf{x}$. Here $\mathrm{d}W^1, \mathrm{d}W^2$ are independent standard white noises, and the constant $\gamma > 0$ is the collision rate.

The construction of a solution $\mathbf{X}$ to (22) can be provided starting from approximations $\{{}^n\mathbf{X} : n \geq 1\}$, where ${}^n\mathbf{X} = ({}^nX^1, {}^nX^2)$ has the property that on small time periods $\bigl[\frac{j}{n}, \frac{j+1}{n}\bigr)$, $j \geq 0$, given ${}^n\mathbf{X}_{j/n}$ the two subpopulations ${}^nX^1, {}^nX^2$ behave as *independent* one-dimensional catalytic super-Brownian motions with frozen, smoothed and truncated branching rate functions (catalyst) $\gamma\bigl(S_{1/n}{}^nX^2_{i/n} \wedge n\bigr)$ and $\gamma\bigl(S_{1/n}{}^nX^1_{i/n} \wedge n\bigr)$, respectively. (This partly relies on ideas of [28].)

Uniqueness however is based on Mytnik's self-duality property we next want to deal with. We first mention that analogously to Theorem 3.2 the mutually catalytic super-Brownian motion in $\mathbb{R}$ can also be defined as a $\bigl(\mathcal{C}^+_{\text{rap}}\bigr)^2$-valued process ([15]). Here $\mathcal{C}_{\text{rap}} = \mathcal{C}_{\text{rap}}(\mathbb{R})$ is the set of all *rapidly decreasing* continuous functions f defined on $\mathbb{R}$, that is continuous functions such that $|f|_{-\lambda} < \infty$ for all $\lambda > 0$ [recall (21)], topologized by

$$\mathrm{d}^{\mathcal{C}}_{\mathrm{rap}}(f,g) \;:=\; \sum_{n=1}^{\infty} 2^{-n}\left(|f-g|_{-n}\wedge 1\right), \qquad f,g\in\mathcal{C}_{\mathrm{rap}}.$$

Introduce the pairing

$$\langle f,\tilde{f}\rangle \;:=\; \int_{\mathbb{R}} \mathrm{d}b\; f(b)\tilde{f}(b), \qquad f\in\mathcal{C}^{+}_{\mathrm{tem}},\quad \tilde{f}\in\mathcal{C}^{+}_{\mathrm{rap}}.$$

Moreover, write A for the orthonormal matrix $2^{-1/2}\begin{pmatrix}1 & 1\\ 1 & -1\end{pmatrix}$ and introduce the duality function

$$\Psi(\mathbf{x},\widetilde{\mathbf{x}}) \;:=\; \exp\left[-\left\langle(\mathbf{x}A)^1,(\widetilde{\mathbf{x}}A)^1\right\rangle + i\left\langle(\mathbf{x}A)^2,(\widetilde{\mathbf{x}}A)^2\right\rangle\right],$$

$\mathbf{x}\in\left(\mathcal{C}^{+}_{\mathrm{tem}}\right)^2$, $\widetilde{\mathbf{x}}\in\left(\mathcal{C}^{+}_{\mathrm{rap}}\right)^2$. From [26] we now quote the following result, implying uniqueness in the martingale problem related to (22) by a standard procedure.

Theorem 3.3 (Self-duality). *Let* $\mathbf{X}$ *and* $\widetilde{\mathbf{X}}$ *be independent mutually catalytic super-Brownian motions in* $\mathbb{R}$ *with* $\mathbf{X}_0\in\left(\mathcal{C}^{+}_{\mathrm{tem}}\right)^2$ *and* $\widetilde{\mathbf{X}}_0\in\left(\mathcal{C}^{+}_{\mathrm{rap}}\right)^2$. *Then the following self-duality formula holds*

$$E\Psi(\mathbf{X}_t,\widetilde{\mathbf{X}}_0) \;=\; E\Psi(\mathbf{X}_0,\widetilde{\mathbf{X}}_t), \qquad t\geq 0.$$

This identity is actually simple to verify by using Itô's formula.

3.4. Long-term behavior of the mutually catalytic SBM in $\mathbb{R}$

Recall that in $d=1$, for the catalyst ϱ we have

$$\begin{aligned} \|\varrho_T\| &\xrightarrow[T\uparrow\infty]{\mathcal{L}} 0 \quad\text{if}\quad \|\varrho_0\|<\infty,\\ \varrho_T &\xrightarrow[T\uparrow\infty]{\mathcal{L}} 0 \quad\text{if}\quad \varrho_0=\ell, \end{aligned} \tag{23}$$

[formulas (14) and (9)/(10)]. On the other hand, given ϱ with $\varrho_0=\ell$, for the (one-dimensional) reactant X^{ϱ} we have

$$\|X^{\varrho}_T\| \xrightarrow[T\uparrow\infty]{\mathcal{L}} \|X^{\varrho}_{\infty}\| \quad\text{if}\quad \|X^{\varrho}_0\|<\infty, \tag{24}$$

where $\|X^{\varrho}_{\infty}\|$ has full expectation and is non-trivial if $X^{\varrho}_0\neq 0$ [Theorem 2.7 (a)], whereas

$$X^{\varrho}_T \xrightarrow[T\uparrow\infty]{\mathcal{L}} \ell \quad\text{if}\quad X^{\varrho}_0=\ell \tag{25}$$

(Theorem 2.2). For the mutually catalytic model $\mathbf{X}$ in $d=1$ we have instead the following persistent convergence statements ([15]). Set $\|X^i_T\| := \int_{\mathbb{R}}\mathrm{d}b X^i_T(b)$.

Theorem 3.4 (Global segregation of types). *Recall that* $d=1$.

(a) (Finite $\mathbf{X}_0$): *If* $\mathbf{X}_0\in(\mathcal{C}^{+}_{\mathrm{rap}})^2$, *then*

$$\|\mathbf{X}_T\| \;:=\; \left(\|X^1_T\|,\|X^2_T\|\right) \xrightarrow[T\uparrow\infty]{\mathcal{L}} \|\mathbf{X}_{\infty}\| \;:=\; B_{\tau}$$

where $t \mapsto B_t$ is a Brownian motion in $\mathbb{R}_+^2$ starting from $\|\mathbf{X}_0\|$, and τ is its first hitting time of the boundary of $\mathbb{R}_+^2$. In particular,

$$E\|\mathbf{X}_\infty\| \ = \ E\|\mathbf{X}_0\| \quad (\textit{persistence}),$$

and

$$\|X_\infty^1\| \cdot \|X_\infty^2\| = 0 \ a.s. \quad (\textit{global segregation}).$$

(b) **(Uniform initial states):** *If $\mathbf{X}_0 = \mathbf{c}1$, $\mathbf{c} \in \mathbb{R}_+^2$, instead, with 1 denoting the function constantly 1 on $\mathbb{R}$, then*

$$\mathbf{X}_T \ \xrightarrow[T\uparrow\infty]{\mathcal{L}} \ \mathbf{X}_\infty \ := \ B_\tau 1 \quad (\textit{global segregation}),$$

where B now starts from $\mathbf{c}$.

Note that this global segregation (or non-coexistence) of types can be considered as a "symmetrization" of the assertion on (ϱ, X^ϱ) in the first formula line of (23) and in (24), respectively.

The statement (b) follows from (a) via the self-duality Theorem 3.3. The verification of (a) however requires some efforts. First of all, since $T \mapsto \|X_T^1\| + \|X_T^2\|$ is a uniformly integrable non-negative martingale, it converges as $T \uparrow \infty$. This implies that its square function

$$T \ \mapsto \ A_T \ := \ \gamma \int_0^T \mathrm{d}t \ \langle X_t^1, X_t^2 \rangle$$

converges to a finite value as $T \uparrow \infty$. Then the strategy of proof is that the assumption that $\|X_T^1\| \cdot \|X_T^2\|$ is bounded away from zero as $T \uparrow \infty$ implies that A_T converges to infinity, giving a contradiction. For the involved details, see [15], or [22], where one can find a generalization of the model to more than two types.

3.5. Finite time survival

After describing the behavior in the long run, let us turn to the more delicate question about finite time survival of both types. The answer is actually dependent on the initial state. A $\mathbb{Z}^d$-version of the following theorem is given in [25], whereas the present $\mathbb{R}$-version follows as a special case of [22].

Theorem 3.5 (Finite time behavior). *Let $d = 1$ and $0 < \|\mathbf{X}_0\| < \infty$.*

(a) **(Finite time survival):** *Under certain additional conditions on $\mathbf{X}_0$,*

$$\|X_T^1\| \cdot \|X_T^2\| \ > 0 \quad \textit{for all } T > 0, \ a.s.$$

(b) **(Finite time extinction of a type with high probability):** *Similarly, under a certain additional condition on X_0^1, for each $\varepsilon \in (0,1]$ and $T > 0$, one can choose X_0^2 ($\neq 0$) so small, that*

$$P\big(X_t^2 = 0 \textit{ for all } t \geq T\big) \ \geq \ 1 - \varepsilon.$$

A sufficient condition for the situation in (a) is given if X_0^1 and X_0^2 are separated in different half lines of $\mathbb{R}$ and have there sufficiently large tails at infinity. The proof of (a) is essentially based on a variance estimate and Chebychev's inequality.

The verification of (b) is more involved and relies on the following philosophy. First of all, because 0 is an absorbing state for the subprocess $t \mapsto X_t^2$, it suffices to consider $\mathbf{X}$ on a possibly smaller time interval $[0, T]$. Since initially X^2 is not too large by assumption and serves as the catalyst for X^1, then X^1 should not be very small on $[0, T]$. But since X^1 serves as the catalyst for X^2, then X^2 should have the chance to die by time T. Actually, under the present conditions and on a suitable "individual" time scale, some norm of X_T^2 can be bounded from above by a *supercritical* Feller's branching diffusion which certainly dies by a given time $T > 0$ with positive probability. Finally, by making the initial state of this diffusion sufficiently small, the latter extinction probability can be forced to be at least $1-\varepsilon$.

We mention that for the case of uniform initial states, where by Theorem 3.4 (b) global segregation occurs in the long run, one can nevertheless expect that the predominant type near the origin *oscillates* as time goes on, just as proved in [3] for the $\mathbb{Z}^d$-model, $d \leq 2$.

So far we discussed only a one-dimensional continuum mutually catalytic branching model. Why this restriction to dimension one? Could one perhaps hope for a non-trivial existence of some mutually catalytic measure-valued processes in higher dimensional continua?

3.6. First arguments against a non-trivial existence in $\mathbb{R}^2$

Actually, there are a couple of arguments against non-trivial existence in the higher-dimensional case. First of all, variances blow up in some variance calculations working with approximations. But perhaps more important, non-trivial existence seems to imply some *paradoxes* that we will now explain.

Recall that in the constant medium case, the SBM ϱ in dimensions $d \geq 2$ has singular states. The same is true in the case of a regular medium, that is, if we have a catalyst, described by absolutely continuous measures. Consequently, a regular catalyst leads to a singular reactant (in $d \geq 2$). On the other hand, if the catalyst is singular enough, the reactant may become absolutely continuous states ([7, 8, 24]). Summarizing, in catalytic models of dimension $d \geq 2$, roughly speaking, catalyst and reactant exhibit an "*alternation property*", say. In particular, locally only one of the substances can be absolutely continuous.

What should follow from this for the symmetric, mutually catalytic model in $\mathbb{R}^2$ we want to have? Might such alternation property be reasonable in the symmetric case?

In dimension two, starting with uniform initial states, the continuum model should be *self-similar*, as one can easily check by formal calculations. But intuitively it should also have the same long-term behavior as the $\mathbb{Z}^2$-model ([15]). That is, Theorem 3.4 (b) should be true also in $d = 2$. But by the self-similarity, this global statement about $\mathbf{X}_\infty$ should be turned into a local property of $\mathbf{X}_1$ say:

(i) The limiting uniform states should be turned into *absolutely continuous* measure states $\mathbf{X}_1$, say, which then do *not* have the alternation property discussed above.

(ii) More important, the global segregation should lead to a *local segregation* of types: Locally only one type should be present. In other words, if two types were locally present, then one of the types should immediately kill the other one. On the other hand, the sought-after non-trivial stochastic model, if it existed, would live from the interaction!

How to resolve such paradoxes, other than to conclude, that the higher-dimensional continuum model would degenerate to the heat flow? At least, if the model existed, it should be rather delicate!

These kind of questions attracted several persons, and now there is consensus about the non-trivial existence of the 2-dimensional continuum model, provided that the collision rate γ is sufficiently small.

3.7. Mutually catalytic super-Brownian motion in $\mathbb{R}^2$

The following theorem is proved in [6].

Theorem 3.6 (Non-degenerate existence). *Suppose that $d = 2$, that $\gamma > 0$ is sufficiently small, and that $\mathbf{X}_0 = \mathbf{c}\ell$, $\mathbf{c} = (c^1, c^2) > 0$. Then $\mathbf{X} = (X^1, X^2)$ exists as a pair of non-trivial continuous measure-valued processes, described by a* martingale problem, *involving the collision local time $L_{\mathbf{X}}$. That is, for smooth test functions $\varphi^i \geq 0$,*

$$t \mapsto M_t^i(\varphi^i) := \langle X_t^i, \varphi^i\rangle - c^i \langle \ell, \varphi^i\rangle - \int_0^t \mathrm{d}s \left\langle X_s^i, \frac{1}{2}\Delta\varphi^i\right\rangle,$$

$i = 1, 2$, are orthogonal, square integrable martingales, with square functions

$$t \mapsto \langle\!\langle M^i(\varphi^i)\rangle\!\rangle_t = \gamma \int_{[0,t]\times\mathbb{R}^2} L_{\mathbf{X}}\big(\mathrm{d}(s,b)\big)\, (\varphi^i)^2(b),$$

where $L_{\mathbf{X}}\big(\mathrm{d}(s,b)\big)$ is the collision local time of X^1 and X^2, which is the limit in law as $\varepsilon \downarrow 0$ of the locally finite measures

$$\mathrm{d}s\mathrm{d}b\, \frac{1}{\varepsilon} \int_0^\varepsilon \mathrm{d}r\; X_s^1 * \mathrm{p}_r(b) X_s^2 * \mathrm{p}_r(b)$$

on $\mathbb{R}_+ \times \mathbb{R}^2$.

Note that the existence of the collision local time $L_{\mathbf{X}}$ is an essential part of the martingale problem.

The key tool for the construction of $\mathbf{X}$ are expensive 4th moment estimates. In fact, there is a closed system of equations for the 4th moments of approximating processes, which can successfully be handled.

The restriction to small γ comes from the fact that (formal) moments of order higher than 2 exhibit a phase transition: They are infinite for sufficiently large $\gamma > 0$. (Imagine that too much reaction leads to very large fluctuations in the sense that moments of order three and more explode.)

Since the model exists non-trivially according to this theorem, how can the "paradoxes" mentioned earlier be resolved? In other words, what features the model does have, and how can these be understood?

3.8. Resolving the paradoxes

The following theorem is also taken from [6].

Theorem 3.7 (Local properties). *Consider the process* $\mathbf{X}$ *constructed in the proof of Theorem 3.6, and fix a time point* $t > 0$.

(a) **(Density functions):** *Both states* X_t^i *are absolutely continuous:*

$$X_t^i(\mathrm{d}b) \;=\; X_t^i(b)\mathrm{d}b, \qquad i = 1,2.$$

(b) **(Law of $\mathbf{X}_t(x)$):** *For almost all* $b \in \mathbb{R}^2$, *the law of* $\mathbf{X}_t(b)$ *coincides with the state* B_τ *of a Brownian motion in* $\mathbb{R}_+^2$, *starting from* $\mathbf{c}$, *at its hitting time* τ *of the boundary of* $\mathbb{R}_+^2$. *In particular:*

$$\mathrm{Var}X_t^i(b) \equiv \infty.$$

(c) **(Local segregation of types):** *For almost all* b,

$$X_t^1(b)X_t^2(b) \;=\; 0 \quad a.s. \tag{26}$$

Consequently, the states of $\mathbf{X}$ are absolutely continuous, as we heuristically concluded in the previous subsection. Moreover, the law of the random density in a point can explicitly be described. The infinite variance of the hitting state can also be obtained by calculations starting from the $\mathbb{Z}^2$-model. Finally, the intuitively predicted local segregation property is confirmed.

Note that the law of the random densities $\mathbf{X}_t(b)$ is independent of t and b. This follows from the spatial shift invariance (in law) and the *self-similarity*

$$X_{K^2t}(K\cdot) \overset{\mathcal{L}}{=} X_t(\cdot), \qquad t, K > 0.$$

But if the types are separated as described in (26), how can it be explained that there is still a non-trivial interaction going on? For this purpose, it might be helpful to first consider the following simplified example (see also Figure 2). For measures μ and ν on $\mathbb{R}$, define the *collision measure* $C_{[\mu,\nu]}$ on $\mathbb{R}$ by

$$\int_{\mathbb{R}} C_{[\mu,\nu]}(\mathrm{d}a)\varphi(a) \;=\; \lim_{\varepsilon\downarrow 0}\int_{\mathbb{R}}\mu(\mathrm{d}a)\varphi(a)\int_{\mathbb{R}}\nu(\mathrm{d}b)\;\mathrm{p}_\varepsilon(a-b)$$

provided that the limit exists in $\mathbb{R}_+$ for all continuous functions $\varphi \geq 0$ on $\mathbb{R}$ with compact support (cf. [17]).

Example 3.8 (Non-trivial collision despite segregation). For $\alpha, \beta \in (0,1)$, consider the segregated locally finite measures μ and ν having the density functions

$$f(b) \;:=\; \mathbf{1}_{(-\infty,0)}|b|^{-\alpha} \quad and \quad g(b) \;:=\; \mathbf{1}_{(0,\infty)}|b|^{-\beta}, \qquad b \in \mathbb{R},$$

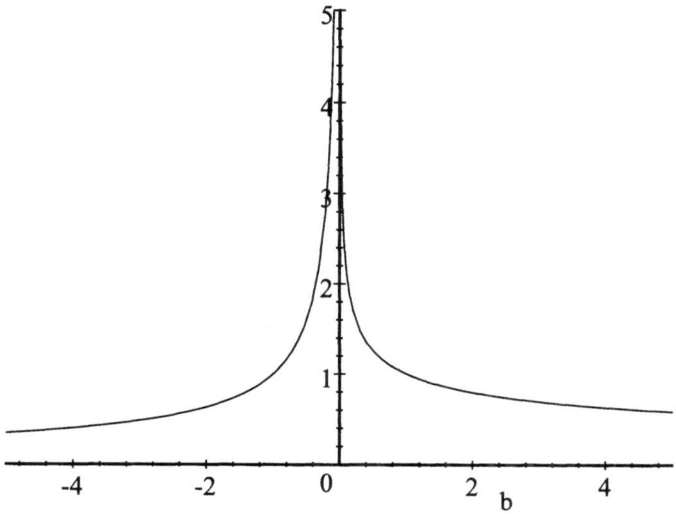

FIGURE 2. Segregated measure densities with $\alpha = \frac{2}{3}$ and $\beta = \frac{1}{3}$.

respectively. Then, for $\varepsilon > 0$, as well as φ continuous and with compact support, by Brownian scaling,

$$\int_{\mathbb{R}} \mu(\mathrm{d}a)\varphi(a) \int_{\mathbb{R}} \nu(\mathrm{d}b)\ \mathrm{p}_\varepsilon(a-b) = \ \varepsilon^{\frac{1}{2}(1-\alpha-\beta)} \int_{\mathbb{R}} \mathrm{d}a\ f(a)\varphi\big(\sqrt{\varepsilon}b\big) \int_{\mathbb{R}} \mathrm{d}b\ g(b)\ \mathrm{p}_1(a-b).$$

But as $\varepsilon \downarrow 0$, the latter integral expression converges to $\varphi(0)c$ with

$$0 \ < \ c \ := \ \int_{\mathbb{R}} \mathrm{d}a\ f(a) \int_{\mathbb{R}} \mathrm{d}b\ g(b)\mathrm{p}_1(a-b) \ < \ \infty. \tag{27}$$

Consequently, as $\varepsilon \downarrow 0$, for $\alpha+\beta > 1$ the approximated collision measures explode, whereas for $\alpha + \beta < 1$ they will finally disappear, as suggested by the informal expression $f(b)g(b)\mathrm{d}b \equiv 0$ for the collision measure. But in the *critical* case $\alpha+\beta = 1$ (as in Figure 2), the collision measure $C_{[\mu,\nu]} = c\delta_0$ exists *non-trivially*, with the constant c [from (27)] depending on the approximation procedure. In simple terms, contact at a common boundary of well-quantified segregated absolutely continuous measures makes possible a proper collision.

Note that lifting up μ, ν to measure-valued paths $t \mapsto \mu_t$ and $t \mapsto \nu_t$ which are constant in time, analogous statements hold for the collision local time $L_{[\mu,\nu]}$. In particular, for critical exponents α, β, the collision local time $L_{[\mu,\nu]}$ exists and is *non-trivial*, namely

$$L_{[\mu,\nu]}\big(\mathrm{d}(s,b)\big) \ = \ \mathrm{d}s\ c\delta_0(\mathrm{d}b) \quad \text{with} \ \ 0 < c < \infty \text{ from (27)},$$

despite the paths μ, ν are separated.

The last example illuminates that the local segregation property (26) is not in contradiction with the existence of a non-trivial collision local time $L_{\mathbf{X}}$ as in

Theorem 3.6, thus resolving an apparent paradox. Consequently, due to the large fluctuations at the interface of the two types (remember the hot spots mentioned in Subsection 1.4), the locally segregated "sexes" actually interact non-trivially only by contact, keeping the mutually catalytic process $\mathbf{X}$ in $\mathbb{R}^2$ stochastically alive.

For the two-dimensional continuum mutually catalytic branching model in the case of finite measure states we refer to [5], and for uniqueness questions in the defining martingale problem to [14].

Acknowledgment. We are grateful to Achim Klenke allowing us to show his simulation picture, and to Jean-François Delmas for a careful reading of the original manuscript. We are also grateful to an anonymous referee for pointing out some queries.

References

[1] M. T. Barlow, S. N. Evans and E. A. Perkins, *Collision local times and measure-valued processes*, Canad. J. Math., **43 (5)** (1991), 897–938.

[2] J. T. Cox and D. Griffeath, *Diffusive clustering in the two dimensional voter model*, Ann. Probab., **14** (1986), 347–370.

[3] J. T. Cox and A. Klenke, *Recurrence and ergodicity of interacting particle systems*, Probab. Theory Related Fields, **116 (2)** (2000), 239–255.

[4] D. A. Dawson, *Measure-valued Markov processes*, In P. L. Hennequin (Ed), *École d'été de probabilités de Saint Flour XXI – 1991*, volume 1541 of Lecture Notes in Mathematics, Springer-Verlag, 1993, 1–260.

[5] D. A. Dawson, A. M. Etheridge, K. Fleischmann, L. Mytnik, E. A. Perkins and J. Xiong, *Mutually catalytic branching in the plane: Finite measure states*, WIAS Berlin, Preprint No. 615, 2000.

[6] D. A. Dawson, A. M. Etheridge, K. Fleischmann, L. Mytnik, E. A. Perkins and J. Xiong, *Mutually catalytic branching in the plane: Infinite measure states*, WIAS Berlin, Preprint No. 633, 2001.

[7] J.-F. Delmas, *Super-mouvement brownien avec catalyse*, Stochastics and Stochastics Reports, **58** (1996), 303–347.

[8] D. A. Dawson and K. Fleischmann, *Super-Brownian motions in higher dimensions with absolutely continuous measure states*, Journ. Theoret. Probab., **8 (1)** (1995), 179–206.

[9] D. A. Dawson and K. Fleischmann, *A continuous super-Brownian motion in a super-Brownian medium*, Journ. Theoret. Probab., **10 (1)** (1997), 213–276.

[10] D. A. Dawson and K. Fleischmann, *Longtime behavior of a branching process controlled by branching catalysts*, Stoch. Process. Appl., **71 (2)** (1997), 241–257.

[11] D. A. Dawson and K. Fleischmann, *Catalytic and mutually catalytic branching*, in *Infinite Dimensional Stochastic Analysis*, Royal Netherlands Academy of Arts and Sciences, 2000, 145–170.

[12] J.-F. Delmas and K. Fleischmann, *On the hot spots of a catalytic super-Brownian motion*, WIAS Berlin, Preprint No. 450 (1998), to appear in Probab. Theory Relat. Fields, 2001.

[13] D. A. Dawson, K. Fleischmann and C. Mueller, *Finite time extinction of superprocesses with catalysts*, Ann. Probab., **28 (2)** (2000), 603–642.

[14] D. A. Dawson, K. Fleischmann, L. Mytnik, E. A. Perkins and J. Xiong, *Mutually catalytic branching in the plane: Uniqueness*, WIAS Berlin, Preprint No. 641, 2001.

[15] D. A. Dawson and E. A. Perkins, *Long-time behavior and coexistence in a mutually catalytic branching model*, Ann. Probab., **26 (3)** (1998), 1088–1138.

[16] A. M. Etheridge and K. Fleischmann, *Persistence of a two-dimensional super-Brownian motion in a catalytic medium*, Probab. Theory Relat. Fields, **110 (1)** (1998), 1–12.

[17] S. N. Evans and E. Perkins, *Absolute continuity results for superprocesses with some applications*, Trans. Amer. Math. Soc., **325 (2)** (1991), 661–681.

[18] S. N. Evans and E. A. Perkins, *Measure-valued branching diffusions with singular interactions*, Canad. J. Math., **46 (1)** (1994), 120–168.

[19] K. Fleischmann and A. Klenke, *Smooth density field of catalytic super-Brownian motion*, Ann. Appl. Probab., **9 (2)** (1999), 298–318.

[20] K. Fleischmann and A. Klenke, *The biodiversity of catalytic super-Brownian motion*, Ann. Appl. Probab., **10 (4)** (2001), 1121–1136.

[21] K. Fleischmann and C. Mueller, *Finite time extinction of catalytic branching processes*, in Luis G. Gorostiza and B. Gail Ivanoff (Eds), *Stochastic Models*, volume 26 of CMS Conference Proceedings, Amer. Math. Soc., Providence, 2000, 125–139.

[22] K. Fleischmann and J. Xiong, *A cyclically catalytic super-Brownian motion*, WIAS Berlin, Preprint No. 528 (1999), to appear in Ann. Probab., 2001.

[23] A. Klenke, *A review on spatial catalytic branching*, in Luis G. Gorostiza and B. Gail Ivanoff (Eds), *Stochastic Models*, volume 26 of CMS Conference Proceedings, Amer. Math. Soc., Providence, 2000, 245–263.

[24] A. Klenke, *Absolute continuity of catalytic measure-valued branching processes*, Stoch. Proc. Appl., **89 (2)** (2000), 227–237.

[25] C. Mueller and E. A. Perkins, *Extinction for two parabolic stochastic PDE's on the lattice*, Ann. Inst. H. Poincaré Probab. Statist., **36 (3)** (2000), 301-338.

[26] L. Mytnik, *Uniqueness for a mutually catalytic branching model*, Probab. Theory Related Fields, **112 (2)** (1998), 245–253.

[27] L. Pagliaro and D. L. Taylor, *Aldolase exists in both the fluid and solid phases of cytoplasm*, J. Cell Biology, **107** (1998), 981–999.

[28] T. Shiga, *Two contrasting properties of solutions for one-dimensional stochastic partial differential equations*, Can. J. Math., **46 (2)** (1994), 415–437.

[29] T. Shiga and A. Shimizu, *Infinite-dimensional stochastic differential equations and their applications*, J. Mat. Kyoto Univ., **20** (1980), 395–416.

The Fields Institute
222 College Street
Toronto, Ontario
Canada M5T
E-mail address: ddawson@fields.utoronto.ca

Weierstrass Institute for Applied Analysis and Stochastics
Mohrenstr. 39
D–10117 Berlin
Germany
E-mail address: fleischmann@wias-berlin.de

Progress in Probability, Vol. 52, 111–120

Sticky Particles, Scalar Conservation Law and Pressureless Gas Equations

A. Dermoune, H. Machrafi, O. Moutsinga

Abstract. Let $\mathbf{P}_0$ be a probability measure on $\mathbb{R}$ and u_0 be a bounded continuous map from $\mathbb{R}$ to $\mathbb{R}$. Let $(\mathbf{P}_0^{(n)} : n \geq 1)$ be a sequence of finite probabilities which converges weakly to $\mathbf{P}_0$. For each n let $\varphi(\cdot, \cdot, P_0^{(n)}, u_0)$ be the forward flow map associated to sticky particles with initial distribution of masses given by $\mathbf{P}_0^{(n)}$ and initial velocity u_0. We prove that the sequence $(\varphi(\cdot, \cdot, P_0^{(n)}, u_0) : n \geq 1)$ converges uniformly on every compact set of $\mathbb{R}_+ \times \mathbb{R}$ to the map $\varphi(\cdot, \cdot, \mathbf{P}_0, u_0)$ given via the generalized variational principle. We show that there exists a measurable map $u(\cdot, \cdot)$ from $\mathbb{R}_+ \times \mathbb{R}$ to $\mathbb{R}$ such that $dt \otimes \mathbf{P}_0$ a.e., $\frac{\partial}{\partial t}\varphi(t, x, \mathbf{P}_0, u_0) = u(t, \varphi(t, x, \mathbf{P}_0, u_0))$. We show also that the family $(\mathbf{P}_t = \mathbf{P}_0 \circ \varphi^{-1}(t, \cdot), u(t, \cdot) : t \geq 0)$ is a weak solution of the pressureless gas equations, and that the function $M(t, x) = P_0(y \in \mathbb{R} : \varphi(t, y, \mathbf{P}_0, u_0) \leq x), (t, x) \in \mathbb{R}_+ \times \mathbb{R}$ is the unique entropy solution of

$$\partial_t M + \partial_x (A(M)) = 0, \quad M(0, x) = \mathbf{P}_0(-\infty, x], \quad \forall x \in \mathbb{R},$$

where the flux function $A(m) = \int_0^m u_0(M_0^{-1}(m'))dm'$ and $M_0^{-1}(m) = \inf\{x : m < M_0(x)\}$.

1. Introduction and the main results

Definition 1.1 (Sticky particles). *Let u_0 be a map from $\mathbb{R}$ to $\mathbb{R}$. We consider n particles on the real line with initial locations $\{x_i : 1 \leq i \leq n\}$, velocities $\{u_0(x_i) : 1 \leq i \leq n\}$, and masses $\{m_i : 1 \leq i \leq n\}$. The particles move with constant velocities unless they collide. At collisions, the colliding particles stick and form a new massive particle. The mass and velocity of this new particle are given by the laws of conservation of mass and momentum.*

Definition 1.2 (The partition ξ_t associated with n sticky particles). *We consider n sticky particles with initial locations $\{x_i : 1 \leq i \leq n\}$. At each time $t > 0$, we denote by $\xi_t := \{G_1, \dots, G_{n(t)}\}$ the partition of $\{x_i : 1 \leq i \leq n\}$, defined by the ordered groups $G_1, \dots, G_{n(t)}$, so that each group of particles is glued to a single one before or at time t, and different groups are at different location at time t.*

2000 *Mathematics Subject Classification.* 60F17, 60K40.
Key words and phrases. Pressureless gas equations, generalized variational principle, non-linear stochastic differential equation.

Definition 1.3. *Let $(P_t : t \geq 0)$ be a family of probability measures on $\mathbb{R}$, weakly continuous with respect to t. For each $t \geq 0$, let $I_t = u(t,\cdot)P_t$ a measure absolutely continuous with respect to P_t. $(P_t, I_t, u(t,\cdot) : t \geq 0)$ is a weak solution of the pressureless gas equations*

$$\begin{cases} \frac{\partial P_t}{\partial t} + \frac{\partial I_t}{\partial x} = 0 \\ \frac{\partial I_t}{\partial t} + \frac{\partial (uI_t)}{\partial x} = 0 \\ P_t \to P_0, uP_t \to u_0 P_0 \quad \textit{weakly}, \end{cases}$$

if, for any $f \in C_0^1(\mathbb{R})$, the space of C^1-functions on $\mathbb{R}$ with compact support, and $0 < t_1 < t_2$,

$$\int f(x)dP_{t_2}(x) - \int f(x)dP_{t_1}(x) = \int_{t_1}^{t_2} \int f'(x)dI_t(x)dt,$$

and

$$\int f(x)dI_{t_2}(x) - \int f(x)dI_{t_1}(x) = \int_{t_1}^{t_2} \int f'(x)u(t,x)dI_t(x)dt.$$

E, Sinai, Rykov in [5] have constructed a weak solution for the pressureless gas equations using sticky particles. The method used in [5] is based on the construction of a partition of $\mathbb{R}$ using the initial data as following.

Generalized variational principle (GVP): $y \in \mathbb{R}$ is the left endpoint of an element of the partition iff for any $y_1 < y < y_2$, the following holds:

$$\frac{\int_{[y_1,y)} \eta + tu_0(\eta)dP_0(\eta)}{P_0([y_1,y))} < \frac{\int_{[y,y_2]} \eta + tu_0(\eta)dP_0(\eta)}{P_0([y,y_2])}.$$

The (GVP) for the right endpoint of an element of the partition is defined by a similar inequality. If $P_0 = \sum_i m_i \delta(x_i - x)$, then we will see in Section 2 that for all $t \geq 0$, the partition given by the (GVP) coincides with the partition ξ_t given in Definition 1.2. So, there is no ambiguity if ξ_t also denotes the partition given by the (GVP).

Having $(\xi_t : t \geq 0)$, E, Sinai, Rykov in [5] have defined the forward flow map associated to the sticky particles as following:

$$\varphi(t,x,P_0,u_0) = \frac{\int_{[y_l(x,t),y_r(x,t)]} \eta + tu_0(\eta)dP_0(\eta)}{P_0([y_l(x,t),y_r(x,t)])}, \tag{1}$$

where $[y_l(x,t), y_r(x,t)]$ is the unique element of ξ_t which contains x. In the sequel $\varphi(t,x,P_0,u_0)$ is denoted by $\varphi(t,x)$, if there is no ambiguity.

In [1] Y. Brenier and E. Grenier have got a link between sticky particles and the following scalar conservation law

$$\partial_t M + \partial_x(A(M)) = 0, \quad M_0(x) = P_0(-\infty,x] \; \forall x \in \mathbb{R}, \tag{2}$$

here the flux function $A(m) = \int_0^m u_0(M_0^{-1}(m'))dm'$, where $M_0^{-1}(m) = \inf\{x : m < M_0(x)\}$.

In [2, 3, 4] A. Dermoune has shown that there is a connection between sticky particles, pressureless gas equations with initial values $(u_0, \mathbf{P}_0)$ and the following nonlinear stochastic differential equation

$$dX_t = \mathbf{E}[u_0(X_0) \,|\, X_t]dt, \quad \text{law}(X_0) = \mathbf{P}_0. \tag{3}$$

The aim of this work is to give a new formulation of the generalized variational principle (GVP) which improves and gives in a precise manner these connections. This new formulation of the (GVP) works also in multi-dimensional case. The following theorem gives the main results.

Theorem 1.4. *Let $\mathbf{P}_0$ be a probability measure and u_0 a continuous bounded function.*

1) *Let $(\varphi_t : t \geq 0)$ be a family of continuous and non-decreasing maps from $\mathbb{R}$ to $\mathbb{R}$ such that:*
   i) *$\varphi_0(x) = x$ for all $x \in \mathbb{R}$,*
   ii) *for all $0 < t_1 < t_2$, and $x \in \mathbb{R}$, $\varphi_{t_1}^{-1}(\varphi_{t_1}(x)) \subset \varphi_{t_2}^{-1}(\varphi_{t_2}(x))$,*
   iii) *for all $t > 0$ and $x \in Supp(\mathbf{P}_0)$,*

$$\varphi_t(x) = \frac{\int_{C(t,x)} \eta + tu_0(\eta) d\mathbf{P}_0(\eta)}{\mathbf{P}_0(C(t,x))},$$

   where $C(t,x) = \varphi_t^{-1}(\varphi_t(x))$. Then φ coincides on $Supp(\mathbf{P}_0)$ with the map $\varphi(\cdot,\cdot,\mathbf{P}_0,u_0)$ given in (1).

2) *Let $(\mathbf{P}_0^{(n)} : n \geq 1)$ be a sequence of finite probabilities on $\mathbb{R}$ which converges weakly to $\mathbf{P}_0$. For each n we consider the forward flow $\varphi(t,x,\mathbf{P}_0^{(n)},u_0)$ given by (1) associated with $P_0^{(n)}$ and the velocity u_0. Then the sequence $(\varphi(\cdot,\cdot,\mathbf{P}_0^{(n)},u_0) : n \geq 1)$ converges uniformly in every compact set of $\mathbb{R}_+ \times \mathbb{R}$ to $\varphi(\cdot,\cdot,P_0,u_0)$.*

3) *There exists a measurable function u on $\mathbb{R}_+ \times \mathbb{R}$ such that $dt \otimes \mathbf{P}_0$ a.e.,*

$$\frac{\partial}{\partial t}\varphi(t,x,\mathbf{P}_0,u_0) = u(t,\varphi(t,x,\mathbf{P}_0,u_0)).$$

4) *Let X_0 be a random variable with probability distribution $\mathbf{P}_0$. Then the process $(X_t := \varphi(t,X_0,\mathbf{P}_0,u_0) : t \geq 0)$ is solution of the nonlinear stochastic differential equation (3). The family $(P_t = law(X_t), u(t,\cdot)P_t, u(t,\cdot) : t \geq 0)$ is a weak solution of the pressureless gas equations.*

From Part 2 of the latter theorem we get the following propagation of chaos for sticky particles.

Corollary 1.5. *Let $(X_i : i \geq 1)$ be an independent copies of the random variable X_0 with probability distribution $\mathbf{P}_0$. For all $k \geq 1$, $(\varphi(t,X_i,\frac{1}{n}\sum_{j=1}^n \delta_{X_j},u_0), i = 1,\ldots,k)$ converges in distribution as $n \to +\infty$ to k independent copies of the random variable $\varphi(t,X_0,\mathbf{P}_0,u_0)$.*

Corollary 1.6. *Suppose that the probability measure* $\mathbf{P}_0$ *has a compact support. Then the function* $M(t,x) = \mathbf{P}(X_t \leq x)$ *is the entropy solution of* (2).

The plan of this paper is the following. In Section 2, we recall the main results concerning sticky particles, and in Section 3 we give the proof of Theorem 1.4 and Corollary 1.6.

2. Detailed study of sticky particles

It is instructive to consider first the case of a finite collection of particles. We consider n sticky particles, and we use the notations of (1.1). The initial positions and masses can be described by the measure $P_0 = \sum_{i=1}^{n} m_i \delta(x - x_i)$. We can suppose that P_0 is a probability measure, and $x_1 < \cdots < x_n$. We denote by $x_i(t)$ the position of the ith particle at time t. We define a map $(t,x) \to \varphi(t,x)$ from $\mathbb{R}_+ \times \mathbb{R}$ to $\mathbb{R}$ as following:

$$\varphi(t,x) = x_i(t) \quad \text{if} \quad x = x_i,$$

and $\varphi(t,x)$ is defined by linear interpolation if $x_i \leq x < x_{i+1}$.

The location x_i can be seen as a realization of a random variable X_0, defined on some probability space $(\Omega, \mathbf{F}, \mu)$, with the distribution P_0 given by $\mu(X_0 = x_i) = P_0(\{x_i\}) = m_i$. The center of mass, at time t, of group of particles belonging to a subset G of $\mathbb{R}$, is given by

$$C(G,t) = \mathbf{E}[X_0 + tu_0(X_0) \,|\, X_0 \in G]. \tag{4}$$

It is a linear function of t. If G is a group of particles glued to a single one before or at time t, then from the conservation of mass and momentum, the location at time t of this group is given by (4).

The following lemmas are due to [5].

Lemma 2.1. *Let* G_1 *and* G_2 *be two neighboring groups of particles such that* $C(G_1,t) < C(G_2,t)$ *for* $t < \tau$, *and* $C(G_1,\tau) = C(G_2,\tau)$. *Then for* $t > \tau$,

$$C(G_2,t) < C(G_1 \cup G_2, t) < C(G_1,t).$$

Proof. Since both $C(G_1,t)$ and $C(G_2,t)$ are linear function of t, we have for $t > \tau$,

$$C(G_1,t) > C(G_2,t).$$

We have for $\alpha = \frac{\mathbf{P}(X_0 \in G_1)}{\mathbf{P}(X_0 \in G_1 \cup G_2)}$,

$$\begin{aligned} C(G_1 \cup G_2, t) &= \mathbf{E}[X_0 + tu_0(X_0) \,|\, X_0 \in G_1 \cup G_2] \\ &= \alpha C(G_1,t) + (1-\alpha) C(G_2,t). \end{aligned}$$

The latter equality achieves the proof. □

Lemma 2.2. *Let* $G = \{x_i : j' \leq i \leq j''\}$ *be an element of the partition* ξ_t *defined in 1.2. Let* i *such that* $j' \leq i \leq j''$, $I_1 = \{x_{j'}, \ldots, x_i\}$, *and* $I_2 = \{x_{i+1}, \ldots, x_{j''}\}$. *Then* $C(I_1,t) \geq C(I_2,t)$.

Proof. Assume to the contrary that $C(I_1, t) < C(I_2, t)$. Since $C(G, t) = \alpha C(I_1, t) + (1 - \alpha)C(I_2, t)$ for some $\alpha \in (0, 1)$, we have

$$C(I_1, t) < C(G, t). \tag{5}$$

Let us consider the evolution of the set of particles I_1. Let τ_1 be the time when the set I_1 is hit from the right by a particle or a cluster of particles $J(\tau_1) \subset I_2$, we add them to our set. We have from Lemma 2.1

$$C(I_1 \cup J(\tau_1), s) \leq C(I_1, s) \quad \forall s \geq \tau_1. \tag{6}$$

Now, we consider the evolution of set of particles $I_1(\tau_1) := I_1(\tau_0) \cup J(\tau_1)$, where $\tau_0 = 0$ and $I_1(0) := I_1$. By induction we get a sequence of times $(\tau_k : k \geq 0)$, and sets $(I_1(\tau_{k+1}) = I_1(\tau_k) \cup J(\tau_{k+1}) : k \geq 0)$, where $J(\tau_{k+1}) \subset I_2 \setminus I_1(\tau_k)$, such that

$$C(I_1(\tau_{k+1}), s) \leq C(I_1(\tau_k), s) \quad \forall s \geq \tau_{k+1}.$$

From these inequalities, and from (6) we have for all $k \geq 0$,

$$C(I_1(\tau_{k+1}), s) \leq C(I_1, s) \quad \forall s \geq \tau_{k+1}. \tag{7}$$

From the assumption of the Lemma the group of particles $G = I_1 \cup I_2$ are glued to a single one before or at time t. Then necessary $I_1(\tau_{k+1}) = I_1 \cup I_2$ for some $\tau_{k+1} \leq t$. Hence we have from (7)

$$C(G, t) < C(I_1, t),$$

contradicting (5). □

Corollary 2.3. *Let $G = \{x_{i'}, \ldots, x_{j''}\}$ be an element of the partition ξ_t defined in 1.2. Then*

$$x_{j''} + tu_0(x_{j''}) \leq C(G, t) \leq x_{i'} + tu_0(x_{i'}).$$

Proof. We have from Lemma 2.2

$$C(\{x_i : i' < i \leq j''\}, t) \leq x_{i'} + tu_0(x_{i'}),$$

and

$$x_{j''} + tu_0(x_{j''}) \leq C(\{x_i : i' \leq i < j''\}, t).$$

From these inequalities and from Lemma 2.1, we derive

$$C(G, t) \leq x_{i'} + tu_0(x_{i'}),$$

and

$$x_{j''} + tu_0(x_{j''}) \leq C(G, t).$$

□

Remark 2.4. The lemmas above work also for all probability measure P_0 on $\mathbb{R}$.

Lemma 2.5. *A particle x is the left endpoint, respectively the right endpoint, of an element of the partition ξ_t defined in 1.2, if and only if*

$$\max_{y<x} C([y,x),t) < \min_{z\geq x} C([x,z],t), \tag{8}$$

respectively

$$\max_{y<x} C([y,x],t) < \min_{z\geq x} C((x,z],t). \tag{9}$$

Proof. The proof of both cases are similar. Let x be a particle satisfying (8), and belonging to the group $G = \{x_i, \dots, x_j\} \in \xi_t$ at time t. Assume that $x_i < x$. From 2.2 we have

$$C([x_i,x),t) \geq C([x,x_j],t),$$

which is absurd.

Assume now that x is the left endpoint of an element of ξ_t. For any $y < x < z$, we want to show that $C([y,x),t) < C([x,z],t)$. Let $I_1, \dots, I_l$ be consecutive elements of ξ_t to the left of x, and $y \in I_1 = \{x_i, i_1 \leq i \leq i_2\}$. Let $J_1, \dots, J_r$ be the consecutive elements of ξ_t to the right of x, and $x \in J_1 = \{x, \dots, x'\}$, and $z \in J_r = \{x_i, j_1 \leq i \leq j_2\}$.

We have first

$$C(I_1,t) < C(I_2,t) < \cdots < C(J_1,t) < \cdots < C(J_r,t).$$

From Lemma 2.1 and 2.2, we have

$$C((y,x_{i_2}],t) < C(I_1,t) < C([x_{i_1},y],t),$$

and

$$C((z,x_{j_2}],t) < C(J_r,t) < C([x_{j_1},z),t).$$

Since

$$C([y,x),t) = \alpha_1 C([y,x^0_{i_2}],t) + \alpha_2 C(I_2,t) + \cdots + \alpha_l C(I_l,t),$$

and

$$C([x,z],t) = \beta_1 C([x,x'],t) + \beta_2 C(J_2,t) + \cdots + \beta_r C([x^0_{j_1},z],t),$$

where $\sum \alpha_i = \sum \beta_i = 1$, and $\alpha_i \geq 0, \beta_i \geq 0$, we must have

$$C([y,x),t) < C([x,z],t).$$

□

Properties of the forward flow $\varphi(\cdot,\cdot)$. We recall that $P_0 = \sum_i m_i \delta(x - x_i)$. Let $I_1, \dots, I_j, \dots$ be the successive group of particles glued to a single one before or at time t. For $x \in I_j$, we have

$$\varphi(t,x) = \mathbf{E}[X_0 + tu_0(X_0) \,|\, X_0 \in I_j],$$

which implies that $\varphi(t,\cdot)$ coincides on $Supp(P_0)$ with the map $\varphi(t,\cdot,P_0,u_0)$ given by the (GVP). For all $t \geq 0$, the map $x \in \mathbb{R} \to \varphi(t,x)$ is non-decreasing. The map $t \in \mathbb{R}_+ \to \varphi(t,x)$, for $x \in \mathbb{R}$, is locally Lipschitz continuous, and satisfies the following property:

$$\varphi(t,X_0) = \mathbf{E}[X_0 + tu_0(X_0) \,|\, \varphi(t,X_0)\,]. \tag{10}$$

Theorem 2.6. *For all $a < b$ and $T > 0$, the set $\{x, \varphi(t,x) \in (a,b)\}$ is uniformly bounded with respect to the class of probabilities P_0 supported by finite sets, and $t \in [0,T]$.*

Proof. Let $x_{\min}(t) = \min\{x \in Supp(P_0), \varphi(t,x) \in (a,b)\}$, and $x_{\max}(t) = \max\{x \in Supp(P_0), \varphi(t,x) \in (a,b)\}$. Obviously $x_{\min}(t)$ (respectively $x_{\max}(t)$) has to be the left endpoint (respectively the right endpoint) of an element I_j in the partition ξ_t. From Corollary 2.3, we have

$$x_{\min}(t) + tu_0(x_{\min}(t)) \geq \varphi(t, x_{\min}(t)) := C(I_j, t) \geq a.$$

Now, using the fact that $\frac{u_0(x)}{x} \to 0$, as $|x| \to +\infty$, we get that $x_{\min}(t)$ is uniformly bounded from below, with respect to $t \in [0,T]$ and the class of probabilities P_0 supported by finite sets. Similarly, from Corollary 2.3 we have

$$x_{\max}(t) + tu_0(x_{\max}(t)) \leq \varphi(t, x_{\max}(t)) \leq b.$$

Again the hypothesis under u_0 yields an upper bound. □

Theorem 2.7.

(1) *Let I_j be an element of the partition ξ_t, and $x_l = \min I_j$, and $x_r = \max I_j$, then*

$$C(\{x_l\}, s) = C(\{x_r\}, s), \quad \text{for some } s \leq t.$$

If $a < b$, and $T > 0$, then:

(2) *the set $\varphi(t, [a,b])$ is uniformly bounded with respect to the distribution of X_0, and $t \in [0,T]$,*

(3) *the set $\frac{\partial \varphi(t,[a,b])}{\partial t}$, defined dt a.e., is uniformly bounded with respect to the class of probabilities P_0 supported by finite sets, and $t \in [0,T]$.*

Proof. Part one. From Corollary 2.3 we have $C(\{x_l\}, t) \geq \varphi(t,x) \geq C(\{x_r\}, t)$. Since $x_l \leq x_r$ we have

$$C(\{x_l\}, s) = C(\{x_r\}, s), \quad \text{for some } s \in (0,t].$$

Part two. Let $a < b$ and $T > 0$. Let $x \in [a,b]$ and I_j be the element of ξ_t which contains x. Let x_l and x_r be respectively the left and right endpoints of I_j. It is easy to see that $\sup x_l \leq b$ and $a \leq \inf x_r$, where inf and sup are taken under $x \in [a,b]$, P_0 belonging to the class of probabilities supported by finite sets and $t \in [0,T]$. Now from Part one, and the fact that

$$\lim_{|y| \to \infty} \frac{u_0(y)}{y} = 0,$$

we derive that $\varphi(t,x)$ is uniformly bounded with respect to the distribution of X_0 and $t \in [0,T]$.

Part three. For $x \in [a,b]$, we have from (10)

$$\frac{\partial \varphi(t,x)}{\partial t} = \mathbf{E}[u_0(X_0) \,|\, \varphi(t,X_0) = \varphi(t,x)] \quad dt \text{ a.e.}$$

It follows from Part three, and Theorem 2.6, that

$$|\frac{\partial\varphi(t,x)}{\partial t}| \leq \max_{y\in K} |u_0(y)|,$$

where K is some compact set which depends on a, b, T, and u_0. □

3. Proof of Theorem 1.4

The proof of the first assertion is intuitively obvious. Let $t > 0$ and $x \in Supp(\mathbf{P}_0)$. We set $\varphi_t^{-1}(\varphi(t,x)) := \{y : \varphi(t,x) = \varphi(t,y)\} = [\alpha, \beta]$. We can show, using Remark 2.4, that α (respectively β) is a left endpoint (respectively right endpoint) of an element of the partition ξ_t. It follows that φ is the forward flow map given by the generalized variational principle. Which achieves the proof of the first assertion of Theorem 1.4.

Let us now prove the second assertion of Theorem 1.4. First we prove that $(\varphi(\cdot,\cdot,P_0^{(n)},u_0) : n \geq 1)$ is relatively compact in $C([0,T] \times [a,b])$ for all $T > 0$, $a < b$.

Suppose that there exist $T > 0$, and $[a,b] \subset \mathbb{R}$ such that the sequence $(\varphi(\cdot,\cdot,P_0^{(n)},u_0) : n \geq 1)$ is not relatively compact on $[0,T] \times [a,b]$. Then there exist $\varepsilon > 0$ and a sequences $\delta_n \to 0+$, $((t_n,x_n),(s_n,y_n) : n \geq 1)$ belonging to $[0,T] \times [a,b]$ such that $|t_n - s_n| + |x_n - y_n| \leq \delta_n$ but

$$|\varphi(t_n,x_n,P_0^{(n)},u_0) - \varphi(s_n,y_n,P_0^{(n)},u_0)| \geq \varepsilon.$$

We can suppose that $t_n \to t, s_n \to s, x_n \to x, y_n \to y$, and as $\delta_n \to 0$ then $t = s$ and $x = y$. We can also suppose that $x_n \leq y_n$ for all n. By the triangular inequality we have

$$|\varphi(t_n,x_n,P_0^{(n)},u_0) - \varphi(s_n,y_n,P_0^{(n)},u_0)| \leq |\varphi(t_n,x_n,P_0^{(n)},u_0) - \varphi(t,x_n,P_0^{(n)},u_0)|$$

$$+|\varphi(t,x_n,P_0^{(n)},u_0) - \varphi(t,y_n,P_0^{(n)},u_0)| + |\varphi(t,y_n,P_0^{(n)},u_0) - \varphi(s_n,y_n,P_0^{(n)},u_0)|.$$

Using the fact that $t \in [0,T] \to \varphi(t,x,P_0^{(n)},u_0)$ is uniformly Lipschitz-continuous, with respect to $P_0^{(n)}$ and $x \in [a,b]$, then necessarily for n large

$$|\varphi(t,x_n,P_0^{(n)},u_0) - \varphi(t,y_n,P_0^{(n)},u_0)| \geq \varepsilon/2.$$

Let $\{x_{i_1}^{(n)},\ldots,x_{i_k}^{(n)}\}$ (respectively $\{y_{j_1}^{(n)},\ldots,y_{j_m}^{(n)}\}$) be the set of ordered particles located in $\varphi(t,x_n,P_0^{(n)},u_0)$ (respectively $\varphi(t,y_n,P_0^{(n)},u_0)$) at time t. The condition $x_n \leq y_n$ implies that $x_{i_k}^{(n)} \leq y_{j_1}^{(n)}$. From Corollary 2.3, and the fact that the map $x \to \varphi(t,x,P_0^{(n)},u_0)$ is non-decreasing, we have

$$x_{i_k}^{(n)} + tu_0(x_{i_k}^{(n)}) \leq \varphi(t,x_n,P_0^{(n)},u_0)$$

$$\leq \varphi(t,y_n,P_0^{(n)},u_0) \leq y_{j_1}^{(n)} + tu_0(y_{j_1}^{(n)}).$$

From $x_n \leq x_{i_k} \leq y_{j_1} \leq y_n$, $x_n - y_n \to 0$ and the continuity of u_0, we have

$$\varphi(t,x_n,P_0^{(n)},u_0) - \varphi(t,y_n,P_0^{(n)},u_0) \to 0.$$

We conclude that the hypothesis "$\varphi(\cdot,\cdot,P_0^{(n)},u_0)$ is not relatively compact on $C([0,T]\times[a,b])$" is absurd.

Now we prove that the sequence $(\varphi(\cdot,\cdot,\mathbf{P}_0^{(n)},u_0) : n\geq 1)$ has a unique limit with respect to the uniform convergence on every compact set.

Let φ be a limit of a subsequence $(\varphi(\cdot,\cdot,\mathbf{P}_0^{(n(k))},u_0) : k\geq 1)$. We can suppose that $n(k)=k$ for all k. We set $X_t=\varphi(t,X_0)$, and $X_t^{(n)}=\varphi(t,X_0^{(n)},\mathbf{P}_0^{(n)},u_0)$, where X_0 is a random variable with probability distribution P_0 and $X_0^{(n)}$ be a finite random variable with probability distribution $\mathbf{P}_0^{(n)}$ which converges almost surely to X_0.

From (10), we have

$$X_t^{(n)}=\mathbf{E}[X_0^{(n)}+tu_0(X_0^{(n)})\mid X_t^{(n)}],\quad \forall t\geq 0. \tag{11}$$

Let f be a continuous function with compact support. Then for all $t\geq 0$, the sequence of random variables $f(X_t^{(n)})X_t^{(n)}$ is uniformly bounded, and converges $a.s.$ to $f(X_t)X_t$. By the dominated convergence theorem we have

$$\lim_{n\to+\infty}\mathbf{E}[f(X_t^{(n)})X_t^{(n)}]=\mathbf{E}[f(X_t)X_t].$$

From Theorem 2.6, we can show that the sequence of random variables $f(X_t^{(n)})[X_0^{(n)}+tu_0(X_0^{(n)})]$ is uniformly bounded. From that and from (11) the sequence

$$\mathbf{E}[f(X_t^{(n)})X_t^{(n)}]=\mathbf{E}[f(X_t^{(n)})(X_0^{(n)}+tu_0(X_0^{(n)}))]$$

converges as $n\to+\infty$ to

$$\mathbf{E}[f(X_t)X_t]=\mathbf{E}[f(X_t)(X_0+tu_0(X_0))].$$

The latter equality holds for all continuous function f with compact support. Now, we derive that

$$X_t=\mathbf{E}[X_0+tu_0(X_0)\mid X_t].$$

This equality implies that for all $x\in\mathbb{R}$, $\varphi(t,\cdot)$ is constant on

$$\varphi^{-1}(t,\varphi(t,x))=\{y:\lim_{n\to+\infty}\varphi(t,y,\mathbf{P}_0^{(n)},u_0)=\lim_{n\to+\infty}\varphi(t,x,\mathbf{P}_0^{(n)},u_0)\},$$

and it is the position of the center of mass of $\varphi^{-1}(t,\varphi(t,x))$. This fact implies that

$$\varphi^{-1}(t,\varphi(t,x))\subset\varphi^{-1}(t+s,\varphi(t+s,x))$$

for all $s\geq 0$, and $\varphi(t,\cdot)$ is clearly non-decreasing for all $t\geq 0$. From the first assertion of Theorem 1.4, we conclude that $\varphi(\cdot,\cdot)=\varphi(\cdot,\cdot,\mathbf{P}_0,u_0)$. Which achieves the proof of the assertion two of Theorem 1.4.

Now, we prove assertions three and four of Theorem 1.4. From the second assertion of Theorem 1.4, we have for all $t,s>0$,

$$\varphi(t,x,\mathbf{P}_0,u_0)=\varphi(t,y,\mathbf{P}_0,u_0)\Rightarrow\varphi(t+s,x,\mathbf{P}_0,u_0)=\varphi(t+s,y,\mathbf{P}_0,u_0).$$

This fact implies that X_{t+s} is $\sigma(X_t)$-measurable. Using the same proof as in [2] we show that

$$X_t = X_0 + \int_0^t \mathbf{E}[u_0(X_0)\,|\,X_s]ds, \quad \forall t \geq 0,$$

and the family $(P_t, u(t,\cdot) = \mathbf{E}[u_0(X_0)\,|\,X_t = \cdot])$ is a weak solution of pressureless gas equations. Which achieves the proof of Theorem 1.4.

The proof of Corollary 1.6 is based on the second assertion of Theorem 1.4 and the following result of Y. Brenier, E. Grenier [1].

Theorem 3.1. Y. Brenier, E. Grenier *Let n sticky particles, with initial weight m_i, position x_i and velocity $u_0(x_i)$, where u_0 is a fixed given bounded continuous function on $\mathbb{R}$, and the total mass is normalized $\sum_i m_i = 1$. Assume that the $(x_i : i)$ belong to a fixed compact and $\sum_i m_i\delta(x - x_i) \to P_0$ weakly when $n \to \infty$. Then, for each $t \geq 0$,*

$$\sum_i m_i\delta(x - x_i(t)) \to \partial_x M(t,x),$$

weakly, where M is the entropy solution of (2).

References

[1] Y. Brenier and E. Grenier, *Sticky particles and scalar conservation laws*, SIAM J. Num. Anal., **35** (1998), 2317–2328.

[2] A. Dermoune, *Probabilistic interpretation for system of conservation law arising in adhesion particle dynamics*, C.R.A.S. Paris Sér. I Math., **326** (1998), 595–599.

[3] A. Dermoune, *Probabilistic interpretation of sticky particles model*, The Annals of Probability, **27 (3)** (1999), 1357–1367.

[4] A. Dermoune, *Sticky particle model and propagation of chaos*, Nonlinear Anal., **45** (2001), 529–541.

[5] W. E, Yu. G. Rykov, and Ya. G. Sinai, *Generalized variational principles, global weak solutions and behavior with random initial data for systems of conservation laws arising in adhesion particle dynamics*, Commun. Math. Phys., **177** (1996), 349–380.

Laboratoire de Statistique et Probabilités
Université de Sciences et Technologies de Lille
59655 Villeneuve d'Ascq cedex
France
E-mail address: `Azzouz.Dermoune@univ-lille1.fr`

Progress in Probability, Vol. 52, 121–132

Affine Short Rate Models

D. Filipović

Abstract. We give a complete characterization of affine term structure (ATS) models based on a general non-negative Markov short rate process. This applies to the classical Cox–Ingersoll–Ross (CIR) model but includes as well short rate processes with jumps. We provide a link to the theory of branching processes and show how CBI-processes naturally enter the field of term structure modelling. Using Markov semigroup theory we exploit the full structure behind an ATS and provide a deeper understanding of some well-known properties of the CIR model. As a byproduct we get that any conservative CBI-process is a semimartingale.

1. Introduction

In valuing interest rate sensitive instruments in an arbitrage-free environment one is mostly interested in models which provide analytical expressions which are computationally well tractable. Models with an affine term structure (ATS), see (1) below, have been found convenient both by academics and practitioners. Prominent among them are the Vasicek [12] and Cox–Ingersoll–Ross (CIR) [3] short rate models where the short rates $r = (r_t)_{t \geq 0}$ follow a particular type of diffusion process. The time t price of a default-free zero coupon bond with maturity T is given by the expression

$$P(t,T) = e^{-A(T-t)-B(T-t)r_t} \tag{1}$$

where A and B are deterministic functions. These are determined by a system of Riccati equations containing the drift and diffusion coefficients of r.

Conversely, it has been shown in [2] and [5] that any diffusion (and hence *continuous*) short rate process r providing an ATS is necessarily of Vasicek or CIR type. Their results are based on Itô calculus.

However, interest rates do not evolve continuously in time. The above continuous short rate models should be seen as simplified descriptions of the real behavior of the interest rates. Coherent models include jumps.

2000 *Mathematics Subject Classification.* Primary: 91B28; Secondary: 60J85.
Key words and phrases. Affine term structure models, CBI-processes, infinitely decomposable processes, Markovian short rates.
Support from Crédit Suisse is gratefully acknowledged.

In this paper we exploit the full structure behind an ATS. We characterize the class of *all* (including *non-continuous*) *non-negative* Markov short rate processes r which imply an ATS model. Indeed, we start with any time homogeneous Markov process r on $E := [0, \infty)$ which satisfies the natural condition that $\int_0^t r_s \, ds$ exists and is finite a.s. for all t. Assuming an ATS, we then show – using Markov semigroup theory – that r is necessarily a *Feller process* with generator

$$\begin{aligned} \mathcal{A}f(x) &= \alpha x f''(x) + (b' + \beta x) f'(x) \\ &\quad + \int_{E^\circ} \Big(f(x+y) - f(x) - f'(x)(1 \wedge y) \Big) \Big(m(dy) + x\mu(dy) \Big) \end{aligned} \tag{2}$$

where $E^\circ := (0, \infty)$ and the coefficients will be specified below (see Theorem 5.6). Such processes are known as *CBI-processes*[1] (continuous state branching processes with immigration) and have been well studied in [8] and [11].

Conversely, any operator $\mathcal{A}$ of the above form generates in a unique way a conservative Feller process r on E which provides an ATS model. Thus, we establish a one-to-one correspondence between conservative CBI-processes (describing branching phenomena) and Markov short rate processes implying an ATS.

We identify the *pricing semigroup*

$$Q_t f(x) = \int_E f(y) \, q_t(x, dy) := \mathbb{E}\Big[e^{-\int_0^t r_s \, ds} f(r_t) \mid r_0 = x \Big]$$

in terms of the Laplace transform of its transition function $q_t(x, dy)$. Thus, by Laplace inversion we get (in principle) analytic expressions for all european claims $f(r_t)$. This explains a known feature of the CIR model.

We restate the result in [11] that any conservative CBI-process is *infinitely decomposable* and vice versa. This considerably extends and clarifies another well-known fact behind the CIR model, namely the additivity property of the squared Bessel processes

$$\mathbb{Q}_x^\delta * \mathbb{Q}_{x'}^{\delta'} = \mathbb{Q}_{x+x'}^{\delta+\delta'},$$

see [9, Chapt. XI.1]. As a byproduct we obtain that any conservative CBI – and hence any infinitely decomposable – process is a semimartingale.

The outline of the paper is as follows. In Section 2 we introduce the canonical setting for the short rate process r. In Section 3 we give the exact definition of an ATS and provide a remarkable regularity result: the pricing semigroup (Q_t) is a fortiori Feller. The main results from [8] and [11] are restated in Section 4. There we also specify what an infinitely decomposable Markov process is. Section 5 contains the announced results which are derived by repeated application of the Feynman–Kac formula. Based on these fundamental results we present in Section 6 an extension of the classical CIR model by including jumps. We derive a closed form expression for the bond prices $P(t, T)$.

---

[1]Here is a slight abuse of the notation used in [8]: By a CBI-process we always mean a *stochastically continuous* CBI-process.

Remark 1.1 (On notation). We basically follow the terminology of Revuz and Yor [9].

Let $\mathcal{E}$ denote the Borel σ-algebra on E and $b\mathcal{E}$ the space of bounded measurable functions on E. We write $C_0(E)$ for the space of continuous functions on E vanishing at infinity.

Both $b\mathcal{E}$ and $C_0(E)$ are Banach spaces for the norm $\|f\| := \sup_{x\in E} |f(x)|$.

2. Markovian short rates

For the following definitions we refer to [9, Chapt. III.1] and [6, Chapt. I–II].

Let $r = (r_t)_{t\geq 0}$ be a homogeneous Markov process, given by a transition function $p_t(x, dy)$ on $(E, \mathcal{E})$. We will assume that $r_t(\omega) = \omega(t)$ is realized as coordinate process on the canonical space[2] $\Omega = E^{[0,\infty)}$, $\mathcal{F} = \mathcal{E}^{[0,\infty)}$ with its natural filtration $(\mathcal{F}_t)_{t\geq 0}$. Thus, we are given a family of probability measures $(\mathbb{P}_x)_{x\in E}$ on $(\Omega, \mathcal{F})$ with $\mathbb{P}_x[r_0 = x] = 1$, such that r is Markov with respect to each $\mathbb{P}_x$.

The corresponding semigroup (P_t) on $b\mathcal{E}$, see [6, Chapt. II.1], is given by

$$P_t f(x) = \int_E f(y) p_t(x, dy) = \mathbb{E}_x[f(r_t)], \quad x \in E, \quad f \in b\mathcal{E}.$$

Here $\mathbb{E}_x$ denotes expectation with respect to $\mathbb{P}_x$. The Markov property reads as

$$\mathbb{E}_x[f(r_{t+s}) \mid \mathcal{F}_t] = \mathbb{E}_x[f(r_s \circ \theta_t) \mid \mathcal{F}_t] = \mathbb{E}_{r_t}[f(r_s)] = P_s f(r_t), \quad \mathbb{P}_x\text{-a.s.} \quad \forall f \in b\mathcal{E},$$

on $\{r_t < \infty\}$, where the shift operators $\theta_t : \Omega \to \Omega$ are defined by $(\theta_t\omega)(s) = \omega(t+s)$.

Notice that this is a fairly general setup for a time homogeneous Markov process in E. In order that $\int_0^t r_s \, ds$ is well defined we assume that r is progressively measurable and

$$\int_0^t r_s \, ds < \infty, \quad \mathbb{P}_x\text{-a.s.} \quad \forall t < \infty. \tag{3}$$

This implies that r is conservative, which means $p_t(x, E) = 1$ for all t and x.

It is clear that (P_t) is a positive contraction semigroup on $b\mathcal{E}$. Its (infinitesimal) generator $\mathcal{A}$ is defined for all $f \in b\mathcal{E}$ such that

$$\mathcal{A}f = \lim_{t\to 0} \frac{P_t f - f}{t}$$

exists in $b\mathcal{E}$. The domain of $\mathcal{A}$ is denoted by $\mathcal{D}(\mathcal{A})$. We write $b\mathcal{E}_0$ for the set of functions f in $b\mathcal{E}$ for which $\lim_{t\to 0} \|P_t f - f\| = 0$. It is well known that the resolvent R_λ of the operator $\mathcal{A}$ is a one-to-one mapping from $b\mathcal{E}_0$ onto $\mathcal{D}(\mathcal{A})$. Moreover $R_\lambda(b\mathcal{E}) \subset b\mathcal{E}_0$, see [6, Chapt. I.2].

We use the following notation, which differs from the terminology in [6].

---

[2]With the convention that E is replaced by its one-point compactification if r is not conservative (see the remark after the proof of Proposition (1.7) in [9, Chapt. III.1])

Definition 2.1. *A positive contraction semigroup* (Q_t) *on* $b\mathcal{E}$ *is called a* Feller semigroup *if*

$$Q_t f \in C_0(E) \quad \text{and} \quad \lim_{t\to 0} \|Q_t f - f\| = 0, \quad \forall f \in C_0(E).$$

The corresponding Markov process is called a Feller process.

We suppose each $\mathbb{P}_x$ is a risk neutral measure. That is, if $r_0 = x$, any contingent claim $h \in L^1(\Omega, \mathcal{F}_T, \mathbb{P}_x)$ maturing at time T has price

$$\mathbb{E}_x\left[e^{-\int_t^T r_s\,ds} h \mid \mathcal{F}_t\right] \tag{4}$$

at time $t \leq T$. In particular, the price of a zero coupon bond is $\mathbb{P}_x$-a.s.

$$P(t,T) = \mathbb{E}_x\left[e^{-\int_t^T r_s\,ds} \mid \mathcal{F}_t\right] = \mathbb{E}_x\left[e^{-\int_0^{T-t} r_s\,ds} \circ \theta_t \mid \mathcal{F}_t\right] = \Pi(T-t, r_t) \tag{5}$$

where $\Pi(u,x) := \mathbb{E}_x\left[e^{-\int_0^u r_s\,ds}\right]$. By assumption (3) we have that $\Pi(u,x)$ is a strictly positive non-increasing continuous function in u.

3. Affine term structure

The simplest non-trivial relation (5) is exponential-affine.

Definition 3.1. *A short rate model with*

$$\Pi(u,x) = e^{-A(u)-B(u)x}, \tag{6}$$

for some functions $A(u)$ *and* $B(u)$, *is said to possess an* affine term structure (ATS).

Notice that $A(u)$ and $B(u)$ a fortiori are non-negative and non-decreasing continuous functions with $A(0) = B(0) = 0$.

An immediate consequence of (5) is the following relation

$$\begin{aligned} e^{-A(T)-B(T)x} &= P(0,T) = \mathbb{E}_x\left[e^{-\int_0^t r_s\,ds} P(t,T)\right] \\ &= e^{-A(T-t)}\mathbb{E}_x\left[e^{-\int_0^t r_s\,ds} e^{-B(T-t)r_t}\right], \quad \forall t \leq T. \end{aligned} \tag{7}$$

Without loss of generality we may thus assume that $B(u) > 0$ for $u > 0$, since otherwise we are in the trivial case $B \equiv 0$. Accordingly, the image of B

$$\mathcal{I} := \{B(u) \mid u \geq 0\}$$

contains an open interval, hence is a set of uniqueness for analytic functions in E°.

It is well known that property (5) excludes arbitrage possibilities. And clearly it imposes structural properties on A, B and r. It turns out that r is necessarily a very special type of Feller process and that A and B satisfy a system of generalized Riccati equations which are given by the infinitesimal generator of r.

Definition 3.2. *With regard to* (4) *we call*

$$Q_t f(x) := \mathbb{E}_x\Big[e^{-\int_0^t r_s\,ds} f(r_t)\Big], \quad f \in b\mathcal{E}, \tag{8}$$

the pricing semigroup.

The naming pricing *semigroup* will be legitimated in Lemma 3.3 below. Write $f_\lambda(x) := e^{-\lambda x}$ and define $\Lambda := \{f_\lambda \mid \lambda > 0\}$. By the Stone–Weierstrass theorem the linear hull $\mathcal{L}(\Lambda)$ of Λ is dense in $C_0(E)$. Equality (7) reads now, for $\lambda = B(u)$,

$$Q_t f_{B(u)}(x) = e^{-(A(t+u)-A(u))-B(t+u)x}. \tag{9}$$

This structural property actually holds for all $\lambda \geq 0$.

Lemma 3.3. (Q_t) *is a Feller semigroup. Moreover, for each* $t \geq 0$ *we have*

$$Q_t f_\lambda(x) = e^{-\phi(t,\lambda)-\psi(t,\lambda)x}, \quad \forall \lambda \geq 0, \tag{10}$$

for some functions $\phi(t,\,.\,)$ *and* $\psi(t,\,.\,)$ *which are analytic in* $(0,\infty)$ *and satisfy*

$$\phi(t, B(u)) = A(t+u) - A(u), \quad \psi(t, B(u)) = B(t+u). \tag{11}$$

Proof. Clearly $Q_0 f = f$ and by the Markov property of r

$$Q_{t+s} f(x) = \mathbb{E}_x\Big[e^{-\int_0^{t+s} r_u\,du} f(r_{t+s})\Big] = \mathbb{E}_x\Big[e^{-\int_0^t r_u\,du}\Big(e^{-\int_0^s r_u\,du}\circ\theta_t\Big) f(r_s\circ\theta_t)\Big]$$
$$= \mathbb{E}_x\Big[e^{-\int_0^t r_u\,du} Q_s f(r_t)\Big] = Q_t Q_s f(x), \quad \text{for all } f \in b\mathcal{E}.$$

Moreover $0 \leq Q_t 1 \leq 1$, whence (Q_t) is a positive contraction semigroup on $b\mathcal{E}$.

Let $q_t(x,dy)$ denote the corresponding transition function, see [6, Chapt. II, Theorem 2.1]. For t and x fixed we have

$$Q_t f_\lambda(x) = \int_E e^{-\lambda y} q_t(x,dy)$$

which is the Laplace transform of $q_t(x,dy)$ and hence analytic in $\lambda > 0$. By (9)

$$Q_t f_\lambda(x)\, Q_t f_\lambda(y) = Q_t f_\lambda(0)\, Q_t f_\lambda(x+y), \quad \forall \lambda \in \mathcal{I}. \tag{12}$$

But the product of two analytic functions is again analytic, hence equality (12) holds for all $\lambda \geq 0$ (notice that $0 \in \mathcal{I}$). Now fix $t, \lambda \geq 0$ and define $g(x) := \frac{Q_t f_\lambda(x)}{Q_t f_\lambda(0)}$. The function g is measurable, positive, bounded and satisfies the functional equation $g(x)g(y) = g(x+y)$. Hence there exist a non-negative number $\psi(t,\lambda)$ such that $g(x) = e^{-\psi(t,\lambda)x}$, see [1, Theorem 1.1.8.]. Clearly, we can write $Q_t f_\lambda(0) = e^{-\phi(t,\lambda)}$ for some $\phi(t,\lambda) \geq 0$, and thus (10) and (11) hold.

It remains to prove the Feller property of (Q_t). Fix $x \in E$. First notice that every sequence $(q_{t_n}(x,dy))$ with $t_n \to 0$ contains a subsequence $(q_{t_{n_k}}(x,dy))$ which converges weakly to a measure $\mu(x,dy)$ on $\overline{E} := [0,\infty]$, the one-point compactification of E. This implies

$$\lim_{k\to\infty} Q_{t_{n_k}} f_\lambda(x) = \int_{\overline{E}} e^{-\lambda y}\mu(x,dy) = \tilde{\mu}(x,\lambda), \quad \forall \lambda > 0$$

where $\tilde{\mu}(x,\lambda)$ denotes the Laplace transform of $\mu(x,dy)$. On the other hand

$$\lim_{t\to 0} Q_t f_\lambda(x) = f_\lambda(x), \quad \forall \lambda \in \mathcal{I} \tag{13}$$

by continuity of A and B. Uniqueness of the weak limit and analyticity of the Laplace transform yield $\tilde{\mu}(x,\lambda) = f_\lambda(x)$ and (13) holds for all $\lambda \geq 0$.

Notice that $Q_t f_{\lambda'} \leq Q_t f_\lambda$ for $\lambda' \geq \lambda$ and $Q_t f_\lambda \in C_0(E)$ for $\lambda \in \mathcal{I} \setminus \{0\}$. From (13) it now follows easily that $\lim_{t\to 0} \|Q_t f_\lambda - f_\lambda\| = 0$ for all $\lambda > 0$. Since $\mathcal{L}(\Lambda)$ is dense in $C_0(E)$, the proof is complete. □

4. CBI-processes

Feller semigroups with property (10) have been well studied. They appear as limits of Galton–Watson branching processes with immigration. We refer to [8] and [11] for background and further reference. Following their terminology we define

Definition 4.1. *A Feller process on E whose semigroup satisfies* (10) *is called a* CBI-process[3] *(continuous state branching process with immigration). Its semigroup is denoted as* CBI-semigroup *and the Laplace exponents ϕ and ψ as* CBI-exponents.

Kawazu and Watanabe [8] provide a complete characterization of the generator of the CBI-semigroup (Q_t) and of its CBI-exponents $\phi(t,\lambda)$ and $\psi(t,\lambda)$. For the convenience of the reader we restate their main results.

Theorem 4.2. *Let (Q_t) be a CBI-semigroup with generator $\mathcal{B}$. Then $\mathcal{L}(\Lambda) \subset \mathcal{D}(\mathcal{B})$ and there exist real numbers*

$$\alpha \geq 0, \quad b \geq 0, \quad \beta, \quad c \geq 0, \quad \gamma \geq 0 \tag{14}$$

and non-negative Borel measures $m(dy)$ and $\mu(dy)$ on E° satisfying

$$\int_{E^\circ} (1 \wedge y)\, m(dy) + \int_{E^\circ} (1 \wedge y^2)\, \mu(dy) < \infty, \tag{15}$$

such that for $f \in \mathcal{L}(\Lambda)$

$$\begin{aligned} \mathcal{B}f(x) &= \alpha x f''(x) + (b' + \beta x) f'(x) - (c + \gamma x) f(x) \\ &\quad + \int_{E^\circ} \Big(f(x+y) - f(x) - f'(x)(1 \wedge y)\Big)\Big(m(dy) + x\mu(dy)\Big) \end{aligned} \tag{16}$$

where $b' := b + \int_{E^\circ} (1 \wedge y)\, m(dy)$. Moreover, if we set

$$R(\lambda) := -\alpha\lambda^2 + \beta\lambda + \gamma - \int_{E^\circ} \Big(e^{-\lambda y} - 1 + \lambda(1 \wedge y)\Big)\, \mu(dy)$$

$$F(\lambda) := c + b\lambda + \int_{E^\circ} \Big(1 - e^{-\lambda y}\Big)\, m(dy),$$

---

[3]In the notation of [8] this is in fact a *stochastically continuous* CBI-process.

then $\psi(t,\lambda)$ *and* $\phi(t,\lambda)$ *solve, for* $\lambda \geq 0$

$$\frac{\partial \psi}{\partial t}(t,\lambda) = R(\psi(t,\lambda)), \quad \psi(0,\lambda) = \lambda \tag{17}$$

$$\phi(t,\lambda) = \int_0^t F(\psi(s,\lambda))\, ds. \tag{18}$$

Conversely we have

Theorem 4.3. *Let* α, b, β, c *and* γ *be real numbers and* $m(dy)$, $\mu(dy)$ *non-negative Borel measures on* E° *such that* (14) *and* (15) *hold. Then* $\mathcal{B}$ *given by* (16) *on* $\mathcal{L}(\Lambda)$ *extends uniquely to the generator of a CBI-semigroup* (Q_t) *with CBI-exponents given by* (17) *and* (18).

Note that $R(\lambda)$ might fail to be Lipschitz continuous in $\lambda = 0$. However, we have the following

Theorem 4.4. *The CBI-semigroup* (Q_t) *from Theorem* 4.2 *is conservative if and only if* $\gamma = c = 0$ *and*

$$\int_0^\epsilon \frac{1}{R^+(\lambda)}\, d\lambda = \infty, \quad \forall \epsilon > 0. \tag{19}$$

Let $D(E)$ be the subset of Ω consisting of all cadlag paths. Denote by $(\mathbb{Q}_x)_{x\in E}$ a family of probability measures on $D(E)$. The following definition is due to [11].

Definition 4.5. *The family* $(\mathbb{Q}_x)_{x\in E}$ *is called* infinitely decomposable *if for all* $n \in \mathbb{N}$ *there exists a family* $(\mathbb{Q}_x^{(n)})_{x\in E}$ *of probability measures on* $D(E)$ *satisfying*

$$\mathbb{Q}_{x_1+\cdots+x_n} = \mathbb{Q}_{x_1}^{(n)} * \cdots * \mathbb{Q}_{x_n}^{(n)}, \quad \forall x_i \in E.$$

If $(\mathbb{Q}_x)_{x\in E}$ *corresponds to a Markov process, we call the Markov process itself* infinitely decomposable.

In [11] it is shown that every infinitely decomposable homogeneous Markov process having the strong Markov property and cadlag paths is a conservative CBI-process and vice versa.

5. The short rates as CBI-processes

So far we have shown that the pricing semigroup (Q_t) given by (8) is necessarily a CBI-semigroup if r induces an ATS. We shall prove that also r has to be a CBI-process in this case. Therefore we apply the Feynman–Kac formula and characterize the generator $\mathcal{A}$ of (P_t), the semigroup related to r.

Introduce the linear operator V on $b\mathcal{E}$ by $Vf(x) := xf(x)$. It is not hard to see that V is an unbounded closed operator on $b\mathcal{E}$. We denote the resolvent of $\mathcal{B}$ by $R_\lambda^{\mathcal{B}}$. We shall use two versions of the Feynman–Kac formula.

Lemma 5.1 (Feynman–Kac I).

$$R_\lambda^{\mathcal{B}} g = R_\lambda(g - V(R_\lambda^{\mathcal{B}} g))$$

for all $g \in b\mathcal{E}$ with $R_\lambda^{\mathcal{B}} g \in \mathcal{D}(V)$.

Proof. See [10, p. 273], taking into account assumption (3). □

We can now prove the following particular version.

Lemma 5.2 (Feynman–Kac II). *We have $\mathcal{L}(\Lambda) \subset \mathcal{D}(\mathcal{B}) \cap \mathcal{D}(\mathcal{A}) \cap \mathcal{D}(V)$ and*

$$\mathcal{B}f = \mathcal{A}f - Vf$$

for $f \in \mathcal{L}(\Lambda)$.

Proof. From Theorem 4.2 we know that $\mathcal{L}(\Lambda) \subset \mathcal{D}(\mathcal{B})$. Moreover, it is clear that $\mathcal{L}(\Lambda) \subset \mathcal{D}(V)$. It remains to show $\mathcal{L}(\Lambda) \subset \mathcal{D}(\mathcal{A})$.

Let $f \in \mathcal{L}(\Lambda)$. Then there exists a unique $g \in C_0(E)$ with $R_\lambda^{\mathcal{B}} g = f$. Since $R_\lambda^{\mathcal{B}} g \in \mathcal{D}(V)$ Lemma 5.1 applies and $f = R_\lambda(g - Vf) \in R_\lambda(b\mathcal{E}) \subset b\mathcal{E}_0$. Thus $\mathcal{L}(\Lambda) \subset b\mathcal{E}_0$. Since $b\mathcal{E}_0$ is closed therefore

$$C_0(E) = \overline{\mathcal{L}(\Lambda)} \subset b\mathcal{E}_0.$$

Consequently, we have $R_\lambda(C_0(E)) \subset R_\lambda(b\mathcal{E}_0) = \mathcal{D}(\mathcal{A})$. Take f and g as above. Clearly $g - Vf \in C_0(E)$, whence $f = R_\lambda(g - Vf) \in \mathcal{D}(\mathcal{A})$. We have thus shown that $\mathcal{L}(\Lambda) \subset \mathcal{D}(\mathcal{A})$.

Let f and g be as above. From Lemma 5.1 we have $R_\lambda g = f + R_\lambda(Vf)$. Since $f \in \mathcal{D}(\mathcal{A})$ and $Vf \in C_0(E) \subset b\mathcal{E}_0$ we can solve this equation and get $g = (\lambda - \mathcal{A})f + Vf$. On the other hand $g = (\lambda - \mathcal{B})f$. We conclude

$$(\lambda - \mathcal{B})f = (\lambda - \mathcal{A})f + Vf$$

and the lemma follows. □

As a consequence of this lemma, the generator $\mathcal{A}$ of (P_t) can be represented on $\mathcal{L}(\Lambda)$ by the right-hand side of equality (16) with γ replaced by $\tilde{\gamma} = \gamma - 1$. It is not a priori clear whether $\tilde{\gamma}$ is negative or not.

Lemma 5.3. *We have $\tilde{\gamma} \geq 0$.*

Proof. Apply the positive maximum principle to $f(x) = e^{-\lambda x} - e^{-2\lambda x} \in \mathcal{L}(\Lambda)$ and let λ tend to zero. □

In view of Theorem 4.3 we conclude that there exists a unique CBI-semigroup (P_t') related to $\mathcal{A}$ restricted to $\mathcal{L}(\Lambda)$. The situation is as follows.

Lemma 5.4 (Uniqueness). *Let (P_t) and (P_t') be two contraction semigroups on $b\mathcal{E}$ with generators $\mathcal{A}$ and $\mathcal{A}'$. Assume that $\mathcal{A} = \mathcal{A}'$ on a set $D \subset \mathcal{D}(\mathcal{A}) \cap \mathcal{D}(\mathcal{A}')$ with $P_t'(D) \subset D$. Then $(P_t) = (P_t')$ on the closure $\overline{D}$.*

Proof. Take $g \in D$. Then $P_s(P_{t-s}' g)$ is differentiable in $s \in (0,t)$. It is easily seen that $\frac{d}{ds} P_s(P_{t-s}' g) = P_s(\mathcal{A} P_{t-s}' g) - P_s(\mathcal{A}' P_{t-s}' g) = 0$. Hence by continuity $P_s(P_{t-s}' g) = P_t' g$ for all $s \leq t$ and thus $P_t g = P_t' g$. □

The CBI-semigroup property yields $P_t'(\mathcal{L}(\Lambda)) \subset \mathcal{L}(\Lambda)$. Thus Lemma 5.4 applies with $D = \mathcal{L}(\Lambda)$ and therefore $(P_t) = (P_t')$ on $C_0(E)$. We have thus shown that r is a CBI-process. Since r is conservative we have $c = \tilde{\gamma} = 0$, by Theorem 4.4.

Conversely, let now r be a conservative CBI-process on E with semigroup (P_t) and generator $\mathcal{A}$. We assume that r is realized on the canonical space. Moreover, we take the cadlag version, see [9, Chapt. III.2]. Then assumption (3) is satisfied. The right-hand side of (16) gives the form of $\mathcal{A}$ on $\mathcal{L}(\Lambda)$. By Theorem 4.4 we have $c = \gamma = 0$.

Define Q_t as in (8). That (Q_t) is a positive contraction semigroup can be shown as in the first part of the proof of Lemma 3.3. Now again we will use Feynman–Kac formula to identify the generator $\mathcal{B}$ of (Q_t). Let $R_\lambda^{\mathcal{B}}$ denote the resolvent of $\mathcal{B}$.

Lemma 5.5 (Feynman–Kac III).

$$R_\lambda g = R_\lambda^{\mathcal{B}}(g + V(R_\lambda g))$$

for all $g \in b\mathcal{E}$ with $R_\lambda g \in \mathcal{D}(V)$.

Proof. Analogous to Lemma 5.1. □

If one exchanges the roles of $\mathcal{A}$ and $\mathcal{B}$ in the proof of Lemma 5.2 and uses Lemma 5.5 instead of Lemma 5.1 one gets $\mathcal{L}(\Lambda) \subset \mathcal{D}(\mathcal{B})$ and

$$\mathcal{B}f = \mathcal{A}f - Vf, \quad \text{for } f \in \mathcal{L}(\Lambda).$$

In view of Theorem 4.3 the above constructed operator $\mathcal{B}$ extends in a unique way to the generator $\mathcal{B}'$ of a CBI-semigroup (Q_t'). Applying Lemma 5.4 yields $(Q_t) = (Q_t')$. As a consequence the implied term structure is affine.

Summarizing we have shown the following main theorem.

Theorem 5.6 (Main theorem). *Let r be a homogeneous Markov process in E as given at the beginning of Section 2. The following conditions are equivalent.*

1. *r provides an ATS.*
2. *r is a CBI-process uniquely characterized by its generator on $\mathcal{L}(\Lambda)$ given by (2) where $b' := b + \int_{E^\circ}(1 \wedge y)\, m(dy)$, for some real numbers $\alpha \geq 0$, $b \geq 0$ and β and non-negative Borel measures $m(dy)$ and $\mu(dy)$ on E° satisfying (15), and such that (19) holds for*

$$R(\lambda) := -\alpha\lambda^2 + \beta\lambda + 1 - \int_{E^\circ} \left(e^{-\lambda y} - 1 + \lambda(1 \wedge y)\right) \mu(dy). \tag{20}$$

3. *(Q_t) given by (8) is a CBI-semigroup whose CBI-exponents $\phi(t,\lambda)$ and $\psi(t,\lambda)$ satisfy the system of equations (17) and (18), where $R(\lambda)$ is given in (20) and*

$$F(\lambda) := b\lambda + \int_{E^\circ} \left(1 - e^{-\lambda y}\right) m(dy) \tag{21}$$

with α, b, β, $m(dy)$ and $\mu(dy)$ as in condition 2).

4. *r is an infinitely decomposable Feller process.*

The functions $A(u)$ and $B(u)$ appearing in the term structure (6) are given by $A(u) = \phi(u,0)$ and $B(u) = \psi(u,0)$, see (11). Hence condition 3) of Theorem 5.6 yields

Corollary 5.7. *The functions A and B satisfy the generalized Riccati equations*

$$B'(u) = R(B(u)), \quad B(0) = 0 \tag{22}$$

$$A(u) = \int_0^u F(B(s))\, ds, \tag{23}$$

where R and F are defined in (20) and (21).

Consider the trivial decomposition

$$P(t,T) = e^{\int_0^t r_s\, ds}\Big(e^{-\int_0^t r_s\, ds} P(t,T)\Big), \quad 0 \le t \le T,$$

into a product of a finite variation process and a martingale. It follows that $(P(t,T))_{0\le t\le T}$ is a semimartingale. Solving equality (5) for r_t yields

$$r_t = -\frac{\log P(t,T) + A(T-t)}{B(T-t)}.$$

Consequently, r is a semimartingale. As a byproduct we thus obtained the following result.

Corollary 5.8. *Any infinitely decomposable conservative Feller process in E is a semimartingale.*

6. An example

The following example extends the classical CIR model by including jumps. Yet we derive a closed form expression for the bond prices. We choose the parameters according to Theorem 5.6 as follows,

$$\alpha, b, \beta \ge 0, \quad m(dy) = pe^{-qy}\, dy, \quad \mu = 0,$$

for some $p \ge 0$ and $q > 0$. The generator of r is then

$$\mathcal{A}f(x) = \alpha x f''(x) + (b - \beta x) f'(x) + p \int_{E^\circ} \Big(f(x+y) - f(x)\Big) e^{-qy}\, dy,$$

see (2). Thus r can be realized as a jump-diffusion process. Its jump size and inter-arrival times are exponentially distributed with parameters q and p/q, respectively. Now (20) and (21) become

$$R(\lambda) = -\alpha\lambda^2 - \beta\lambda + 1$$

$$F(\lambda) = b\lambda + \frac{p\lambda}{q\lambda + q^2}.$$

The solutions of (22) and (23) accordingly are given in closed form,

$$B(u) = \frac{L_1(u)}{L_2(u)}$$

$$A(u) = -\frac{b}{\alpha}\log\left(\frac{L_3(u)}{L_2(u)}\right) - \frac{p}{1+\beta q - \alpha q^2}\log\left(e^{-u/q}\frac{L_1(u)+qL_2(u)}{qL_3(u)}\right)$$

where

$$L_1(u) := 2(e^{\gamma u} - 1)$$
$$L_2(u) := \gamma(e^{\gamma u} + 1) + \beta(e^{\gamma u} - 1)$$
$$L_3(u) := 2\gamma e^{(\gamma+\beta)u/2}$$

and $\gamma := \sqrt{\beta^2 + 4\alpha}$. In fact, we assume that $\alpha > 0$ and $q \neq (\beta + \gamma)/(2\alpha)$, such that the above expression for $A(u)$ is well defined. Notice that $p = 0$ simply yields the classical CIR model.

In summary, we obtain an explicit expression for the bond prices $P(t,T)$ in terms of r_t and the real parameters α, b, β, p, q (see (1)). This allows in principle to calibrate the model by fitting the today's term structure, $P(0,T)$, which is observed on the market. It is shown in [7] that the present model also admits a closed form expression for the price of a European call option on a bond. Duffie and Gârleanu [4] use this setup for the modelling of intensity based default.

References

[1] N. H. Bingham, C. M. Goldie, and J. L. Teugels, *Regular variation*, Cambridge University Press, Cambridge, 1987.

[2] R. G. Brown and S. M. Schaefer, *Interest rate volatility and the shape of the term structure*, Phil. Trans. R. Soc. Lond. A **347** (1994), 563–576.

[3] J. Cox, J. Ingersoll, and S. Ross, *A theory of the term structure of interest rates*, Econometrica **53** (1985), 385–408.

[4] D. Duffie and N. Gârleanu, *Risk and valuation of collateralized debt obligations*, Financial Analysts Journal **57** (2001), no. 1, 41–59.

[5] D. Duffie and R. Kan, *A yield-factor model of interest rates*, Math. Finance **6** (1996), 379–406.

[6] E. B. Dynkin, *Markov processes*, vol. I, Springer-Verlag, Berlin, 1965.

[7] D. Filipović, *A general characterization of one factor affine term structure models*, Finance and Stochast. **5** (3) (2001), 389–412.

[8] K. Kawazu and S. Watanabe, *Branching processes with immigration and related limit theorems*, Theory Probab. Appl. **16** (1971), 36–54.

[9] D. Revuz and M. Yor, *Continuous martingales and Brownian motion*, Grundlehren der mathematischen Wissenschaften, vol. 293, Springer-Verlag, Berlin-Heidelberg-New York, 1994.

[10] L. C. G. Rogers and D. Williams, *Diffusions, Markov processes, and martingales. Vol. 1*, second ed., John Wiley & Sons Ltd., Chichester, 1994, Foundations.

[11] T. Shiga and S. Watanabe, *Bessel diffusions as a one-parameter family of diffusion processes*, Z. Wahrscheinlichkeitstheorie verw. Geb. **27** (1973), 37–46.

[12] O. Vasicek, *An equilibrium characterization of the term structure*, J. Finan. Econom. **5** (1977), 177–188.

Department of Mathematics
ETH
CH-8092 Zürich
Switzerland
E-mail address: filipo@math.ethz.ch

Progress in Probability, Vol. 52, 133–152

A Filtered EM Algorithm for Parameter Estimation in Linear Filtering

P. Florchinger

Abstract. The aim of this paper is to compute finite-dimensional filters for integrals and stochastic integrals of moments of the state of partially observed linear Gaussian systems when the signal process and the observation noise are possibly correlated. These filters are used with the expectation maximization (EM) algorithm in order to obtain maximum likelihood estimation of the system parameters.

1. Introduction

The Kalman filter, which is widely used by engineers and economists deals with partially observed linear Gaussian systems in the form

$$\begin{aligned} dx_t &= A_t x_t dt + B_t dw_t + E_t dy_t \\ dy_t &= C_t x_t dt + D_t dv_t \end{aligned}$$

where $x_t \in \mathbb{R}^n$ is the unobserved signal and $y_t \in \mathbb{R}^p$ is the observation.

The Kalman filter states that the conditional density of the signal process x_t given the observation history up to time t is a Gaussian random variable the variance of which turns out to be deterministic and the solution of a Riccati equation.

Nevertheless, in order to apply the Kalman filter one needs to know explicitly the system coefficients. For linear filtering systems with independent noises (i.e. $E_t \equiv 0$) a maximum likelihood estimation of the system coefficients via the expectation maximization (EM) algorithm has been studied in [2], [3] and [6].

However, to make use of the EM algorithm, one has to compute filtered estimates of quantities such as

$$\int_0^t x_s (dx_s)^\star, \quad \int_0^t x_s (dy_s)^\star \quad \text{and} \quad \int_0^t x_s x_s^\star ds.$$

In [3] and [2] filtered estimates of such integrals are computed by means of smoothing techniques. The main problem when using this technique is that it requires

---

2000 *Mathematics Subject Classification.* 60G35, 62M20, 93E10, 93E11.
Key words and phrases. Linear filtering, expectation maximization algorithm, parameter estimation.

large memory in numerical implementation. The main result of [6] provides finite-dimensional filters to compute (the components of) the above integrals. Furthermore, the technique developed in [6] permits to design finite-dimensional filters for integrals and stochastic integrals of moments of all orders of the state process.

The purpose of this paper is to extend the results of [6] to linear Gaussian systems when the signal process and the observation noise are correlated.

This paper is divided in three sections and is organized as follows. In Section 2, we introduce the class of linear Gaussian systems we are dealing with in this paper and we compute, by means of the "reference probability measure" technique, the Zakai equation solved by the unnormalized conditional density of x_t given the observation history up to time t. In Section 3, we derive the finite-dimensional filters that we need to obtain filtered estimates of quantities such as

$$\int_0^t x_s(dx_s)^\star, \quad \int_0^t x_s(dy_s)^\star, \quad \int_0^t x_s x_s^\star ds \quad \text{and} \quad \int_0^t (dx_s)^\star y_s.$$

In Section 4, the finite-dimensional filters of Section 3 are used to compute maximum likelihood estimation of the system parameters through a filter-based EM algorithm.

2. Problem setting

In this paper, we make use of the linear Gaussian model of Kalman filtering when the signal process and the observation noise are possibly correlated (see [1] for example).

Let $(\Omega, \mathcal{F}, P)$ be a complete probability space and $(\mathcal{F}_t)_{t\geq 0}$ be a complete right-continuous filtration.

Consider the signal process $(x_t)_{t\geq 0}$ with values in $\mathbb{R}^n$, solution of the stochastic differential equation written in the sense of Itô,

$$x_t = x_0 + \int_0^t A_s x_s ds + \int_0^t B_s dw_s + \int_0^t E_s dy_s \tag{1}$$

and the observation process $(y_t)_{t\geq 0}$ with values in $\mathbb{R}^p$, solution of the stochastic differential equation

$$y_t = \int_0^t C_s x_s ds + \int_0^t D_s dv_s \tag{2}$$

where

1. $x_0 \in \mathbb{R}^n$ is a random variable with the Gaussian density $p_0(x)$ with mean m_0 and variance Σ_0.
2. $(w_t)_{t\geq 0}$ and $(v_t)_{t\geq 0}$ are independent Wiener processes with values in $\mathbb{R}^m$ and $\mathbb{R}^p$, respectively. Moreover, the processes $(w_t)_{t\geq 0}$ and $(v_t)_{t\geq 0}$ and x_0 are assumed to be mutually independent.
3. The matrix functions $A_t \in \mathcal{M}_{n\times n}(\mathbb{R})$, $B_t \in \mathcal{M}_{n\times m}(\mathbb{R})$, $E_t \in \mathcal{M}_{n\times p}(\mathbb{R})$, $C_t \in \mathcal{M}_{p\times n}(\mathbb{R})$ and $D_t \in \mathcal{M}_{p\times p}(\mathbb{R})$ are measurable functions of the time variable t and D_t is a positive definite matrix.

As is usual in filtering theory, define for any function $\phi \in C_0(\mathbb{R}^n;\mathbb{R})$ (the set of functions mapping $\mathbb{R}^n$ into $\mathbb{R}$ with compact support) the filter associated with the system (1)–(2) by

$$\pi_t(\phi) = E\left[\phi(x_t) \;/\; \mathcal{Y}_t\right]$$

where $\mathcal{Y}_t$ is the filtration defined by $\mathcal{Y}_t = \sigma\left(y_s \;/\; 0 \le s \le t\right)$.

For any $t \ge 0$, let Z_t be the exponential

$$Z_t = \exp\left(\int_0^t (C_s x_s)^\star (D_s^{-1})^\star D_s^{-1} dy_s - \frac{1}{2}\int_0^t (C_s x_s)^\star (D_s^{-1})^\star D_s^{-1} C_s x_s ds\right).$$

That is, $(Z_t)_{t\ge 0}$ is the stochastic process solution of the stochastic differential equation

$$dZ_t = Z_t (C_t x_t)^\star (D_t^{-1})^\star D_t^{-1} dy_t \tag{3}$$

with $Z_0 = 1$.

Then, denoting by $\bar{P}$ the probability measure defined by the Radon-Nikodym derivative

$$\left.\frac{d\bar{P}}{dP}\right|_{\mathcal{F}_t} = Z_t^{-1}$$

one can deduce from Girsanov's theorem (see [5] for example) that, under the probability measure $\bar{P}$, the stochastic process $(y_t)_{t\ge 0}$ is a Wiener process, independent of $(w_t)_{t\ge 0}$ and x_0, with variance D_t.

Furthermore, defining, for any function $\phi \in C_0(\mathbb{R}^n;\mathbb{R})$, the unnormalized filter associated to the system (1)–(2) by

$$\rho_t(\phi) = \bar{E}\left[\phi(x_t) Z_t \;/\; \mathcal{Y}_t\right]$$

one has, for any $t \ge 0$, the well-known Kallianpur-Striebel formula

$$\pi_t(\phi) = \frac{\rho_t(\phi)}{\rho_t(1)}.$$

Therefore, if the measure-valued process $(\rho_t(\phi))_{t\ge 0}$ has a density $p_t(x)$ with respect to the Lebesgue measure on $\mathbb{R}^n$, one can prove that the process $(p_t(x))_{t\ge 0}$ solves a stochastic partial differential equation: the Zakai equation.

Theorem 2.1. *The stochastic process $(p_t(x))_{t\ge 0}$ solves the following stochastic partial differential equation*

$$\begin{aligned} p_t(x) = p_0(x) &- \int_0^t \operatorname{div}\left(A_s x p_s(x)\right) ds - \int_0^t \operatorname{div}\left(E_s C_s x p_s(x)\right) ds \\ &+ \frac{1}{2}\int_0^t \operatorname{Tr}\left(\nabla^2 p_s(x) B_s B_s^\star\right) ds + \frac{1}{2}\int_0^t \operatorname{Tr}\left(\nabla^2 p_s(x) E_s D_s D_s^\star E_s^\star\right) ds \\ &+ \int_0^t p_s(x)(C_s x)^\star (D_s^{-1})^\star D_s^{-1} dy_s - \sum_{i=1}^p \int_0^t \operatorname{div}\left(p_s(x) E_s^i\right) dy_s^i \end{aligned} \tag{4}$$

where E_t^i, $1 \le i \le p$, is the vector corresponding to the ith line of the matrix E_t.

Proof. For any function $\phi \in C_0^2(\mathbb{R}^n;\mathbb{R})$, Itô's formula implies that for any $t \geq 0$,

$$\begin{aligned}\phi(x_t)Z_t = \phi(x_0) &+ \int_0^t Z_s(\nabla\phi(x_s))^\star A_s x_s ds + \int_0^t Z_s(\nabla\phi(x_s))^\star B_s dw_s \\ &+ \int_0^t Z_s(\nabla\phi(x_s))^\star E_s dy_s + \frac{1}{2}\int_0^t Z_s \mathrm{Tr}\left(\nabla^2\phi(x_s)B_sB_s^\star\right) ds \\ &+ \frac{1}{2}\int_0^t Z_s \mathrm{Tr}\left(\nabla^2\phi(x_s)E_sD_sD_s^\star E_s^\star\right) ds \\ &+ \int_0^t Z_s\phi(x_s)(C_sx_s)^\star(D_s^{-1})^\star D_s^{-1}dy_s + \int_0^t Z_s(\nabla\phi(x_s))^\star E_sC_sx_sds.\end{aligned}$$

Conditioning each side of the previous equality on $\mathcal{Y}_t$, taking into account that the integral with respect to w_t is a local martingale and using a stochastic Fubini theorem (see [9] or [10] for example), one has

$$\begin{aligned}\rho_t(\phi) = \rho_0(\phi) &+ \int_0^t \rho_s\left((\nabla\phi)^\star A_s x\right) ds + \sum_{i=1}^p \int_0^t \rho_s\left((\nabla\phi)^\star E_s^i\right) dy_s^i \\ &+ \frac{1}{2}\int_0^t \rho_s\left(\mathrm{Tr}\left(\nabla^2\phi B_sB_s^\star\right)\right) ds + \frac{1}{2}\int_0^t \rho_s\left(\mathrm{Tr}\left(\nabla^2\phi E_sD_sD_s^\star E_s^\star\right)\right) ds \\ &+ \int_0^t \rho_s\left((\nabla\phi)^\star E_sC_s\right) ds + \int_0^t \rho_s\left(\phi(C_sx)^\star(D_s^{-1})^\star D_s^{-1}\right) dy_s.\end{aligned}$$

Therefore, equality (4) is obtained by integrating by parts each term of the previous equality as in [5]. □

Since the stochastic differential system (1)–(2) is linear Gaussian, the density $p_t(x)$, solution of the stochastic partial differential equation (4), can be expressed in the following simple form (see [1] for example).

Proposition 2.2. *The density $p_t(x)$, solution of the stochastic partial differential equation (4), is given by*

$$p_t(x) = \eta_t \frac{1}{(2\pi)^{n/2}|\Sigma_t|^{1/2}} \exp\left(-\frac{1}{2}(x-m_t)^\star\Sigma_t^{-1}(x-m_t)\right) \tag{5}$$

where η_t is the normalizing factor given by

$$\eta_t = (2\pi)^{m/2}\exp\left(\int_0^t m_s^\star C_s^\star(D_s^{-1})^\star D_s^{-1}dy_s - \frac{1}{2}\int_0^t m_s^\star C_s^\star(D_s^{-1})^\star D_s^{-1}C_sm_sds\right)$$

and $(m_t)_{t\geq 0}$ and $(\Sigma_t)_{t\geq 0}$ are given as the solution of the stochastic differential equation

$$m_t = m_0 + \int_0^t A_sm_sds + \int_0^t E_sdy_s + \int_0^t \Sigma_sC_s^\star(D_s^{-1})^\star D_s^{-1}(dy_s - C_sm_sds)$$

and the deterministic Riccati equation

$$\dot{\Sigma}_t = A_t^\star\Sigma_t + \Sigma_tA_t + B_tB_t^\star - \Sigma_tC_t^\star(D_t^{-1})^\star D_t^{-1}C_t\Sigma_t.$$

Remark 2.3. The variance Σ_t is deterministic and can be computed off-line.

3. Finite-dimensional filters

In this section, we make use of the following notations introduced by Elliott and Krishnamurthy [6].

For any $i, j \in \{1, \dots, n\}$ denote by e_i and e_j unit vectors in $\mathbb{R}^n$ with 1 in the ith and jth positions, respectively.

Then, for any $i, j \in \{1, \dots, n\}$, let $\left(H_t^{ij}\right)_{t \geq 0}$ and $\left(L_t^{ij}\right)_{t \geq 0}$ be the stochastic processes defined by

$$H_t^{ij} = \int_0^t x_s^\star e_i e_j^\star x_s ds \tag{6}$$

and

$$L_t^{ij} = \int_0^t x_s^\star e_i e_j^\star dx_s. \tag{7}$$

On the other hand, for any $j \in \{1, \dots, p\}$ denote by f_j a unit vector in $\mathbb{R}^p$ with 1 in the jth position and for any $i \in \{1, \dots, n\}$, let $\left(J_t^{ij}\right)_{t \geq 0}$ and $\left(K_t^{ji}\right)_{t \geq 0}$ be the stochastic processes defined by

$$J_t^{ij} = \int_0^t x_s^\star e_i f_j^\star dy_s \tag{8}$$

and

$$K_t^{ji} = \int_0^t y_s^\star f_j e_i^\star dx_s. \tag{9}$$

To estimate the system parameters using the EM algorithm, one has to compute

$$E\left[H_t^{ij} \;/\; \mathcal{Y}_t\right], \quad E\left[L_t^{ij} \;/\; \mathcal{Y}_t\right], \quad E\left[J_t^{ij} \;/\; \mathcal{Y}_t\right], \quad E\left[K_t^{ji} \;/\; \mathcal{Y}_t\right].$$

Following the method initiated in [6], we describe these quantities in terms of a finite number of statistics.

First, we define a measure-valued process associated with the process $\left(H_t^{ij}\right)_{t \geq 0}$ as follows.

Definition 3.1. *The measure-valued process defined for any function $\phi \in C_0(\mathbb{R}^n; \mathbb{R})$ by $\bar{E}\left[Z_t H_t^{ij} \phi(x_t) \;/\; \mathcal{Y}_t\right]$ has a density $\beta_t^{ij}(x)$ with respect to the Lebesgue measure on $\mathbb{R}^n$ so that*

$$\bar{E}\left[Z_t H_t^{ij} \phi(x_t) \;/\; \mathcal{Y}_t\right] = \int_{\mathbb{R}^n} \phi(x) \beta_t^{ij}(x) dx. \tag{10}$$

Remark 3.2. The existence of the density $\beta_t^{ij}(x)$ has been established by means of stochastic partial differential equations techniques by Pardoux [8].

Proposition 3.3. *The density $\beta_t^{ij}(x)$ defined by (10) solves the stochastic partial differential equation*

$$\begin{aligned}\beta_t^{ij}(x) = &-\int_0^t \operatorname{div}\left(\beta_s^{ij}(x)A_s x\right) ds - \int_0^t \operatorname{div}\left(\beta_s^{ij}(x)E_s C_s x\right) ds \\ &+\frac{1}{2}\int_0^t \operatorname{Tr}\left(\nabla^2\beta_s^{ij}(x)B_sB_s^\star\right) ds + \frac{1}{2}\int_0^t \operatorname{Tr}\left(\nabla^2\beta_s^{ij}(x)E_sD_sD_s^\star E_s^\star\right) ds \\ &+\int_0^t \beta_s^{ij}(x)(C_s x)^\star (D_s^{-1})^\star D_s^{-1} dy_s - \sum_{k=1}^p \int_0^t \operatorname{div}\left(\beta_s^{ij}(x)E_s^k\right) dy_s^k \\ &+\int_0^t p_s(x)x^\star e_i e_j^\star x ds. \qquad (11)\end{aligned}$$

Proof. Let ϕ be a "test function" in $C_0^2(\mathbb{R}^n;\mathbb{R})$. By application of Itô's formula to the process $\left(Z_t H_t^{ij}\phi(x_t)\right)_{t\geq 0}$ one gets

$$\begin{aligned}Z_t H_t^{ij}\phi(x_t) = &\int_0^t Z_s H_s^{ij}(\nabla\phi(x_s))^\star A_s x_s ds + \int_0^t Z_s H_s^{ij}(\nabla\phi(x_s))^\star B_s dw_s \\ &+\int_0^t Z_s H_s^{ij}(\nabla\phi(x_s))^\star E_s dy_s + \frac{1}{2}\int_0^t Z_s H_s^{ij}\operatorname{Tr}\left(\nabla^2\phi(x_s)B_sB_s^\star\right) ds \\ &+\frac{1}{2}\int_0^t Z_s H_s^{ij}\operatorname{Tr}\left(\nabla^2\phi(x_s)E_sD_sD_s^\star E_s^\star\right) ds \\ &+\int_0^t Z_s H_s^{ij}\phi(x_s)(C_s x_s)^\star (D_s^{-1})^\star D_s^{-1} dy_s \\ &+\int_0^t Z_s H_s^{ij}(\nabla\phi(x_s))^\star E_s C_s x_s ds + \int_0^t Z_s\phi(x_s)x_s^\star e_i e_j^\star x_s ds.\end{aligned}$$

Conditioning each side of the previous equality on $\mathcal{Y}_t$ and using a stochastic Fubini theorem, one has

$$\begin{aligned}\bar{E}\left[Z_t H_t^{ij}\phi(x_t) \;/\; \mathcal{Y}_t\right] = &\int_0^t \bar{E}\left[Z_s H_s^{ij}(\nabla\phi(x_s))^\star A_s x_s \;/\; \mathcal{Y}_s\right] ds \\ &+\int_0^t \bar{E}\left[Z_s H_s^{ij}(\nabla\phi(x_s))^\star E_s \;/\; \mathcal{Y}_s\right] dy_s \\ &+\frac{1}{2}\int_0^t \bar{E}\left[Z_s H_s^{ij}\operatorname{Tr}\left(\nabla^2\phi(x_s)B_sB_s^\star\right) \;/\; \mathcal{Y}_s\right] ds \\ &+\frac{1}{2}\int_0^t \bar{E}\left[Z_s H_s^{ij}\operatorname{Tr}\left(\nabla^2\phi(x_s)E_sD_sD_s^\star E_s^\star\right) \;/\; \mathcal{Y}_s\right] ds \\ &+\int_0^t \bar{E}\left[Z_s H_s^{ij}\phi(x_s)(C_s x_s)^\star (D_s^{-1})^\star D_s^{-1} \;/\; \mathcal{Y}_s\right] dy_s\end{aligned}$$

$$+\int_0^t \bar{E}\left[Z_s H_s^{ij}(\nabla\phi(x_s))^\star E_s C_s x_s \;/\; \mathcal{Y}_s\right] ds$$

$$+\int_0^t \bar{E}\left[Z_s \phi(x_s) x_s^\star e_i e_j^\star x_s \;/\; \mathcal{Y}_s\right] ds.$$

Then, the latter equality reads in terms of the densities $p_t(x)$ and $\beta_t^{ij}(x)$,

$$\begin{aligned}\int_{\mathbb{R}^n}\phi(x)\beta_t^{ij}(x)dx &= \int_0^t\int_{\mathbb{R}^n}(\nabla\phi(x))^\star A_s x\beta_s^{ij}(x)dxds\\ &+\int_0^t\int_{\mathbb{R}^n}(\nabla\phi(x))^\star E_s C_s x\beta_s^{ij}(x)dxds\\ &+\frac{1}{2}\int_0^t\int_{\mathbb{R}^n}\mathrm{Tr}\left(\nabla^2\phi(x)B_sB_s^\star\right)\beta_s^{ij}(x)dxds\\ &+\frac{1}{2}\int_0^t\int_{\mathbb{R}^n}\mathrm{Tr}\left(\nabla^2\phi(x)E_sD_sD_s^\star E_s^\star\right)\beta_s^{ij}(x)dxds\\ &+\int_0^t\int_{\mathbb{R}^n}\phi(x)(C_sx)^\star(D_s^{-1})^\star D_s^{-1}\beta_s^{ij}(x)dxdy_s\\ &+\int_0^t\int_{\mathbb{R}^n}(\nabla\phi(x))^\star E_s\beta_s^{ij}(x)dxdy_s\\ &+\int_0^t\int_{\mathbb{R}^n}\phi(x)x^\star e_i e_j^\star x p_s(x)dxds.\end{aligned}$$

Equality (11) is obtained by integrating by parts each term in the latter equation and by noticing that the equality holds for all "test function" ϕ. □

In the following result, we prove that the measure $\beta_t^{ij}(x)$ can be described in terms of a finite number of statistics.

Theorem 3.4. *The density $\beta_t^{ij}(x)$ is given by*

$$\beta_t^{ij}(x) = \left(a_t^{ij} + x^\star b_t^{ij} + x^\star c_t^{ij} x\right) p_t(x) \tag{12}$$

where the processes $a_t^{ij} \in \mathbb{R}$, $b_t^{ij} \in \mathbb{R}^n$ and $c_t^{ij} \in \mathcal{S}_{n\times n}(\mathbb{R})$ (the set of real-valued symmetric $n \times n$ matrices) are solutions of the following stochastic differential equations

$$\begin{aligned}da_t^{ij} &= \mathrm{Tr}\left(c_t^{ij}B_tB_t^\star\right)dt + \mathrm{Tr}\left(c_t^{ij}E_tD_tD_t^\star E_t^\star\right)dt\\ &\quad + b_t^{ij\star}B_tB_t^\star\Sigma_t^{-1}m_tdt + b_t^{ij\star}E_tD_tD_t^\star E_t^\star\Sigma_t^{-1}m_tdt \qquad (13)\\ &\quad - b_t^{ij\star}E_tdy_t\\ a_0^{ij} &= 0\end{aligned}$$

$$\begin{aligned}
db_t^{ij} &= -\left(A_t^\star + C_t^\star E_t^\star + \Sigma_t^{-1}(B_tB_t^\star + E_tD_tD_t^\star E_t^\star)\right) b_t^{ij}\,dt \\
&\quad +2c_t^{ij}\left(B_tB_t^\star + E_tD_tD_t^\star E_t^\star\right)\Sigma_t^{-1}m_t dt \\
&\quad -2c_t^{ij}E_t dy_t
\end{aligned} \tag{14}$$

$$b_0^{ij} = 0$$

$$\begin{aligned}
\frac{dc_t^{ij}}{dt} &= -\left(A_t^\star + C_t^\star E_t^\star + \Sigma_t^{-1}(B_tB_t^\star + E_tD_tD_t^\star E_t^\star)\right) c_t^{ij} \\
&\quad -c_t^{ij}\left(A_t + E_tC_t + (B_tB_t^\star + E_tD_tD_t^\star E_t^\star)\Sigma_t^{-1}\right) \\
&\quad +\frac{1}{2}(e_je_i^\star + e_ie_j^\star)
\end{aligned} \tag{15}$$

$$c_0^{ij} = 0$$

Proof. Let $\bar{\beta}_t(x)$ be given by

$$\bar{\beta}_t(x) = (a_t + x^\star b_t + x^\star c_t x)p_t(x) \tag{16}$$

where $a_t \in \mathbb{R}$, $b_t \in \mathbb{R}^n$ and $c_t \in \mathcal{S}_{n\times n}(\mathbb{R})$.

Then, as noted in Remark 3.2, if $\bar{\beta}_t(x)$ is a solution of the stochastic partial differential equation (11), it is unique.

Taking into account the fact that $p_t(x)$ solves the Zakai equation (4) yields,

$$\begin{aligned}
d\bar{\beta}_t(x) = (da_t + x^\star db_t + x^\star dc_t x)p_t(x) \\
- (a_t + x^\star b_t + x^\star c_t x)\operatorname{div}(A_t x p_t(x))dt \\
- (a_t + x^\star b_t + x^\star c_t x)\operatorname{div}(E_tC_t x p_t(x))dt \\
+ \frac{1}{2}(a_t + x^\star b_t + x^\star c_t x)\operatorname{Tr}\left(\nabla^2 p_t(x)B_tB_t^\star\right)dt \\
+ \frac{1}{2}(a_t + x^\star b_t + x^\star c_t x)\operatorname{Tr}\left(\nabla^2 p_t(x)E_tD_tD_t^\star E_t^\star\right)dt \\
+ (a_t + x^\star b_t + x^\star c_t x)p_t(x)(C_t x)^\star (D_t^{-1})^\star D_t^{-1} dy_t \\
- \sum_{k=1}^{p}(a_t + x^\star b_t + x^\star c_t x)\operatorname{div}\left(p_t(x)E_t^k\right)dy_t^k.
\end{aligned}$$

Also, if we substitute the expression of $\bar{\beta}_t(x)$ given by (16) in the differential form of the right-hand side of (11) we get

$$\begin{aligned}
&- (b_t + 2c_t x)^\star A_t x p_t(x)dt - (a_t + x^\star b_t + x^\star c_t x)\operatorname{div}(A_t x p_t(x))dt \\
&- (b_t + 2c_t x)^\star E_tC_t x p_t(x)dt - (a_t + x^\star b_t + x^\star c_t x)\operatorname{div}(E_tC_t x p_t(x))dt \\
&+ p_t(x)\operatorname{Tr}\left(c_tB_tB_t^\star\right)dt - (b_t + 2c_t x)^\star B_tB_t^\star\Sigma_t^{-1}(x - m_t)p_t(x)dt \\
&+ \frac{1}{2}(a_t + x^\star b_t + x^\star c_t x)\operatorname{Tr}\left(\nabla^2 p_t(x)B_tB_t^\star\right)dt
\end{aligned}$$

$$
\begin{aligned}
&+ p_t(x)\mathrm{Tr}\left(c_t E_t D_t D_t^\star E_t^\star\right) dt - (b_t + 2c_t x)^\star E_t D_t D_t^\star E_t^\star \Sigma_t^{-1}(x - m_t) p_t(x) dt \\
&\quad + \frac{1}{2}(a_t + x^\star b_t + x^\star c_t x)\mathrm{Tr}\left(\nabla^2 p_t(x) E_t D_t D_t^\star E_t^\star\right) dt \\
&\quad + (a_t + x^\star b_t + x^\star c_t x) p_t(x) (C_t x)^\star (D_t^{-1})^\star D_t^{-1} dy_t \\
&- (b_t + 2c_t x)^\star p_t(x) E_t dy_t - \sum_{k=1}^{p} (a_t + x^\star b_t + x^\star c_t x)\mathrm{div}\left(p_t(x) E_t^k\right) dy_t^k \\
&\qquad + p_t(x) x^\star e_i e_j^\star x dt.
\end{aligned}
$$

Therefore, by equating the coefficients of x, $x^\star$ and the constants in the last two expressions, we see that $\bar{\beta}_t(x)$ given by (16) solves the stochastic partial differential equation (11) provided (13), (14) and (15) hold. □

Now, as in [6], we explicitly solve the stochastic differential equation (14) and the ordinary differential equation (15).

Let $G_t \in \mathcal{M}_{n\times n}(\mathbb{R})$ be the matrix solution of the ordinary differential equation

$$
\frac{dG_t}{dt} = -\left(A_t^\star + C_t^\star E_t^\star + \Sigma_t^{-1}(B_t B_t^\star + E_t D_t D_t^\star E_t^\star)\right) G_t
$$

with $G_0 = I$.

Then, the following representation result is proved easily by using variation of constants.

Proposition 3.5. *The stochastic processes $(b_t^{ij})_{t\geq 0}$ and $(c_t^{ij})_{t\geq 0}$ solutions of (14) and (15) are given by*

$$
b_t^{ij} = 2G_t\left(\int_0^t G_s^{-1} c_s^{ij}\left(B_s B_s^\star + E_s D_s D_s^\star E_s^\star\right)\Sigma_s^{-1} m_s ds - \int_0^t G_s^{-1} c_s^{ij} E_s dy_s\right)
$$

and

$$
c_t^{ij} = \frac{1}{2} G_t\Big(\int_0^t G_s^{-1}(e_j e_i^\star + e_i e_j^\star)(G_s^{-1})^\star ds\Big) G_t^\star.
$$

Remark 3.6. The matrix G_t is deterministic, invertible and can be computed offline.

Now, we define a measure-valued process associated with the process $(J_t^{ij})_{t\geq 0}$ by means of a procedure similar to that exposed above.

Definition 3.7. *The measure-valued process defined for any function $\phi \in C_0(\mathbb{R}^n;\mathbb{R})$ by $\bar{E}\left[Z_t J_t^{ij}\phi(x_t) \;/\; \mathcal{Y}_t\right]$ has a density $\gamma_t^{ij}(x)$ with respect to the Lebesgue measure on $\mathbb{R}^n$ so that*

$$
\bar{E}\left[Z_t J_t^{ij}\phi(x_t) \;/\; \mathcal{Y}_t\right] = \int_{\mathbb{R}^n} \phi(x)\gamma_t^{ij}(x) dx. \tag{17}
$$

Remark 3.8. The existence of the density $\gamma_t^{ij}(x)$ has been established by means of stochastic partial differential equations techniques by Pardoux [8].

Proposition 3.9. *The density $\gamma_t^{ij}(x)$ defined by (17) solves the stochastic partial differential equation*

$$\begin{aligned}\gamma_t^{ij}(x) = &-\int_0^t \operatorname{div}\left(\gamma_s^{ij}(x)A_s x\right) ds - \int_0^t \operatorname{div}\left(\gamma_s^{ij}(x)E_s C_s x\right) ds \\ &+\frac{1}{2}\int_0^t \operatorname{Tr}\left(\nabla^2\gamma_s^{ij}(x)B_sB_s^\star\right) ds + \frac{1}{2}\int_0^t \operatorname{Tr}\left(\nabla^2\gamma_s^{ij}(x)E_sD_sD_s^\star E_s^\star\right) ds \\ &+\int_0^t \gamma_s^{ij}(x)(C_s x)^\star (D_s^{-1})^\star D_s^{-1} dy_s - \sum_{k=1}^{p}\int_0^t \operatorname{div}\left(\gamma_s^{ij}(x)E_s^k\right) dy_s^k \\ &+\int_0^t p_s(x)x^\star e_i f_j^\star dy_s + \int_0^t p_s(x)x^\star C_s^\star f_j e_i^\star x ds \\ &-\int_0^t \operatorname{div}\left(p_s(x)E_sD_sD_s^\star f_j e_i^\star x\right) ds \end{aligned} \tag{18}$$

Proof. The product $Z_t J_t^{ij}\phi(x_t)$ is calculated by means of Itô's formula and each side of the resulting formula is conditioned on $\mathcal{Y}_t$. The stochastic partial differential equation (18) is obtained after integration by parts. □

Arguing as before, the density $\gamma_t^{ij}(x)$ can be described as follows.

Theorem 3.10. *The density $\gamma_t^{ij}(x)$ is given by*

$$\gamma_t^{ij}(x) = \left(\bar{a}_t^{ij} + x^\star \bar{b}_t^{ij} + x^\star \bar{c}_t^{ij} x\right) p_t(x) \tag{19}$$

where the processes $\bar{a}_t^{ij} \in \mathbb{R}$, $\bar{b}_t^{ij} \in \mathbb{R}^n$ and $\bar{c}_t^{ij} \in \mathcal{S}_{n\times n}(\mathbb{R})$ are solutions of the following stochastic differential equations

$$\begin{aligned} d\bar{a}_t^{ij} &= \operatorname{Tr}\left(\bar{c}_t^{ij} B_t B_t^\star\right) dt + \bar{b}_t^{ij\star} B_t B_t^\star \Sigma_t^{-1} m_t dt \\ &\quad +\operatorname{Tr}\left(\bar{c}_t^{ij} E_t D_t D_t^\star E_t^\star\right) dt + \bar{b}_t^{ij\star} E_t D_t D_t^\star E_t^\star \Sigma_t^{-1} m_t dt \\ &\quad -\operatorname{Tr}\left(E_t D_t D_t^\star f_j e_i^\star\right) dt - \bar{b}_t^{ij\star} E_t dy_t \end{aligned} \tag{20}$$

$$\bar{a}_0^{ij} = 0$$

$$\begin{aligned} d\bar{b}_t^{ij} &= -\left(A_t^\star + C_t^\star E_t^\star + \Sigma_t^{-1}(B_t B_t^\star + E_t D_t D_t^\star E_t^\star)\right) \bar{b}_t^{ij} dt \\ &\quad +2\bar{c}_t^{ij}\left(B_t B_t^\star + E_t D_t D_t^\star E_t^\star\right) \Sigma_t^{-1} m_t dt \\ &\quad -e_i f_j^\star D_t D_t^\star E_t^\star \Sigma_t^{-1} m_t dt + e_i f_j^\star dy_t - 2\bar{c}_t^{ij} E_t dy_t \end{aligned} \tag{21}$$

$$\bar{b}_0^{ij} = 0$$

$$\begin{aligned}\frac{d\bar{c}_t^{ij}}{dt} &= -\left(A_t^\star + C_t^\star E_t^\star + \Sigma_t^{-1}(B_tB_t^\star + E_tD_tD_t^\star E_t^\star)\right)\bar{c}_t^{ij} \\ &\quad -\bar{c}_t^{ij}\left(A_t + E_tC_t + (B_tB_t^\star + E_tD_tD_t^\star E_t^\star)\Sigma_t^{-1}\right) \qquad (22)\\ &\quad +\frac{1}{2}\left(e_if_j^\star C_t + e_if_j^\star D_tD_t^\star E_t^\star\Sigma_t^{-1} + C_t^\star f_je_i^\star + \Sigma_t^{-1}E_tD_tD_t^\star f_je_i^\star\right)\\ \bar{c}_0^{ij} &= 0\end{aligned}$$

Proof. Recalling that the density $p_t(x)$ solves the Zakai equation (4) one can prove by means of computations similar to those exposed in the proof of Theorem 3.4 that $\gamma_t^{ij}(x)$ given by (19) is solution of (18) provided that (20), (21) and (22) hold. □

Furthermore, by using variation of constants technique, one gets the following representation result for the processes $(\bar{b}_t^{ij})_{t\geq 0}$ and $(\bar{c}_t^{ij})_{t\geq 0}$.

Proposition 3.11. *The stochastic processes* $(\bar{b}_t^{ij})_{t\geq 0}$ *and* $(\bar{c}_t^{ij})_{t\geq 0}$ *solutions of* (21) *and* (22) *are given by*

$$\begin{aligned}\bar{b}_t^{ij} = G_t\Bigg(2\int_0^t G_s^{-1}\bar{c}_s^{ij}\left(B_sB_s^\star + E_sD_sD_s^\star E_s^\star\right)\Sigma_s^{-1}m_s ds \\ -\int_0^t G_s^{-1}e_if_j^\star D_sD_s^\star E_s^\star\Sigma_s^{-1}m_s ds \\ +\int_0^t G_s^{-1}e_if_j^\star dy_s - 2\int_0^t G_s^{-1}\bar{c}_s^{ij}E_s dy_s\Bigg)\end{aligned}$$

and

$$\begin{aligned}\bar{c}_t^{ij} = \frac{1}{2}G_t\Big(\int_0^t G_s^{-1}\big(e_if_j^\star C_s + e_if_j^\star D_sD_s^\star E_s^\star\Sigma_s^{-1} \\ +C_s^\star f_je_i^\star + \Sigma_s^{-1}E_sD_sD_s^\star f_je_i^\star\big)(G_s^{-1})^\star ds\Big)G_t^\star.\end{aligned}$$

Now, we study the stochastic process $\left(L_t^{ij}\right)_{t\geq 0}$ by using a method similar to that used previously.

Definition 3.12. *The measure-valued process defined for any function* $\phi\in C_0(\mathbb{R}^n;\mathbb{R})$ *by* $\bar{E}\left[Z_tL_t^{ij}\phi(x_t)\ /\ \mathcal{Y}_t\right]$ *has a density* $\lambda_t^{ij}(x)$ *with respect to the Lebesgue measure on* $\mathbb{R}^n$ *so that*

$$\bar{E}\left[Z_tL_t^{ij}\phi(x_t)\ /\ \mathcal{Y}_t\right] = \int_{\mathbb{R}^n}\phi(x)\lambda_t^{ij}(x)dx. \qquad (23)$$

Remark 3.13. The existence of the density $\lambda_t^{ij}(x)$ has been established by means of stochastic partial differential equations techniques by Pardoux [8].

Arguing as in the proof of Proposition 3.3, we get the following result.

Proposition 3.14. *The density* $\lambda_t^{ij}(x)$ *defined by* (23) *solves the stochastic partial differential equation*

$$\begin{aligned}
\lambda_t^{ij}(x) = &- \int_0^t \operatorname{div}\left(\lambda_s^{ij}(x) A_s x\right) ds - \int_0^t \operatorname{div}\left(\lambda_s^{ij}(x) E_s C_s x\right) ds \\
&+ \frac{1}{2}\int_0^t \operatorname{Tr}\left(\nabla^2 \lambda_s^{ij}(x) B_s B_s^\star\right) ds + \frac{1}{2}\int_0^t \operatorname{Tr}\left(\nabla^2 \lambda_s^{ij}(x) E_s D_s D_s^\star E_s^\star\right) ds \\
&+ \int_0^t \lambda_s^{ij}(x)(C_s x)^\star (D_s^{-1})^\star D_s^{-1} dy_s - \sum_{k=1}^p \int_0^t \operatorname{div}\left(\lambda_s^{ij}(x) E_s^k\right) dy_s^k \\
&+ \int_0^t p_s(x) x^\star e_i e_j^\star A_s x ds + \int_0^t p_s(x) x^\star e_i e_j^\star E_s dy_s \\
&+ \int_0^t p_s(x) x^\star e_i e_j^\star E_s C_s x ds - \int_0^t \operatorname{div}\left(p_s(x) B_s B_s^\star e_j e_i^\star x\right) ds \\
&- \int_0^t \operatorname{div}\left(p_s(x) E_s D_s D_s^\star E_s^\star e_j e_i^\star x\right) ds. \qquad (24)
\end{aligned}$$

Furthermore, if we proceed similarly as above, we can prove that the measure $\lambda_t^{ij}(x)$ can be described in terms of a finite number of statistics.

Theorem 3.15. *The density* $\lambda_t^{ij}(x)$ *is given by*

$$\lambda_t^{ij}(x) = \left(\hat{a}_t^{ij} + x^\star \hat{b}_t^{ij} + x^\star \hat{c}_t^{ij} x\right) p_t(x) \qquad (25)$$

where the processes $\hat{a}_t^{ij} \in \mathbb{R}$, $\hat{b}_t^{ij} \in \mathbb{R}^n$ *and* $\hat{c}_t^{ij} \in \mathcal{S}_{n\times n}(\mathbb{R})$ *are solutions of the following stochastic differential equations*

$$\begin{aligned}
d\hat{a}_t^{ij} &= \operatorname{Tr}\left(\hat{c}_t^{ij} B_t B_t^\star\right) dt + \operatorname{Tr}\left(\hat{c}_t^{ij} E_t D_t D_t^\star E_t^\star\right) dt \\
&\quad + \hat{b}_t^{ij\star}\left(B_t B_t^\star + E_t D_t D_t^\star E_t^\star\right) \Sigma_t^{-1} m_t dt \qquad (26)\\
&\quad - \operatorname{Tr}\left(e_i e_j^\star B_t B_t^\star\right) dt - \operatorname{Tr}\left(e_i e_j^\star E_t D_t D_t^\star E_t^\star\right) dt \\
&\quad - \hat{b}_t^{ij\star} E_t dy_t \\
a_0^{ij} &= 0 \\
d\hat{b}_t^{ij} &= -\left(A_t^\star + C_t^\star E_t^\star + \Sigma_t^{-1}(B_t B_t^\star + E_t D_t D_t^\star E_t^\star)\right) \hat{b}_t^{ij} dt \\
&\quad + 2\hat{c}_t^{ij}\left(B_t B_t^\star + E_t D_t D_t^\star E_t^\star\right) \Sigma_t^{-1} m_t dt \qquad (27)\\
&\quad - e_i e_j^\star (B_t B_t^\star + E_t D_t D_t^\star E_t^\star) \Sigma_t^{-1} m_t dt \\
&\quad + \left(e_i e_j^\star - 2\hat{c}_t^{ij}\right) E_t dy_t \\
b_0^{ij} &= 0
\end{aligned}$$

$$\begin{aligned}\frac{d\hat{c}_t^{ij}}{dt} &= -\left(A_t^\star + C_t^\star E_t^\star + \Sigma_t^{-1}(B_tB_t^\star + E_tD_tD_t^\star E_t^\star)\right)\hat{c}_t^{ij} \\ &\quad -\hat{c}_t^{ij}\left(A_t + E_tC_t + (B_tB_t^\star + E_tD_tD_t^\star E_t^\star)\Sigma_t^{-1}\right) \\ &\quad +\frac{1}{2}\left(e_ie_j^\star\left(A_t + E_tC_t + (B_tB_t^\star + E_tD_tD_t^\star E_t^\star)\Sigma_t^{-1}\right)\right. \\ &\qquad \left. + \left(A_t^\star + C_t^\star E_t^\star + \Sigma_t^{-1}(B_tB_t^\star + E_tD_tD_t^\star E_t^\star)\right)e_je_i^\star\right) \\ \hat{c}_0^{ij} &= 0\end{aligned} \tag{28}$$

The last two equations can be solved by means of a variation of constants technique.

Proposition 3.16. *The stochastic processes* $(\hat{b}_t^{ij})_{t\geq 0}$ *and* $(\hat{c}_t^{ij})_{t\geq 0}$ *solutions of* (27) *and* (28) *are given by*

$$\begin{aligned}\hat{b}_t^{ij} = G_t\Bigg(2\int_0^t & G_s^{-1}\hat{c}_s^{ij}\left(B_sB_s^\star + E_sD_sD_s^\star E_s^\star\right)\Sigma_s^{-1}m_s ds \\ &- \int_0^t G_s^{-1}\left(e_ie_j^\star(B_sB_s^\star + E_sE_s^\star)\right)\Sigma_s^{-1}m_s ds \\ &\qquad + \int_0^t G_s^{-1}\left(e_ie_j^\star - 2\hat{c}_s^{ij}\right)E_s dy_s\Bigg)\end{aligned}$$

and

$$\begin{aligned}\hat{c}_t^{ij} = \frac{1}{2}G_t\Bigg(\int_0^t & G_s^{-1}\left(e_ie_j^\star\left(A_s + E_sC_s + (B_sB_s^\star + E_sE_s^\star)\Sigma_s^{-1}\right)\right. \\ &\left. + \left(A_s^\star + C_s^\star E_s^\star + \Sigma_s^{-1}(B_sB_s^\star + E_sE_s^\star)\right)e_je_i^\star\right)(G_s^{-1})^\star ds\Bigg)G_t^\star.\end{aligned}$$

Now, we proceed similarly to study the stochastic process $(K_t^{ji})_{t\geq 0}$ omitting details.

Definition 3.17. *The measure-valued process defined for any function* $\phi\in C_0(\mathbb{R}^n;\mathbb{R})$ *by* $\bar{E}\left[Z_tK_t^{ji}\phi(x_t)\ /\ \mathcal{Y}_t\right]$ *has a density* $\zeta_t^{ji}(x)$ *with respect to the Lebesgue measure on* $\mathbb{R}^n$ *so that*

$$\bar{E}\left[Z_tK_t^{ji}\phi(x_t)\ /\ \mathcal{Y}_t\right] = \int_{\mathbb{R}^n}\phi(x)\zeta_t^{ji}(x)dx. \tag{29}$$

Remark 3.18. The existence of the density $\zeta_t^{ji}(x)$ has been established by means of stochastic partial differential equations techniques by Pardoux [8].

Proposition 3.19. *The density $\zeta_t^{ji}(x)$ defined by (29) solves the stochastic partial differential equation*

$$\begin{aligned}\zeta_t^{ji}(x) = &-\int_0^t \operatorname{div}\left(\zeta_s^{ji}(x)A_s x\right) ds - \int_0^t \operatorname{div}\left(\zeta_s^{ji}(x)E_s C_s x\right) ds \\ &+\frac{1}{2}\int_0^t \operatorname{Tr}\left(\nabla^2\zeta_s^{ji}(x)B_sB_s^\star\right) ds + \frac{1}{2}\int_0^t \operatorname{Tr}\left(\nabla^2\zeta_s^{ji}(x)E_sD_sD_s^\star E_s^\star\right) ds \\ &+\int_0^t \zeta_s^{ji}(x)(C_s x)^\star (D_s^{-1})^\star D_s^{-1} dy_s - \sum_{k=1}^{p}\int_0^t \operatorname{div}\left(\zeta_s^{ji}(x)E_s^k\right) dy_s^k \\ &+\int_0^t p_s(x)y_s^\star f_j e_i^\star A_s x dt + \int_0^t p_s(x)y_s^\star f_j e_i^\star E_s dy_s \\ &+\int_0^t p_s(x)y_s^\star f_j e_i^\star E_s C_s x dt - \int_0^t \operatorname{div}\left(p_s(x)B_sB_s^\star e_i f_j^\star y_s\right) ds \\ &-\int_0^t \operatorname{div}\left(p_s(x)E_sD_sD_s^\star E_s^\star e_i f_j^\star y_s\right) ds. \qquad (30)\end{aligned}$$

Again, the measure $\zeta_t^{ji}(x)$ can be described in terms of a finite number of statistics.

Theorem 3.20. *The density $\zeta_t^{ji}(x)$ is given by*

$$\zeta_t^{ji}(x) = \left(\tilde{a}_t^{ji} + x^\star \tilde{b}_t^{ji} + x^\star \tilde{c}_t^{ji} x\right) p_t(x) \qquad (31)$$

where the processes $\tilde{a}_t^{ji} \in \mathbb{R}$, $\tilde{b}_t^{ji} \in \mathbb{R}^n$ and $\tilde{c}_t^{ji} \in \mathcal{S}_{n\times n}(\mathbb{R})$ are solutions of the following stochastic differential equations

$$\begin{aligned} d\tilde{a}_t^{ji} &= \operatorname{Tr}\left(\tilde{c}_t^{ji} B_tB_t^\star\right) dt + \operatorname{Tr}\left(\tilde{c}_t^{ji} E_tD_tD_t^\star E_t^\star\right) dt \\ &\quad + \left(\tilde{b}_t^{ji\star} - y_t^\star f_j e_i^\star\right)\left(B_tB_t^\star + E_tD_tD_t^\star E_t^\star\right)\Sigma_t^{-1} m_t dt \qquad (32) \\ &\quad - \left(\tilde{b}_t^{ji\star} - y_t^\star f_j e_i^\star\right) E_t dy_t \\ \tilde{a}_0^{ji} &= 0 \\ d\tilde{b}_t^{ji} &= -\left(A_t^\star + C_t^\star E_t^\star + \Sigma_t^{-1}(B_tB_t^\star + E_tD_tD_t^\star E_t^\star)\right)\tilde{b}_t^{ji} dt \\ &\quad +2\tilde{c}_t^{ji}\left(B_tB_t^\star + E_tD_tD_t^\star E_t^\star\right)\Sigma_t^{-1} m_t dt \qquad (33) \\ &\quad + \left(A_t^\star + C_t^\star E_t^\star + \Sigma_t^{-1}(B_tB_t^\star + E_tD_tD_t^\star E_t^\star)\right) e_i f_j^\star y_t dt \\ &\quad -2\tilde{c}_t^{ji} E_t dy_t \\ \tilde{b}_0^{ji} &= 0 \end{aligned}$$

$$
\begin{aligned}
\frac{d\tilde{c}_t^{ji}}{dt} &= -\left(A_t^\star + C_t^\star E_t^\star + \Sigma_t^{-1}(B_t B_t^\star + E_t D_t D_t^\star E_t^\star)\right)\tilde{c}_t^{ji} \\
&\quad -\tilde{c}_t^{ji}\left(A_t + E_t C_t + (B_t B_t^\star + E_t D_t D_t^\star E_t^\star)\Sigma_t^{-1}\right) \\
\tilde{c}_0^{ji} &= 0
\end{aligned}
\tag{34}
$$

By using variation of constants technique, we obtain explicit solutions to the above equations.

Proposition 3.21. *The stochastic processes* $(\tilde{b}_t^{ji})_{t\geq 0}$ *and* $(\tilde{c}_t^{ji})_{t\geq 0}$ *solutions of* (33) *and* (34) *are given by*

$$
\begin{aligned}
\tilde{b}_t^{ji} = G_t \Bigg(& 2\int_0^t G_s^{-1}\tilde{c}_s^{ji}\left(B_s B_s^\star + E_s D_s D_s^\star E_s^\star\right)\Sigma_s^{-1} m_s ds \\
& + \int_0^t G_s^{-1}\left(A_t^\star + C_t^\star E_t^\star + \Sigma_t^{-1}(B_t B_t^\star + E_t D_t D_t^\star E_t^\star)\right) e_i f_j^\star y_s ds \\
& -2\int_0^t G_s^{-1}\tilde{c}_s^{ji} E_s dy_s \Bigg)
\end{aligned}
$$

and

$$
\tilde{c}_t^{ji} = G_t G_t^\star .
$$

Then, finite-dimensional filters for the stochastic processes

$$
\left(H_t^{ij}\right)_{t\geq 0}, \quad \left(J_t^{ij}\right)_{t\geq 0}, \quad \left(L_t^{ij}\right)_{t\geq 0} \quad \text{and} \quad \left(K_t^{ji}\right)_{t\geq 0}
$$

are obtained by using arguments similar to those invoked in Theorem 3.10 in [6].

Theorem 3.22. *Finite-dimensional filters for the stochastic processes*

$$
\left(H_t^{ij}\right)_{t\geq 0}, \quad \left(J_t^{ij}\right)_{t\geq 0}, \quad \left(L_t^{ij}\right)_{t\geq 0} \quad \textit{and} \quad \left(K_t^{ji}\right)_{t\geq 0}
$$

are given by

$$
\begin{aligned}
E\left[H_t^{ij} \;/\; \mathcal{Y}_t\right] &= a_t^{ij} + m_t^\star b_t^{ij} + \sum_{k,l=1}^{n}\left(c_t^{ij}\right)_{kl}(\Sigma_t)_{kl} + m_t^\star c_t^{ij} m_t \\
E\left[J_t^{ij} \;/\; \mathcal{Y}_t\right] &= \bar{a}_t^{ij} + m_t^\star \bar{b}_t^{ij} + \sum_{k,l=1}^{n}\left(\bar{c}_t^{ij}\right)_{kl}(\Sigma_t)_{kl} + m_t^\star \bar{c}_t^{ij} m_t \\
E\left[L_t^{ij} \;/\; \mathcal{Y}_t\right] &= \hat{a}_t^{ij} + m_t^\star \hat{b}_t^{ij} + \sum_{k,l=1}^{n}\left(\hat{c}_t^{ij}\right)_{kl}(\Sigma_t)_{kl} + m_t^\star \hat{c}_t^{ij} m_t \\
E\left[K_t^{ji} \;/\; \mathcal{Y}_t\right] &= \tilde{a}_t^{ji} + m_t^\star \tilde{b}_t^{ji} + \sum_{k,l=1}^{n}\left(\tilde{c}_t^{ji}\right)_{kl}(\Sigma_t)_{kl} + m_t^\star \tilde{c}_t^{ji} m_t
\end{aligned}
$$

Proof. According to (5) one has, for any $u \in \mathbb{R}^n$,

$$\int_{\mathbb{R}^n} u^\star x p_t(x)dx = (u^\star m_t)\eta_t$$

and for any matrix $M \in \mathcal{M}_{n\times n}(\mathbb{R})$,

$$\int_{\mathbb{R}^n} x^\star M x p_t(x)dx = \left(\sum_{k,l=1}^{n} M_{kl}\left(\Sigma_t\right)_{kl} + m_t^\star M m_t\right)\eta_t.$$

Therefore, by means of the Kallianpur-Striebel formula, one gets

$$\begin{aligned} E\left[H_t^{ij} \;/\; \mathcal{Y}_t\right] &= \frac{\bar{E}\left[Z_t H_t^{ij} \;/\; \mathcal{Y}_t\right]}{\bar{E}\left[Z_t \;/\; \mathcal{Y}_t\right]} \\ &= \frac{1}{\eta_t}\int_{\mathbb{R}^n} \beta_t^{ij}(x)dx \\ &= a_t^{ij} + m_t^\star b_t^{ij} + \sum_{k,l=1}^{n}\left(c_t^{ij}\right)_{kl}\left(\Sigma_t^{ij}\right)_{kl} + m_t^\star c_t^{ij} m_t. \end{aligned}$$

The proofs of the last three estimates are similar. □

4. Filtered EM algorithm for system parameter estimate

In this section, we consider the homogeneous version of the filtering system introduced in Section 2. That is, $(x_t, y_t) \in \mathbb{R}^n \times \mathbb{R}^p$ is the solution of the stochastic differential system

$$dx_t = Ax_t dt + Bdw_t + Edy_t \tag{35}$$

$$dy_t = Cx_t dt + Ddv_t \tag{36}$$

The purpose of this section is to compute, by means of a filtered EM algorithm, maximum likelihood estimates of the system coefficients given the observation history $\mathcal{Y}_t$.

The EM algorithm is an iterative procedure for maximizing a likelihood function in a context of partial information (see [4] for example). It can be described as follows:

Let $(P_\theta)_{\theta\in\Theta}$ be a family of probability measures on the measurable space $(\Omega, \mathcal{F})$, all absolutely continuous with respect to a fixed probability measure P_0 defined on $(\Omega, \mathcal{F})$, and assume that the σ-algebra $\mathcal{Y}_t$ containing the observation history is available.

The log-likelihood function for the estimation of the parameter θ based on the available observation history is defined as

$$\mathcal{L}(\theta) = \log E_0\left[\frac{dP_\theta}{dP_0}/\mathcal{Y}_t\right]$$

where E_0 denotes the expectation under the measure P_0 and the maximum likelihood estimate (MLE) is given by

$$\hat{\theta} \in \operatorname{argmax}_{\theta\in\Theta} \mathcal{L}(\theta).$$

Furthermore, a straightforward application of Jensen's inequality gives

$$\mathcal{L}(\theta) - \mathcal{L}(\theta') \geq E_{\theta'}\left[\log\left(\frac{dP_\theta}{dP_{\theta'}}\right) / \mathcal{Y}_t\right].$$

Then, setting $Q(\theta,\theta') = E_{\theta'}\left[\log\left(\frac{dP_\theta}{dP_{\theta'}}\right) / \mathcal{Y}_t\right]$, it holds that for each value θ' of the parameter, $\mathcal{L}(\theta') + Q(\theta,\theta')$ is a lower bound for the log-likelihood function $\mathcal{L}(\theta)$ with equality if $\theta = \theta'$.

Hence, one gets the following algorithm for computing the maximum likelihood estimate

1. $p = 0$, guess $\hat{\theta}_0$
2. (E-Step) Set $\theta' = \hat{\theta}_p$ and compute $Q(.,\theta')$
3. (M-Step) Find $\hat{\theta}_{p+1} \in \operatorname{argmax}_{\theta\in\Theta} Q(\theta,\theta')$
4. If $\hat{\theta}_p = \hat{\theta}_{p+1}$ then set final estimate to $\hat{\theta}_p$ else repeat from step 2 with $p = p+1$.

4.1. Estimation of the coefficients A and C

In this section, we compute maximum likelihood estimate of the parameter $\theta = (A, C)$ given the observation history $\mathcal{Y}_t$ assuming that the coefficients B, D and E are known.

In the case of the linear Gaussian system (35)–(36) one gets the following result.

Proposition 4.1. *For any θ and $\tilde{\theta}$ in Θ,*

$$\begin{aligned} Q(\theta,\tilde{\theta}) = E&\left[\int_0^t x_s^\star A^\star (BB^\star + EDD^\star E^\star)^\# dx_s \ / \ \mathcal{Y}_t\right] \\ &- \frac{1}{2}E\left[\int_0^t x_s^\star A^\star (BB^\star + EDD^\star E^\star)^\# A x_s ds \ / \ \mathcal{Y}_t\right] \\ &+ E\left[\int_0^t x_s^\star C^\star (DD^\star)^{-1} dy_s - \frac{1}{2}\int_0^t x_s^\star C^\star (DD^\star)^{-1} C x_s ds \ / \ \mathcal{Y}_t\right] \\ &+ E\left[R\left(\tilde{\theta}\right) \ / \ \mathcal{Y}_t\right] \end{aligned} \tag{37}$$

where # denotes the pseudo-inverse and $R\left(\tilde{\theta}\right)$ is a function which does not involve the parameter θ.

Proof. By means of Girsanov's theorem, one can prove easily that the updating of the estimate from $\tilde{A}$ to A is related to the Radon-Nikodym derivative

$$\begin{aligned}\frac{dP(A)}{dP(\tilde{A})} &= \exp\left(\int_0^t x_s^\star (A-\tilde{A})^\star (BB^\star + EDD^\star E^\star)^{\#} dx_s\right) . \\ & \exp\left(-\frac{1}{2}\int_0^t x_s^\star (A-\tilde{A})^\star (BB^\star + EDD^\star E^\star)^{\#} (A-\tilde{A}) x_s ds\right)\end{aligned}$$

which gives

$$\begin{aligned}E\left[\log\left(\frac{dP(A)}{dP(\tilde{A})}\right) \;/\; \mathcal{Y}_t\right] &= E\left[\int_0^t x_s^\star A^\star (BB^\star + EDD^\star E^\star)^{\#} dx_s \;/\; \mathcal{Y}_t\right] \\ & -\frac{1}{2} E\left[\int_0^t x_s^\star A^\star (BB^\star + EDD^\star E^\star)^{\#} A x_s ds \;/\; \mathcal{Y}_t\right] \\ & + E\left[R_1(\tilde{A}) \;/\; \mathcal{Y}_t\right] \qquad (38)\end{aligned}$$

where $R_1(\tilde{A})$ is a function which does not involve A.

The updating of the estimate from $\tilde{C}$ to C is related to the Radon-Nikodym derivative

$$\frac{dP(C)}{dP(\tilde{C})} = \exp\left(\int_0^t x_s^\star C^\star (DD^\star)^{-1} dy_s - \frac{1}{2}\int_0^t x_s^\star C^\star (DD^\star)^{-1} C x_s ds\right)$$

which gives

$$\begin{aligned}E\left[\log\left(\frac{dP(C)}{dP(\tilde{C})}\right) \;/\; \mathcal{Y}_t\right] &= E\left[\int_0^t x_s^\star C^\star (DD^\star)^{-1} dy_s \;/\; \mathcal{Y}_t\right] \\ & -\frac{1}{2} E\left[\int_0^t x_s^\star C^\star (DD^\star)^{-1} C x_s ds \;/\; \mathcal{Y}_t\right] \\ & + E\left[R_2(\tilde{A}) \;/\; \mathcal{Y}_t\right] \qquad (39)\end{aligned}$$

where $R_2(\tilde{C})$ is a function which does not involve C.

Equality (37) is obtained by adding (38) and (39). □

Then, when dealing with the M-step of the algorithm one has to set the derivatives $\dfrac{\partial Q}{\partial \theta} = 0$ which leads to

$$A = E\left[\int_0^t x_s (dx_s)^\star \;/\; \mathcal{Y}_t\right]^\star \left(E\left[\int_0^t x_s x_s^\star ds \;/\; \mathcal{Y}_t\right]\right)^{-1} = \tilde{L}_t^\star \tilde{H}_t^{-1} \qquad (40)$$

and

$$C = E\left[\int_0^t x_s (dy_s)^\star \;/\; \mathcal{Y}_t\right]^\star \left(E\left[\int_0^t x_s x_s^\star ds \;/\; \mathcal{Y}_t\right]\right)^{-1} = \tilde{J}_t^\star \tilde{H}_t^{-1} \qquad (41)$$

where $\tilde{H}_t$ and $\tilde{L}_t$ are the matrices with elements $E\left[H_t^{ij} \ / \ \mathcal{Y}_t\right]$ and $E\left[L_t^{ij} \ / \ \mathcal{Y}_t\right]$, $i,j \in \{1,\dots,n\}$ computed in Theorem 3.22 and $\tilde{J}_t$ is the matrix with elements $E\left[J_t^{ij} \ / \ \mathcal{Y}_t\right]$, $i \in \{1,\dots,n\}$, $j \in \{1,\dots,p\}$ computed in Theorem 3.22.

Hence, (40) and (41) give a filtered-EM algorithm for estimating the parameters A and C.

4.2. Estimation of the matrix D

An application of Itô's formula to the tensor product of y_t with itself gives

$$y_t y_t^\star = \int_0^t y_s (dy_s)^\star + \int_0^t dy_s y_s^\star + DD^\star t.$$

Then, one can evaluate $DD^\star$ in terms of the quadratic variation of the stochastic process $(y_t)_{t\geq 0}$.

4.3. Estimation of the matrix E

By using Itô's formula, one can prove easily that for any $t \geq 0$,

$$x_t y_t^\star = \int_0^t x_s (dy_s)^\star + \int_0^t dx_s y_s^\star + EDD^\star t.$$

Then, conditioning both sides of the previous equality on $\mathcal{Y}_t$, yields

$$m_t y_t^\star = E\left[\int_0^t x_s (dy_s)^\star \ / \ \mathcal{Y}_t\right] + E\left[\int_0^t dx_s y_s^\star \ / \ \mathcal{Y}_t\right] + EDD^\star t. \tag{42}$$

The components of the stochastic integrals in (42) are given by the filtered estimates obtained in Theorem 3.22.

Consequently, we have a procedure leading to an estimate of the matrix $EDD^\star$.

4.4. Estimation of the matrix B

An application of Itô's formula to the tensor product of x_t with itself gives

$$x_t x_t^\star = x_0 x_0^\star + \int_0^t x_s (dx_s)^\star + \int_0^t dx_s x_s^\star + (BB^\star + EDD^\star E^\star)t.$$

Conditioning both sides of the previous equality on $\mathcal{Y}_t$, one gets

$$\Sigma_t = E\left[x_0 x_0^\star \ / \ \mathcal{Y}_t\right] + E\left[\int_0^t x_s (dx_s)^\star \ / \ \mathcal{Y}_t\right] + E\left[\int_0^t dx_s x_s^\star \ / \ \mathcal{Y}_t\right] + (BB^\star + EDD^\star E^\star)t. \tag{43}$$

The components of the stochastic integrals in (43) are given by the filtered estimates obtained in Theorem 3.22 and $E\left[x_0 x_0^\star \ / \ \mathcal{Y}_t\right]$ can be expressed in terms of finite-dimensional statistics (see Theorem 12.11 in [7] for example).

Therefore, we have a procedure leading to an estimate of the matrix $(BB^\star + EDD^\star E^\star)$ and hence of the matrix $BB^\star$.

References

[1] A. Bensoussan, *Stochastic control of partially observable systems*, Cambridge University Press, Cambridge (UK), **1992**.

[2] F. Campillo and F. Le Gland, *MLE for partially observed diffusions: direct maximization vs. the EM algorithm*, Stochastic Processes and Their Applications, **33** (1989), 245–274.

[3] A. Dembo and O. Zeitouni, *Parameter estimation of partially observed continuous-time stochastic process via the EM algorithm*, Stochastic Processes and Their Applications, **23** (1986), 91–113.

[4] A. Dempster, N. Laird, and D. Rubin, *Maximum likelihood estimation from incomplete data via the EM algorithm*, Journal of the Royal Statistical Society, **B 39 (1)** (1977), 1–38.

[5] R. Elliott, *Stochastic calculus and applications*, Applications of Mathematics **18**, Springer Verlag, Berlin, Heidelberg, New York (1982).

[6] R. Elliott and V. Krishnamurthy, *Exact finite-dimensional filters for maximum likelihood parameters estimation of continuous-time linear Gaussian systems*, SIAM Journal on Control and Optimization, **35 (6)** (1997), 1908–1923.

[7] R. Lipster and A.Shiryayev, *Statistics of Random Processes 2*, Springer Verlag, Berlin, Heidelberg, New York, **1977**.

[8] E. Pardoux, *Stochastic partial differential equations and filtering of diffusion processes*, Stochastics **3 (2)** (1979), 127–167.

[9] E. Pardoux, *Filtrage non linéaire et équations aux dérivées partielles stochastiques associées*, in: P.L.Hennequin (ed.), cole d'Eté de Probabilités de Saint Flour XIX, *Lecture Notes in Mathematics* **1464**, Springer Verlag, Berlin, Heidelberg, New York (1991) 67–163.

[10] E. Wong and B. Hajek, *Stochastic Processes in Engineering*, Springer Verlag, Berlin, Heidelberg, New York, **1985**.

23 Allée des Oeillets
F 57160 Moulins les Metz
France
E-mail address: `patrick.florchinger@wanadoo.fr`

Progress in Probability, Vol. 52, 153–171

Instability of a Quantum Particle Induced by a Randomly Varying Spring Coefficient

J. Garnier

Abstract. This paper investigates the evolution of a quantum particle in a harmonic oscillator whose spring coefficient randomly fluctuates around its mean value. The perturbations are small, but they act long enough so that we can solve the problem in the asymptotic framework corresponding to a perturbation amplitude which tends to zero and a perturbation duration which tends to infinity. We describe the effective evolution equation of the state vector which reads as a stochastic partial differential equation. We exhibit a closed-form equation for the transition probabilities, which can be interpreted in terms of a jump process. Using standard probability tools, we are then able to compute explicitly the probabilities for observing the different energy eigenstates and give the exact statistical distribution of the energy of the particle.

1. Introduction

This paper is a contribution to the study of time-dependent perturbations of quantum systems. One can find in the literature a lot of work devoted to special types of perturbations: sudden, adiabatic, periodic, ... [27]. The considered phenomena are described by a Hamiltonian which is the sum $H^0 + H^1$ of a time-independent piece H^0 whose eigenvalue problem has been solved, and of a small time-dependent perturbation. The typical question one asks is the following. If at $t = 0$ the system is in the eigenstate ψ^0 of H^0, what is the probability for it to be observed in a given eigenstate ? Most results that have been obtained follow a scheme in which the answers are computed in a perturbation series in powers of H^1 [21, 27]. We shall apply a new method for obtaining answers to the above questions, which is based on the one hand on some rigorous asymptotic theory and on the other hand on a representation of the evolution of the transition probabilities in terms of a Markov jump process. In this paper we shall focus on a quadratic perturbation of the harmonic oscillator, although the method can be applied to more general situations.

2000 *Mathematics Subject Classification.* 35Q40, 35R60, 81S99, 82D30.
Key words and phrases. Harmonic oscillator, random perturbations, asymptotic analysis, stochastic calculus.

The quantum harmonic oscillator has been extensively studied because a lot of systems close to a stable equilibrium can be described by an oscillator or a collection of decoupled harmonic oscillators [21]. Furthermore modifications of this model have been investigated, handling by the perturbation theory. Indeed, even for this simple model it is exceptional to find closed-form expressions, except for very particular types of perturbations [17]. Nevertheless rigorous results have been obtained for time-dependent perturbations of the harmonic oscillator. Most of them concern periodic driven force [8, 9, 16]. Although the problem is far less understood in the case of random perturbations, one can find some results in the literature devoted to systems with randomly time-dependent external driving force. A general class of quantum systems in Markovian potentials has been treated in detail [25, 26]. Under suitable conditions on the dynamics of the random potential, it is shown in [23] that the spectrum of the quasi-energy operator is continuous. In [7] the authors study the long-time stability of oscillators driven by time-dependent forces originating from dynamical systems with varying degrees of randomness and focus on the asymptotic energy growth. Recently in [13] we have studied the energy density of a charged particle in a harmonic oscillator driven by a time-dependent homogeneous electric field.

In this paper we consider a particle in a harmonic oscillator whose spring coefficient is not constant but randomly fluctuates around its mean value. We aim at studying this problem by a rigorous and non-perturbative method. Our approach is inspired by the works of Papanicolaou and his co-workers about waves in random media [19, 24]. The first step consists in determining the characteristic scales of the problem at hand: mean oscillation frequency of the harmonic oscillator, amplitude, coherence time and duration of the random perturbations. We then study the asymptotic evolution of the state vector in the asymptotic framework based on the separation of these scales. Our main aim is to exhibit the asymptotic regime which corresponds to the case where the amplitudes of the random fluctuations go to zero and the duration of the external perturbation goes to infinity. We then describe explicitly the effective random evolution of the state vector and the probability transitions.

The paper is organized as follows. In Section 2 we review the main features of the harmonic oscillator and we formulate the problem at hand in Section 3. In Section 4 we establish a relation between the Lyapunov exponents of the energy of the quantum harmonic oscillator and that of the classical harmonic oscillator, which has already been studied. In Sections 5–6 we derive effective evolution equations for the modal decomposition and for the state vector of the particle. By exploiting a representation of the evolution of the energy of the particle in terms of a jump process, we are able to give closed-form expressions for the transition probabilities in Section 7. Finally we compare the theoretical results with numerical simulations in Section 8.

2. The harmonic oscillator

We consider the quantum oscillator, that is to say, a particle of mass M whose state vector in the coordinate basis obeys the Schrödinger equation [21]:

$$i\hbar\frac{\partial\psi}{\partial t} = -\frac{\hbar^2}{2M}\frac{\partial^2\psi}{\partial x^2} + \frac{1}{2}M\omega^2 x^2\psi, \tag{1}$$

where ω is the oscillation frequency. In order to transform this equation into a standard and dimensionless form, we multiply the spatial coordinate x by $r_0^{-1} := (M\omega/\hbar)^{1/2}$ and the time t by $t_0^{-1} := \omega$, so that (1) now reads:

$$2i\frac{\partial\psi}{\partial t} = -\frac{\partial^2\psi}{\partial x^2} + x^2\psi. \tag{2}$$

The spectrum of the harmonic oscillator is pure point with state energies $(2p+1)/2$ and corresponding eigenstates [21]:

$$\begin{aligned} f_p(x) &= \frac{1}{\sqrt{2^p\sqrt{\pi}p!}}H_p(x)\,e^{-x^2/2}, \\ H_p(x) &= (-1)^p e^{x^2}\frac{d^p}{dx^p}e^{-x^2}. \end{aligned}$$

The family $(f_p)_{p\in\mathbb{N}}$ is complete in the following sense [15, Prop. 1.5.7].

Proposition 2.1.
1. *The $(f_p)_{p\in\mathbb{N}}$ are an orthonormal basis of $L^2(\mathbb{R},\mathbb{C})$.*
2. *$(t,x)\mapsto e^{-i\frac{2p+1}{2}t}f_p(x)$ is a solution of (2) for any $p\in\mathbb{N}$.*

We define the eigenstate decomposition as the map $\Theta : \psi \in L^2(\mathbb{R},\mathbb{C}) \mapsto (c_p)_{p\in\mathbb{N}}$, where c_p are the coefficients of the expansion of ψ in the basis (f_p):

$$\Theta(\psi)_p := c_p = \int_{\mathbb{R}} f_p(x)\psi(x)dx.$$

By Proposition 2.1, Θ is an isometry from $L^2(\mathbb{R},\mathbb{C})$ onto l^2, the space of all the sequences $(c_p)_{p\in\mathbb{N}}$ from $\mathbb{N}$ into $\mathbb{C}$ which are squared integrable. In view of the fundamental postulates of the quantum mechanics, if ψ is the state vector of the particle, then the measurement of the energy will yield the eigenvalue $(2p+1)/2$ with probability $|\Theta(\psi)_p|^2$.

3. Evolution driven by time-dependent forces

Suppose that the spring coefficient is shaken so that it fluctuates randomly with respect to its mean value. The problem is to compute the manner in which this randomness distributes the population from the initial (low) energy state into other (high) ones. The perturbed equation which governs the evolution of the state vector is then:

$$2i\frac{\partial\psi}{\partial t} = -\frac{\partial^2\psi}{\partial x^2} + x^2(1+\varepsilon m(t))\psi. \tag{3}$$

We assume that the amplitudes of the fluctuations are of order $\varepsilon \ll 1$. The real-valued function m is assumed to be zero-mean, time-stationary and time-ergodic process. More exactly the forthcoming results require that the random process $t \mapsto m(t)$ fulfills the technical mixing condition "m is ϕ-mixing, with $\phi \in L^{1/2}(\mathbb{R}^+)$" (see [20, Section 4-6-2]).

The initial state vector at time $t = 0$ is ψ_0, which corresponds to the decomposition $c(0) = \Theta(\psi_0)$. Since Θ is an isometry, it is equivalent to study the evolution of the expansion of ψ in the family of eigenstates $(f_p)_{p\in\mathbb{N}}$, i.e. the corresponding normalized coefficients c:

$$c_p(t) = \Theta(\psi(t,.))_p e^{i\frac{2p+1}{2}t},$$

$$\psi(t,x) = \sum_{p'=0}^{\infty} c_{p'}(t) e^{-i\frac{2p'+1}{2}t} f_{p'}(x). \tag{4}$$

Substituting the expression (4) into Eq. (3) and integrating with respect to $f_p(x)dx$ we get the equation that governs the evolution of c:

$$\frac{dc_p}{dt} = \varepsilon\frac{i}{4}m(t)\left(\sqrt{\Gamma_p}c_{p-2}e^{-2it} + (2p+1)c_p + \sqrt{\Gamma_{p+2}}c_{p+2}e^{2it}\right),$$

where the coefficient Γ_p is simply:

$$\Gamma_p = p(p-1).$$

The coefficients $\mathbb{E}[|c_p|^2(t)]$ represent the probabilities that the particle driven by the random process $\varepsilon m x^2$ be observed in the state f_p at time t. Equivalently, one can say that the measurement of the energy at time t:

$$E(t) = \frac{1}{2}\int_{\mathbb{R}}\left(\left|\frac{\partial\psi}{\partial x}\right|^2 + x^2|\psi|^2\right)(t,x)dx \tag{5}$$

will yield the eigenvalue $p + 1/2$ with probability $\mathbb{E}[|c_p|^2(t)]$. This implies that the expected value of the energy can be expressed as:

$$\mathbb{E}[E(t)] = \frac{1}{2} + \sum_{p=0}^{\infty} p\mathbb{E}[|c_p|^2(t)]. \tag{6}$$

4. Lyapunov exponents of the energy

We mention in the introduction that there is a close relation between the classical and quantum harmonic oscillators. In particular their energy growths may be related by the following way. As pointed out in [18], a state vector ψ can be characterized by its Wigner transform:

$$\mathcal{W}(x,k) = \frac{1}{\pi}\int dy \psi^*(x+y)\psi(x-y)e^{i2ky}.$$

Although $\mathcal{W}$ is not necessarily positive, it possesses some properties of a local density of state and in particular the following identities hold true:

$$\begin{aligned} \int k^2 \mathcal{W}(x,k)dk &= \left|\frac{\partial \psi}{\partial x}(x)\right|^2, \\ \int \mathcal{W}(x,k)dk &= |\psi(x)|^2. \end{aligned}$$

The Wigner transform $\mathcal{W}(t,x,k)$ of the state vector $\psi(t)$ satisfies the transport equation:

$$\frac{\partial \mathcal{W}}{\partial t} = -k\frac{\partial \mathcal{W}}{\partial x} + x(1+\varepsilon m(t))\frac{\partial \mathcal{W}}{\partial k}$$

which coincides with the classical Liouville equation. The solution can be expressed as:

$$\mathcal{W}(t,x,k) = \mathcal{W}(0, X(-t,x,k), K(-t,x,k)),$$

where (X,K) is the classical harmonic oscillator whose time-dependent Hamiltonian is $H = \frac{K^2}{2} + (1+\varepsilon m(t))\frac{X^2}{2}$:

$$\begin{aligned} \frac{\partial X}{\partial t} &= K(t), & X(0) &= x, \\ \frac{\partial K}{\partial t} &= -(1+\varepsilon m(t))X(t), & K(0) &= k. \end{aligned} \tag{7}$$

$X(-t,x,k), K(-t,x,k)$ are the initial conditions that evolve into (x,k) at time t. The energy of the quantum particle is therefore:

$$\begin{aligned} E(t) &= \frac{1}{2}\int dxdk(x^2+k^2)\mathcal{W}(t,x,k) \\ &= \frac{1}{2}\int dxdk\left(X(t,x,k)^2 + K(t,x,k)^2\right)\mathcal{W}(0,x,k). \end{aligned}$$

Introducing polar coordinates $R(t,\theta_0)$ and $\theta(t,\theta_0)$:

$$\begin{aligned} X(t,x,k) &= \sqrt{x^2+k^2}R(t,\theta_0)\cos(\theta(t,\theta_0)), \\ K(t,x,k) &= \sqrt{x^2+k^2}R(t,\theta_0)\sin(\theta(t,\theta_0)), \\ \theta_0 &= \arctan(k/x), \end{aligned}$$

System (7) is equivalent to:

$$\begin{aligned} \frac{\partial R}{\partial t} &= -R\varepsilon m(t)\frac{\sin(2\theta)}{2}, & R(0,\theta_0) &= 1, \\ \frac{\partial \theta}{\partial t} &= -1-(1+\varepsilon m(t))\frac{1+\cos(2\theta)}{2}, & \theta(0,\theta_0) &= \theta_0. \end{aligned}$$

The energy of the quantum particle then writes:

$$E(t) = \frac{1}{2}\int_0^{2\pi} R(t,\theta_0)^2 p(\theta_0)d\theta_0,$$

where $p(\theta_0)$ is the density with respect to the Lebesgue measure over $[-\pi, \pi]$ of $\arctan(k/x)$ under $\mathcal{W}(0, x, k)$:

$$p(\theta) = \int_0^\infty \mathcal{W}(0, r\cos(\theta), r\sin(\theta)) r dr.$$

If the initial state is the fundamental eigenstate, then $p(\theta_0) \equiv 1/(2\pi)$. Under appropriate assumptions on the law of the process m, [4, Theorem 4] proves that there exists an analytic function $q \mapsto g(q)$ such that:

$$\lim_{t\to\infty} \frac{1}{t} \ln \mathbb{E}\left[R^q(t,\theta)\right] = g(q), \tag{8}$$

$$\lim_{t\to\infty} \frac{1}{t} \ln R(t,\theta) = g'(0) \text{ almost surely.} \tag{9}$$

Moreover the convergence is uniform over $\theta \in [-\pi, \pi]$. In [4] this theorem is stated for piecewise constant process m, various versions exist which yield the same conclusion for many different classes of processes m [2, 6, 12, 28]. Thus the energy of the quantum particle satisfies:

$$\lim_{t\to\infty} \frac{1}{t} \ln \mathbb{E}\left[E(t)\right] = g(2),$$

$$\lim_{t\to\infty} \frac{1}{t} \ln E(t) = 2g'(0) \text{ almost surely.}$$

Nevertheless the Lyapunov exponent of the q-th moment ($q > 1$) requires a little more information than (8). For instance the exponential growth rate of $\mathbb{E}[E(t)^2]$ requires the knowledge of the growth rate of $\mathbb{E}[R^2(t,\theta_0)R^2(t,\theta_1)]$ for all pairs $(\theta_0, \theta_1) \in [-\pi, \pi]^2$. Although the expression of $g(q)$ is very intricate, even for a very simple random process m, it can be expanded as powers of ε [3]:

$$g(2q) = \frac{q^2+q}{2}\alpha_2\varepsilon^2 + O(\varepsilon^3),$$

$$2g'(0) = \frac{1}{2}\alpha_2\varepsilon^2 + O(\varepsilon^3),$$

where α_2 is the parameter:

$$\alpha_2 = \int_0^\infty \cos(2t)\mathbb{E}[m(0)m(t)]dt.$$

The main feature is that the sample and mean Lyapunov exponents of the system are different. More exactly that the mean Lyapunov energy is twice as large as the sample Lyapunov energy. In the following sections we aim at describing more precisely the energy distribution of the quantum particle. Indeed $E(t)$ is the mean energy, where the averaging is with respect to the quantum distribution. Nevertheless the second quantum theory postulate claims that an observation of the energy can only give a value amongst the discrete set of eigenvalues $1/2+\mathbb{N}$. Consequently a precise description of the energy distribution consists in the determination of the probability transitions.

5. Asymptotic behavior of the modal energy distribution

Let us define the process $\bar{c}$ as the solution of the following infinite-dimensional system of linear stochastic differential equations starting from $c(0)$:

$$d\bar{c}_p = \frac{\sqrt{\alpha_2}}{4}\left(\sqrt{\Gamma_{p+2}}\bar{c}_{p+2} - \sqrt{\Gamma_p}\bar{c}_{p-2}\right) \circ dW_t^1 + \frac{i\sqrt{\alpha_2}}{4}\left(\sqrt{\Gamma_{p+2}}\bar{c}_{p+2} + \sqrt{\Gamma_p}\bar{c}_{p-2}\right) \circ dW_t^2 + \frac{i\sqrt{\alpha_0}}{2\sqrt{2}}(2p+1)\bar{c}_p \circ dW_t^3, \quad (10)$$

where W^j, $j = 1, \dots, 3$ are independent standard Brownian motions, $\circ$ stands for the Stratonovich stochastic integral, and α_j is the real parameter given by:

$$\alpha_j = \int_0^\infty \mathbb{E}[m(0)m(t)]\cos(jt)dt,$$

which is proportional to the power spectral density of the process m (nonnegative by the Wiener-Khintchine theorem [22]).

Proposition 5.1.

1. *There exists a unique solution $\bar{c}$ of Equation* (10).
2. *The processes $c(./\varepsilon^2)$ converge in distribution as continuous functions from $[0,\infty)$ into l^2 to the diffusion Markov process $\bar{c}$ solution of* (10) *as $\varepsilon \to 0$.*

Proof. Apply formally the (unique!) theorem of [24]. Take care to separate the real and imaginary parts of the process $c(./\varepsilon^2)$: Denoting $\chi_{2p} := \mathrm{Re}c_p$ and $\chi_{2p+1} := \mathrm{Im}c_p$, the process $\chi^\varepsilon(t) := \chi(t/\varepsilon^2)$ satisfies the linear differential equation:

$$\frac{d\chi^\varepsilon(t)}{dt} = \frac{1}{\varepsilon}m\left(\frac{t}{\varepsilon^2}\right)F\left(\chi^\varepsilon, \frac{t}{\varepsilon^2}\right),$$

where

$$\begin{aligned}
F_{2p}(\chi,h) &= \frac{\cos(2h)}{4}\left(\sqrt{\Gamma_{p+2}}\chi_{2p+5} - \sqrt{\Gamma_p}\chi_{2p-3}\right) \\
&\quad + \frac{\sin(2h)}{4}\left(\sqrt{\Gamma_{p+2}}\chi_{2p+4} + \sqrt{\Gamma_p}\chi_{2p-4}\right) - \frac{2p+1}{4}\chi_{2p+1}, \\
F_{2p+1}(\chi,h) &= \frac{\cos(2h)}{4}\left(\sqrt{\Gamma_{p+2}}\chi_{2p+4} + \sqrt{\Gamma_p}\chi_{2p-4}\right) \\
&\quad + \frac{\sin(2h)}{4}\left(-\sqrt{\Gamma_{p+2}}\chi_{2p+5} + \sqrt{\Gamma_p}\chi_{2p-3}\right) + \frac{2p+1}{4}\chi_{2p}.
\end{aligned}$$

Note that we deal with an infinite-dimensional system while only finite-dimensional systems are addressed in [24]. Nevertheless $c(./\varepsilon^2)$ can be approximated by finite-dimensional processes. The technical developments are proposed in [13] in the case of a perturbation of the form $2xm(t)$ and are very similar in the present case $x^2m(t)$. This technique based on a martingale approach to some limit theorems in the diffusion-approximation regime is now well known and extensively reviewed in literature [19, 20]. The processes $\left(c_p(./\varepsilon^2)\right)_{p\in\mathbb{N}}$ converge in distribution in l^2 to the diffusion process $(\bar{\chi}_{2p}(.) + i\bar{\chi}_{2p+1}(.))_{p\in\mathbb{N}}$ where $\bar{\chi}$ is the diffusion process with infinitesimal generator $\mathcal{L}$:

$$\mathcal{L} = \sum_{i,j=0}^{\infty} a_{ij}(\bar{\chi}) \frac{\partial^2}{\partial \bar{\chi}_i \partial \bar{\chi}_j} + \sum_{j=0}^{\infty} b_j(\bar{\chi}) \frac{\partial}{\partial \bar{\chi}_j}.$$

The diffusion and drift coefficients are:

$$\begin{aligned} a_{i,j}(\bar{\chi}) &= \int_0^{\infty} \mathbb{E}\left[m(0)m(t)\right] \langle F_i(\bar{\chi}, h) F_j(\bar{\chi}, h+t) \rangle_h \, dt, \\ b_j(\bar{\chi}) &= \int_0^{\infty} \sum_{i=0}^{\infty} \mathbb{E}\left[m(0)m(t)\right] \left\langle F_i(\bar{\chi}, h) \frac{\partial F_j}{\partial \chi_i}(\bar{\chi}, h+t) \right\rangle_h dt, \end{aligned}$$

where $\langle . \rangle_h$ stands for an averaging over a period in h. The application of Itô's formula to system (10) then yields that the distribution of $\bar{c}$ is the same as $(\bar{\chi}_{2p}(.) + i\bar{\chi}_{2p+1}(.))_{p\in\mathbb{N}}$. □

There exist also technical conditions on the initial condition $c(0)$ so that the above proposition holds true. These conditions require that the initial sequence $(c(0)_p)_{p\in\mathbb{N}}$ decays fast enough and they are fulfilled in particular if the initial state is a pure eigenstate, i.e. $c(0)_p = \delta_{p,p_0}$ for some p_0. Thus, in order to avoid unnecessary intricate technical developments, we shall assume throughout the paper that the initial state is a pure eigenstate. If the initial state is f_{p_0}, the coefficients

$$C_{p_0,p}(t) := \mathbb{E}[|\bar{c}_p|^2(t)]$$

represent the probabilities that the particle driven by the random process $\varepsilon m x^2$ be observed in the state f_p at time t/ε^2 in the asymptotic framework $\varepsilon \to 0$. Proposition 5.1 is very useful since it allows us to compute efficiently these relevant quantities:

Proposition 5.2. *The family $(C_{p_0,p}(t))_{p\in\mathbb{N}}$ satisfies a closed-form set of ordinary differential equations:*

$$\frac{dC_{p_0,p}}{dt} = \frac{\alpha_2}{8}\Gamma_{p+2}(C_{p_0,p+2} - C_{p_0,p}) + \frac{\alpha_2}{8}\Gamma_p(C_{p_0,p-2} - C_{p_0,p}), \tag{11}$$

starting from $C_{p_0,p}(0) = \delta_{p,p_0}$.

Proof. This results from a direct application of Itô's formula. □

System (11) is one of the most important results of the paper. It shows that the probabilities $C_{p_0,p}$ can be computed theoretically from the coupling coefficients Γ_p, and that their evolutions are self-consistent in the sense that no other relevant quantities appear. In particular the relative phases between the coefficients $(\bar{c}_p)$ of the expansion of the state vector ψ in the basis (f_p) have no importance in the asymptotic evolution of the probability distribution $(C_{p_0,p})$. This statement is not at all obvious, since it is not satisfied by the original equations (3).

6. Asymptotic behavior of the state vector

We denote by $\tilde{\psi}^\varepsilon(x,t)$ the state vector at the regularly spaced instants $2\pi[t/(2\pi\varepsilon^2)]$ defined by:

$$\tilde{\psi}^\varepsilon(t,x) = \psi\left(2\pi\left[\frac{t}{2\pi\varepsilon^2}\right], x\right),$$

where $[\tau]$ stands for the integral part of a real number τ. The process $\tilde{\psi}^\varepsilon$ possesses nice convergence properties in the space of the càd-làg functions $\mathbf{D}$ equipped with the Skorohod topology The following result is a straightforward corollary of Proposition 5.1 since Θ is an isometry from L^2 into l^2.

Proposition 6.1. *$\tilde{\psi}^\varepsilon$ converges in distribution in $\mathbf{D}([0,\infty), L^2)$ to $\tilde{\psi}$ which is the unique solution of:*

$$\begin{aligned} d\tilde{\psi} &= \frac{\sqrt{\alpha_2}}{4}\left(1 + 2x\frac{\partial}{\partial x}\right)\tilde{\psi}\circ dW_t^1 \\ &\quad + \frac{i\sqrt{\alpha_2}}{4}\left(x^2 + \frac{\partial^2}{\partial x^2}\right)\tilde{\psi}\circ dW_t^2 + \frac{i\sqrt{\alpha_0}}{2\sqrt{2}}\left(-x^2 + \frac{\partial^2}{\partial x^2}\right)\tilde{\psi}\circ dW_t^3, \end{aligned} \tag{12}$$

starting from $\tilde{\psi}(0,x) = f_{p_0}(x)$.

Very fortunately the solution of Equation (12) can be written explicitly as:

$$\tilde{\psi}(t,x) = \sqrt{a(t)}e^{-ib(t)x^2} f_{p_0}(a(t)x)e^{-i(2p_0+1)\phi(t)},$$

where the coefficients a, b, and ϕ obey the following system of stochastic differential equations:

$$da = \frac{\sqrt{\alpha_2}}{2}a\circ dW_t^1 + \sqrt{\alpha_2}ab\circ dW_t^2 + 2\sqrt{2}\sqrt{\alpha_0}ab\circ dW_t^3, \tag{13}$$

$$db = \sqrt{\alpha_2}b\circ dW_t^1 - \frac{\sqrt{\alpha_2}}{4}(1+a^4-4b^2)\circ dW_t^2 + \frac{\sqrt{\alpha_0}}{2\sqrt{2}}(1-a^4+4b^2)\circ dW_t^3, \tag{14}$$

$$d\phi = \frac{\sqrt{\alpha_2}}{8}a^2\circ dW_t^2 + \frac{\sqrt{\alpha_0}}{2\sqrt{2}}a^2\circ dW_t^3,$$

starting from $a(0) = 1$, $b(0) = 0$, $\phi(0) = 0$. In particular the spatial distribution of the quantum particle is $|\tilde{\psi}(t,x)|^2 = a(t)|f_{p_0}(a(t)x)|^2$, which is the initial profile rescaled by the factor $a(t)$. Furthermore, the state vector $\tilde{\psi}_\delta^\varepsilon$ in the intermediate planes $\delta + 2\pi[t/(2\pi\varepsilon^2)]$ defined by:

$$\tilde{\psi}_\delta^\varepsilon(t,x) = \psi\left(\delta + 2\pi\left[\frac{t}{2\pi\varepsilon^2}\right], x\right),$$

converges in distribution in $\mathbf{D}([0,\infty), L^2)$ to $\tilde{\psi}_\delta$:

$$\tilde{\psi}_\delta(t,x) = \sum_{p=0}^{\infty}\bar{c}_p(t)e^{-i(p+1/2)\delta}f_p(x).$$

The asymptotic state vector $\tilde{\psi}_\delta$ can be derived from $\tilde{\psi}$ through the following equation, in which t is frozen:

$$2i\frac{\partial\tilde{\psi}_\delta}{\partial\delta} - x^2\tilde{\psi}_\delta + \frac{\partial^2\tilde{\psi}_\delta}{\partial x^2} = 0, \qquad \tilde{\psi}_\delta(t,x)\Big|_{\delta=0} = \tilde{\psi}(t,x).$$

The general solution $\tilde{\psi}_\delta(t,x)$ is found to be:

$$\tilde{\psi}_\delta(t,x) = \sqrt{a(\delta,t)}e^{-ib(\delta,t)x^2} f_{p_0}(a(\delta,t)x)e^{-i(2p_0+1)\phi(\delta,t)},$$

where

$$\begin{aligned} a(\delta,t) &= \frac{a(t)}{\sqrt{(\cos(\delta)+2b(t)\sin(\delta))^2 + a(t)^4\sin(\delta)^2}}, \\ b(\delta,t) &= \frac{b(t)\cos(2\delta) + (4b(t)^2 + a(t)^4 - 1)\frac{\sin(2\delta)}{4}}{(\cos(\delta)+2b(t)\sin(\delta))^2 + a(t)^4\sin(\delta)^2}, \\ \phi(\delta,t) &= \phi(t) - \frac{\delta}{2}. \end{aligned}$$

We can give simple representations of the distributions of the variables $a(\delta,t)$ and $b(\delta,t)$ in terms of two auxiliary Brownian motions.

Proposition 6.2.

1. *The processes $(a(\delta,.), b(\delta,.))$ obey the same distribution as $(a(.), b(.))$.*
2. *For any δ and t, the variables $(a^2(\delta,t), b(\delta,t))$ obey the distributions of:*

$$a^2(t) = \exp\left(\sqrt{\alpha_2}W_t^4 - \frac{\alpha_2}{2}t\right), \tag{15}$$

$$b(t) = \frac{\sqrt{\alpha_2}}{2}\exp\left(\sqrt{\alpha_2}W_t^4 - \frac{\alpha_2}{2}t\right)\int_0^t \exp\left(-\sqrt{\alpha_2}W_{t'}^4 + \frac{\alpha_2}{2}t'\right) dW_{t'}^5, \tag{16}$$

where W^4 and W^5 are independent standard Brownian motions.

Proof. The proof consists in applying Itô's formula for the system (13–14) on the one hand and for the processes defined by Eqs. (15–16) on the other hand. One can then check that both formulations are identical. □

We can deduce relevant features from the representations (15–16). Let us discuss one of them, which concerns the fundamental difference between the long-range mean behavior and long-range typical behavior of $a(t)$. For large t it is well known that the typical value of W_t^4 is of the order of $\sqrt{t}$, so that the typical value of the random variable $a^2(t)$ is of the order of $\exp\left(-\frac{\alpha_2}{2}t \pm \sqrt{\alpha_2 t}\right)$. In this exponential the first term prevails for large t, so that $a^2(t)$ is exponentially small. Nevertheless, the mean value $\mathbb{E}[a^2(t)] = 1$, whatever t. This means that the *mean* value does not correspond at all to a *typical* value that can be reached by the random variable $a^2(t)$. The mean value is actually imposed by very few events which represent large deviations of the Brownian motion. These exceptional events are enhanced by the exponential, which gives $\mathbb{E}[\exp\sqrt{\alpha_2}W_t^4] = \exp(\frac{\alpha_2}{2}t)$. These observations are of course closely related with the fact that the sample and mean Lyapunov exponents of the system at hand are different, as exhibited in Section 4.

Proposition 6.3. *Let us consider the energy $E(t)$ defined by (5). The quantity $E(./\varepsilon^2)$ converges as a continuous function from $[0,\infty)$ into $\mathbb{R}$ to the process $\bar{E}$ given by:*

$$\bar{E}(t) = E_0\left(\frac{a^2(t)}{2} + \frac{1+4b^2(t)}{2a^2(t)}\right), \tag{17}$$

where E_0 is the energy of the particle in the initial state f_{p_0}: $E_0 = p_0 + 1/2$. The moments of the energy satisfy:

$$\mathbb{E}[\bar{E}(t)] = E_0 e^{\alpha_2 t}, \tag{18}$$

$$\mathbb{E}[\bar{E}(t)^2] = E_0^2\left(\frac{2}{3}e^{3\alpha_2 t} + \frac{1}{3}\right), \tag{19}$$

$$\lim_{t\to\infty}\frac{1}{t}\mathbb{E}[\bar{E}(t)^q] = \frac{q^2+q}{2}\alpha_2. \tag{20}$$

Proof. The first moment of $\bar{E}$ can be computed directly since we get from the expression (6) that it satisfies a closed-form differential equation $\frac{d\mathbb{E}[\bar{E}]}{dt} = \alpha_2\mathbb{E}[\bar{E}]$. The expression of $\bar{E}$ in terms of a, b can be obtained from the expression (17). By Itô's formula we get:

$$\frac{d\mathbb{E}[a^n b^q]}{dt} = \alpha_2\left(\frac{n(n-2)}{8} + \frac{nq+q(q-1)}{2}\right)\mathbb{E}[a^n b^q] + \alpha_2\frac{q(q-1)}{8}\mathbb{E}[a^n b^{q-2}],$$

so that we can compute $\mathbb{E}[\bar{E}(t)^q]$ for any q. □

The energy has a very large variance, much larger than the square of its mean, which shows that it has very large fluctuations with respect to its mean value. This can be also put into evidence by comparing the typical value of the energy with its mean value. As pointed out here above, $a^2(t)$ decays as $\exp(-\frac{\alpha_2}{2}t)$ for $t \gg 1$, since $W_t^4 \sim \sqrt{t} \ll t$ with very high probability. Furthermore $b(t)$ obeys a normal distribution with mean 0 and variance 1/2 for $t \gg 1$. By substituting these approximations into Eq. (17), we get that the typical behavior of the energy $\bar{E}$ is for large t:

$$\frac{1}{t}\ln\bar{E}(t) \simeq \frac{\alpha_2}{2} \text{ for } t \gg 1. \tag{21}$$

The typical exponential behavior (21) is different from the exponential behavior of the mean energy (18), and the departure originates from large deviations of the Brownian motion W^4.

7. Interpretation of the limit system in terms of a jump process

We shall show that the transition probabilities $C_{p_0,p}(t)$ can be regarded as the statistical distribution of a jump process. Let N_t be the Markov process with state space $\mathbb{N}$ and infinitesimal generator $\mathcal{K}$:

$$\mathcal{K} = \frac{\alpha_2}{8}\left(\Gamma_{N+2}\nabla_2 + \Gamma_N\nabla_{-2}\right), \tag{22}$$

where $\nabla_{p'} g(N) = g(N+p') - g(N)$. $(N_t)_{t\geq 0}$ is a time-homogeneous jump process defined on a probability space $(\Omega, \mathcal{F}, \mathbf{P})$. The operator $\nabla_{p'}$ corresponds to a jump from n to $n+p'$. The complete construction of the process N_t is standard [11, Section 7, Theorem 33]. There exists a sequence of integer-valued random variables $(\xi_j)_{j\in\mathbb{N}}$ and positive real-valued variables $(\tau_j)_{j\in\mathbb{N}}$ such that:

$$N_t = \xi_n \text{ if } \sum_{j=0}^{n-1} \tau_j \leq t < \sum_{j=0}^{n} \tau_j.$$

The sequence $(\xi_j)_{j\in\mathbb{N}}$ is a Markov chain with stationary transition matrix Q:

$$\mathbf{P}\left(\xi_{n+1} = p + p' / \xi_n = p\right) = Q_{p,p+p'},$$

where $Q_{p,p+p'} = 0$ if $p' \neq \{-2, 2\}$ and:

$$\begin{aligned} Q_{p,p+2} &= \frac{\Gamma_p}{\Gamma_p + \Gamma_{p+2}} = \frac{1}{2} - \frac{2p+1}{2(p^2+p+1)}, \\ Q_{p,p-2} &= \frac{\Gamma_{p+2}}{\Gamma_p + \Gamma_{p+2}} = \frac{1}{2} + \frac{2p+1}{2(p^2+p+1)}. \end{aligned}$$

Given $(\xi_j)_{j\in\mathbb{N}}$, the random variables $(\tau_j)_{j\in\mathbb{N}}$ are conditionally independent and exponentially distributed, with parameters $(q(\xi_j))_{j\in\mathbb{N}}$:

$$q(p) = \frac{\alpha_2}{8}(\Gamma_p + \Gamma_{p+2}) = \frac{\alpha_2}{4}(p^2 + p + 1).$$

We denote by $\mathbf{P}_{t_0,p_0}$ the distribution of the paths $(N_t)_{t\geq t_0}$ starting at time t_0 with the initial condition $N_{t_0} = p_0$. The Markov process has stationary transition probabilities:

$$P_t(p_0, p) = \mathbf{P}_{t',p_0}(N_{t'+t} = p) \text{ independent of } t',$$

and it is reversible since the infinitesimal generator $\mathcal{K}$ is self-adjoint. Furthermore P_t is the unique probabilistic solution (i.e. $\sum_{p'} P_t(p, p+p') = 1$) of the Kolmogorov's forward equation [10, Section X-3]:

$$\frac{\partial P_t}{\partial t}(p_0, p_1) = -q(p_1) P_t(p_0, p_1) + \sum_p q(p) Q_{p,p_1} P_t(p_0, p).$$

In view of the choices of the kernel Q and parameters $q(.)$, we get that the Kolmogorov's forward equation is equivalent to (11), which implies the relation $C_{p_0,p}(t) = \mathbf{P}_{0,p_0}(N_t = p)$. This interpretation of the transition probabilities is very powerful to solve problems and study the system (11) since it allows us to apply existing results on Markov jump processes which can be found in the literature.

Note that we have only showed that (N_t) gives the correct statistical distribution of the measurement of the energy of the particle at a given time t. In terms of probability theory, we have only proved that the one-dimensional distributions of (N_t) and the ones of the energy distribution coincide. We shall see that (N_t) gives the correct statistical distribution of any sequence of measurements. For that purpose we revisit the postulates of quantum mechanics in terms of the jump process N_t.

- *Postulate 1.* "The state vector ψ obeys the Schrödinger equation (3)." The time evolution of the process N_t is governed by the Markovian dynamics described by the infinitesimal generator (22).
- *Postulate 2.* "The measurement of the energy at (normalized) time t will yield one of the eigenvalues $(2p+1)/2$ with probability $C_{p_0,p}(t)$." The jump process $1/2+N_t$ takes values only in the set $1/2+\mathbb{N}$. At time t one will find the value $1/2+p$ with probability $\mathbf{P}_{0,p_0}(N_t=p)$ which is equal to $C_{p_0,p}(t)$.
- *Postulate 3.* "If a measure at time t of the energy gives the result $1/2+p$, then the state of the system will change from $\psi(t)$ to f_p as a result of the measurement." Let us assume that we start from state f_{p_0} at time 0. If we observe the energy of the particle at times t_1 and t_2, then the probability for measuring first the energy $1/2+p_1$ and then $1/2+p_2$ is equal to the probability for observing $1/2+p_1$ at t_1 starting from f_{p_0} multiplied by the probability for observing $1/2+p_2$ at t_2 starting from f_{p_1} at t_1, because the system is in state f_{p_1} just after the measurement at t_1. This product of probabilities also reads as $\mathbf{P}_{0,p_0}(N_{t_1}=p_1)\times\mathbf{P}_{t_1,p_1}(N_{t_2}=p_2)$. From the Markov property of the process N_t, this product is exactly equal to $\mathbf{P}_{0,p_0}(N_{t_1}=p_1,N_{t_2}=p_2)$. Of course this statement can be generalized to any sequence of measurements so that we conclude that if we observe the energies of the particle at times $t_1,\ldots,$ and t_n, then the probability to measure the sequence of energies: $1/2+p_1,\ldots,$ $1/2+p_n$ is exactly $\mathbf{P}_{0,p_0}(N_{t_1}=p_1,\ldots,N_{t_n}=p_n)$. This means that the dynamics of the observations is Markovian and exactly described by the jump process N_t (Markovian means that the future is independent from the past conditionally to the present). More remarkable, this property is essentially equivalent to the Postulate 3 of quantum mechanics. Indeed, if instead of Postulate 3 we assume that the dynamics of the observations is Markovian, then just after a measurement the system depends only on the result of the measurement, which means that just after measuring the energy $1/2+p_1$, the system must be in a state with energy $1/2+p_1$ with probability 1. Since there exists a unique eigenstate f_{p_1} with energy $1/2+p_1$, this means that the system must be in state f_{p_1} with probability 1 (here the non-degeneracy of the system plays a primary role). As a conclusion, the Postulate 3 is exactly the right condition under which we can extent the statement "(N_t) describes the statistical distribution of the measurement of the energy at some given time" to the statement "(N_t) describes the statistical distribution of any sequence of measurements of the energy".

By studying the generating function of the jump process (N_t) the system (11) can be solved in the sense that closed-form expressions can be derived for the probabilities $C_{p_0,p}(t)$. We sketch out the method we use to derive these expressions. We first establish the equation that governs the evolution of the generating function

$u(t,z) := \mathbb{E}\left[z^{N_t}\right]$ of the jump process:

$$\frac{\partial u}{\partial t} = \frac{\alpha_2}{8}(1-z^2)\frac{\partial^2}{\partial z^2}\left((1-z^2)u\right).$$

We then express $v(t,z) := (1-z^2)u(t,z)$ as the expectation of a functional of a diffusion process Z:

$$v(t,z) = \mathbb{E}\left[Z^{p_0}(t)(1-Z^2(t))\right],$$

where the infinitesimal generator of the Markov process Z starting from z is:

$$\mathcal{L}_Z = \frac{\alpha_2}{8}(1-Z^2)^2\frac{\partial^2}{\partial Z^2}.$$

We introduce the auxiliary process $Y_t = \text{argth}(Z_t)$ whose infinitesimal generator is:

$$\mathcal{L}_Y = \frac{\alpha_2}{8}\frac{\partial^2}{\partial Y^2} + \frac{\alpha_2}{4}\tanh(Y)\frac{\partial}{\partial Y}.$$

We can then apply a generalized version [1, Prop.1] of Bougerol's identity which states that, for any t, $\sinh(Y_t)$ has the same distribution as X_t:

$$X_t = \exp\left(\frac{\sqrt{\alpha_2}}{2}B_t^1 + \frac{\alpha_2}{4}t\right)\left(\frac{z}{\sqrt{1-z^2}} + \frac{\sqrt{\alpha_2}}{2}\int_0^t \exp\left(-\frac{\sqrt{\alpha_2}}{2}B_s^1 - \frac{\alpha_2}{4}s\right)dB_s^2\right),$$

where B^1 and B^2 are two independent Brownian motions. We then get the expression:

$$v(t,z) = \mathbb{E}\left[X_t^{p_0}\left(1+X_t^2\right)^{-1-p_0/2}\right]. \tag{23}$$

We finally note that for $z=0$, $u(t,0) = v(t,0) = \mathbb{P}_{0,p_0}(N_t = 0)$. Furthermore, still for $z=0$, the distribution of X_t can be represented as:

$$X_t = \frac{\sqrt{\alpha_2}}{2}\int_0^t \exp\left(\frac{\sqrt{\alpha_2}}{2}B_s^1 + \frac{\alpha_2}{4}s\right)dB_s^2. \tag{24}$$

Substituting into Eq. (23) yields in particular that:

$$C_{0,0}(t) = \mathbf{P}_{0,0}(N_t = 0) = \mathbb{E}\left[\left(1+X_t^2\right)^{-1}\right],$$

with X_t as defined by Eq. (24). Corollary 2.3.3 [5] gives the probability density of X_t (after a straightforward application of Girsanov's formula). Some further calculations then establish:

$$C_{0,0}(t) = \frac{2}{\sqrt{\pi}}\exp\left(-\frac{\alpha_2 t}{8}\right)\int_0^\infty du\frac{e^{-u^2}}{\cosh\left(u\sqrt{\alpha_2 t/2}\right)}.$$

This proves in particular that the decay of the probability $C_{0,0}(t)$ for long t is $\sqrt{2\pi}(\alpha_2 t)^{-1/2}\exp(-\alpha_2 t/8)$. The transition probabilities $C_{0,p}(t)$ in the particular case when the initial state is the fundamental can then be computed by repeated derivations from Eq. (11):

$$C_{0,2p}(t) = \frac{2}{\sqrt{\pi}}\exp\left(-\frac{\alpha_2 t}{8}\right)\int_0^\infty due^{-u^2}\frac{F_{0,2p}(\alpha_2 t, u\sqrt{\alpha_2 t/2})}{\cosh\left(u\sqrt{\alpha_2 t/2}\right)},$$

where the functions $F_{0,2p}$ are:

$$F_{0,0}(\bar{t},\bar{u}) = 1, \tag{25}$$

$$F_{0,2}(\bar{t},\bar{u}) = \frac{1}{2} - 2\tanh(\bar{u})\frac{\bar{u}}{\bar{t}}, \tag{26}$$

$$F_{0,4}(\bar{t},\bar{u}) = \left(\frac{3}{8} - \frac{2}{3}\frac{\bar{u}^2}{\bar{t}^2}\right) + \tanh(\bar{u})\left(\frac{2}{3}\frac{\bar{u}}{\bar{t}^2} - \frac{7}{3}\frac{\bar{u}}{\bar{t}}\right) + \tanh^2(\bar{u})\frac{4}{3}\frac{\bar{u}^2}{\bar{t}^2}, \tag{27}$$

$$F_{0,2p}(\bar{t},\bar{u}) = \frac{4}{\Gamma_{2p}}\left(\frac{\bar{u}}{\bar{t}}\frac{\partial F_{0,2p-2}}{\partial \bar{u}} + 2\frac{\partial F_{0,2p-2}}{\partial \bar{t}}\right) + \left(1 - \frac{1}{\Gamma_{2p}} - \frac{4}{\Gamma_{2p}}\frac{\bar{u}}{\bar{t}}\tanh(\bar{u})\right)F_{0,2p-2} + \frac{\Gamma_{2p-2}}{\Gamma_{2p}}\left(F_{0,2p-2} - F_{0,2p-4}\right). \tag{28}$$

We now examine the general case where the initial state is f_{2p_0}. The jump process is reversible, so $\mathbf{P}_{0,2p_0}(N_t = 2p) = \mathbf{P}_{0,2p}(N_t = 2p_0)$ which reads:

$$C_{2p_0,2p}(t) = C_{2p,2p_0}(t).$$

In particular $C_{2p_0,0}(t) = C_{0,2p_0}(t)$ has been derived here above, and repeated derivations finally establish the expression of $C_{2p_0,2p}(t)$:

$$C_{2p_0,2p}(t) = \frac{2}{\sqrt{\pi}}\exp\left(-\frac{\alpha_2 t}{8}\right)\int_0^\infty du e^{-u^2}\frac{F_{2p_0,2p}(\alpha_2 t, u\sqrt{\alpha_2 t/2})}{\cosh\left(u\sqrt{\alpha_2 t/2}\right)},$$

where the functions $F_{2p_0,2p}$ are:

$$\begin{aligned} F_{2p_0,2p}(\bar{t},\bar{u}) = {} & \frac{4}{\Gamma_{2p}}\left(\frac{\bar{u}}{\bar{t}}\frac{\partial F_{2p_0,2p-2}}{\partial \bar{u}} + 2\frac{\partial F_{2p_0,2p-2}}{\partial \bar{t}}\right) \\ & + \left(1 - \frac{1}{\Gamma_{2p}} - \frac{4}{\Gamma_{2p}}\frac{\bar{u}}{\bar{t}}\tanh(\bar{u})\right)F_{2p_0,2p-2} \\ & + \frac{\Gamma_{2p-2}}{\Gamma_{2p}}\left(F_{2p_0,2p-2} - F_{2p_0,2p-4}\right), \\ F_{2p_0,0}(\bar{t},\bar{u}) = {} & F_{0,2p_0}(\bar{t},\bar{u}). \end{aligned}$$

As an application the functions $F_{2,2p}$ which govern the probabilities that a particle initially in state f_2 be observed in state f_{2p} at normalized time t are:

$$F_{2,0}(\bar{t},\bar{u}) = \frac{1}{2} - 2\tanh(\bar{u})\frac{\bar{u}}{\bar{t}}, \tag{29}$$

$$F_{2,2}(\bar{t},\bar{u}) = \left(\frac{1}{4} - 4\frac{\bar{u}^2}{\bar{t}^2}\right) + \tanh(\bar{u})\left(4\frac{\bar{u}}{\bar{t}^2} - 2\frac{\bar{u}}{\bar{t}}\right) + 8\tanh^2(\bar{u})\frac{\bar{u}^2}{\bar{t}^2}, \tag{30}$$

$$\begin{aligned} F_{2,4}(\bar{t},\bar{u}) = {} & \left(\frac{3}{16} + 4\frac{\bar{u}^2}{\bar{t}^3} - 5\frac{\bar{u}^2}{\bar{t}^2}\right) + \tanh(\bar{u})\left(\frac{20}{3}\frac{\bar{u}^3}{\bar{t}^3} - 4\frac{\bar{u}}{\bar{t}^3} + 5\frac{\bar{u}}{\bar{t}^2} - \frac{23}{12}\frac{\bar{u}}{\bar{t}}\right) \\ & + \tanh^2(\bar{u})\left(10\frac{\bar{u}^2}{\bar{t}^2} - 8\frac{\bar{u}^2}{\bar{t}^3}\right) - 8\tanh^3(\bar{u})\frac{\bar{u}^3}{\bar{t}^3}. \end{aligned} \tag{31}$$

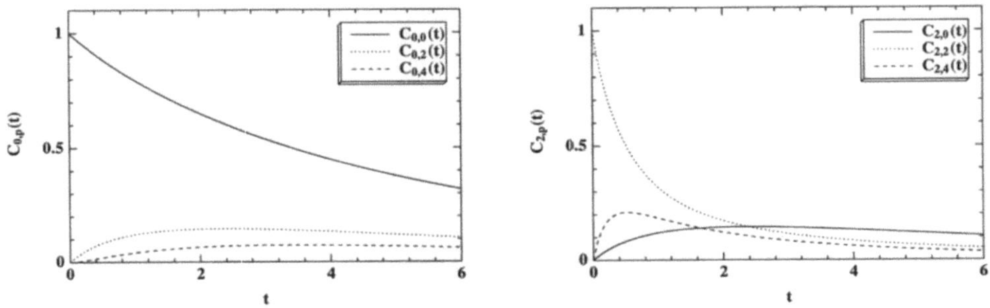

FIGURE 1. Theoretical histograms of the probability distributions $(C_{p_0,p}(t))_{p\in\mathbb{N}}$ of the energy of the particle under time-dependent perturbations with $\alpha_2 = 1$. In the left picture (resp. right picture) the particle is assumed to be at $t = 0$ in the state f_0 (resp. f_2). Only even states have positive probabilities.

8. Numerical simulations

The results in the previous sections are theoretically valid in the limit case $\varepsilon \to 0$, where the amplitude (resp. duration) of the perturbation goes to zero (resp. infinity). In this section we aim at showing that the asymptotic behaviors of the state vector can be easily observed in numerical simulations in the case where the perturbation is small, so that its effect appears after a long time. We use a split-step method to simulate the one-dimensional perturbed linear Schrödinger equations (3). We adopt in this section the following model for the perturbation:

$$m(t) = u_l \text{ if } lt_c \leq t < (l+1)t_c,$$

where $(u_l)_{l=0,\dots,M-1}$ is a sequence of M independent and identically distributed variables, which take the value $-1/2$ or $1/2$ with probability $1/2$. t_c is the so-called coherence time of the random process m. The power spectral density is:

$$\alpha_j = \frac{1}{4}\frac{1-\cos(jt_c)}{j^2 t_c},$$

and $\alpha_0 = t_c/8$. The quantity Mt_c which is equal to the time duration of the perturbation will be chosen so large that we can observe the effect of the small perturbation $\varepsilon m(t)x^2$. We measure the L^2-norm and the energy (5) of the particle that we can compare with the corresponding data of the initial state vector. We present results corresponding to simulations where the initial state at $z = 0$ is the fundamental Gaussian mode $f_0(x)$ or the second mode $f_2(x)$. We have first simulated the homogeneous Schrödinger equation (with $m \equiv 0$) which admits as an exact solution $f_0(x)e^{-it/2}$ and $f_2(x)e^{-5it/2}$ respectively. We can therefore check the accuracy of the numerical method, since we can see that the modulus of the computed solution maintains a very close resemblance to the initial profile (data not shown), while the L^2-norm and the energy are almost constant.

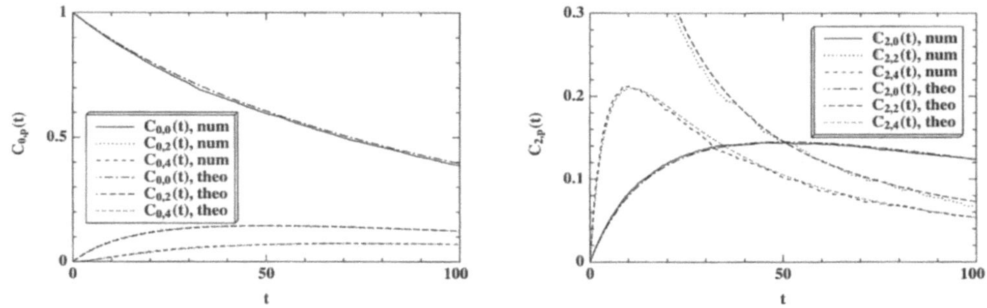

FIGURE 2. Evolution of the modal distribution of the particle computed by simulation of the perturbed Schrödinger equation (3) and averaged over 2000 realizations and comparison with the theoretical formulas given by (25–27) and (29–31). In the left picture (resp. right picture) the initial state vector is the fundamental state f_0 (resp. the state f_2). The values of the parameters are $t_c = 0.4$ ($\alpha_2 \simeq 0.0474$) and $t_{\max} = 100$ ($M = 250$).

The other simulations are carried out with different realizations of the random process m with $\varepsilon = 1$. In Figure 2 we plot the modal distributions of the particle averaged over 2000 realizations and compare them with the theoretical distributions derived in Section 7. It thus appears that the numerical simulations are in very good agreement with the theoretical results. All these observations confirm that our formulas describe with accuracy the evolution of the harmonic oscillator under small perturbations and long times.

9. Conclusion

We have analyzed in this paper the effects of random perturbations of the spring coefficient on the evolution of a quantum particle in a quadratic potential. The precise results we have obtained are a consequence of the particular form of the stable system at hand (quantum harmonic oscillator) and of the considered perturbation. Eigenvalues and eigenstates of this system are explicit and tabulated formulas are available, so that we are able to perform exact calculations and compute closed-form expressions for the eigenstates probabilities. Nevertheless the results demonstrated in this paper can be generalized to a large class of systems. In [14] we study the energy distribution of the quantum harmonic oscillator under a very general class of random time-dependent perturbations which satisfy some spatial growth conditions. Nevertheless the quadratic perturbation we address in this paper did not fulfill these conditions, so a specific approach was required. Second we can also expect generalizations to systems that are different from the harmonic oscillator. Indeed the basic assumption which requires the existence of a complete set of normal eigenstates for the unperturbed Hamiltonian holds true for many systems. The remainder of the study then consists in technical developments to exhibit and analyze the coupling mechanisms between the eigenstates.

References

[1] L. Alili, D. Dufresne and M. Yor, *Sur l'identité de Bougerol pour les fonctionnelles du mouvement Brownien avec shift*, in M. Yor (Ed.), *Exponential functionals and principal values related to Brownian Motion*, Revista Matemática, Iberoamericana, 1997.

[2] L. Arnold, *A formula connecting sample and moment stability of linear stochastic systems*, SIAM J. Appl. Math., **44** (1984), 793–802.

[3] L. Arnold, G. Papanicolaou and V. Wihstutz, *Asymptotic analysis of the Lyapunov exponent and rotation number of the random oscillator and applications*, SIAM J. Appl. Math., **46** (1986), 427–450.

[4] P. H. Baxendale and R. Z. Khaminskii, *Stability index for products of random transformations*, Adv. Appl. Prob., **30** (1998), 968–988.

[5] P. Bougerol, *Exemples de théorèmes locaux sur les groupes résolubles*, Ann. Inst. H. Poincaré, **19** (1983), 369–391.

[6] P. Bougerol and J. Lacroix, *Products of random matrices with applications to Schrödinger operators*, Birkhauser, Boston, **1985**.

[7] L. Bunimovich, H. R. Jauslin, J. L. Lebowitz, A. Pellegrinotti and P. Nielaba, *Diffusive energy growth in classical and quantum driven oscillators*, J. Stat. Phys., **62** (1991), 793–817.

[8] M. Combescure, *The quantum stability problem for time-periodic perturbations of the harmonic oscillator*, Ann. Inst. H. Poincaré A, **47** (1987), 63–83.

[9] V. Enss and K. Veselić, *Bound states and propagating states for time-dependent Hamiltonians*, Ann. Inst. H. Poincaré A, **39** (1983), 159–191.

[10] W. Feller, *Introduction to probability theory and its applications*, Wiley, New York, **1971**.

[11] D. Freedman, *Markov processes*, Holden-Day, San Francisco, **1971**.

[12] H. Furstenberg, *Noncommuting random products*, Trans. Amer. Math. Soc., **108** (1963), 377–428.

[13] J. Garnier, *Asymptotic behavior of the quantum harmonic oscillator driven by a random time-dependent electric field*, J. Stat. Phys., **93** (1998), 211–241.

[14] J. Garnier, *Energy distribution of the quantum harmonic oscillator under random time-dependent perturbations*, Phys. Rev. E, **60** (1999), 3676–3687.

[15] J. Glimm and A. Jaffe, *Quantum physics, a functional integral point of view*, Springer, New York, **1981**.

[16] G. A. Hagedorn, M. Loss and J. Slawny, *Non-stochasticity of time-dependent quadratic Hamiltonians and the spectra of canonical transformations*, J. Phys. A, **19** (1986), 521–531.

[17] R. M. Hervé and M. Hervé, *A mathematical treatment for a perturbation of the one-dimensional harmonic oscillator*, J. Math. Phys., **32** (1991), 956–958.

[18] M. Hillery, R. F. O'Connel, M. O. Scully and E. P. Wigner, *Distribution Functions in Physics: Fundamentals*, Phys. Rep., **106** (1984), 121–167.

[19] W. Kohler and G. Papanicolaou, *Power statistics for wave propagation in one dimension and comparison with transport theory*, J. Math. Phys., **14** (1973), 1733–1745; **15** (1974), 2186–2197.

[20] H. J. Kushner, *Approximation and weak convergence methods for random processes*, MIT Press, Cambridge, **1984**.

[21] A. Messiah, *Quantum Mechanics*, North Holland, Amsterdam, **1962**.

[22] D. Middleton, *Introduction to statistical communication theory*, Mc Graw Hill, New York, **1960**.

[23] M. G. Nerurkar and H. R. Jauslin, *Stability of oscillators driven by ergodic processes*, J. Math. Phys., **35** (1994), 628–645.

[24] G. Papanicolaou and W. Kohler, *Asymptotic theory of mixing stochastic ordinary differential equations*, Comm. Pure Appl. Math., **27** (1974), 641–668.

[25] C. A. Pillet, *Some results on the quantum dynamics of a particle in a Markovian potential*, Commun. Math. Phys., **102** (1985), 237–254.

[26] C. A. Pillet, *Asymptotic completeness for a quantum particle in a Markovian short range potential*, Commun. Math. Phys., **105** (1986), 259–280.

[27] L. I. Schiff, *Quantum Mechanics*, Mac Graw Hill, New York, **1968**.

[28] V. N. Tutubalin, *On limit theorems for products of random matrices*, Theory Prob. Appl., **10** (1965), 15–27.

Centre de Mathématiques Appliquées
Centre National de la Recherche Scientifique Unité Mixte de Recherche 7641
Ecole Polytechnique
91128 Palaiseau Cedex
France
E-mail address: Josselin.Garnier@polytechnique.fr

Progress in Probability, Vol. 52, 173–187

On the Superreplication Approach for European Interest Rates Derivatives

F. Gozzi and T. Vargiolu

Abstract. In this paper we analyse the superreplication approach to stochastic volatility in the case of European interest rates derivatives. We exploit some general results of [13] and [17] to prove that the minimal superstrategy is given by the solution of a nonlinear PDE associated to the model, that is the so-called Black-Scholes-Barenblatt (BSB) equation. In particular we show how this approach applies to the case of caps and floors extending results of [6].

1. Introduction

In this paper we analyse the superreplication approach to stochastic volatility in the case of European interest rates derivatives including also cases when the value function is nonsmooth. More precisely our starting point are some general results of [17] about the characterization of superstrategy for European multiasset derivatives via the regular ($C^{1,2}$) solutions of the associated Black-Scholes-Barenblatt (BSB) equation and of [13] about existence, uniqueness, regularity of the solution of the BSB equation and characterization of superstrategies in the case when the solution is not $C^{1,2}$. We show that such results can be applied in the case of interest rates derivatives with suitable arrangements. In particular we treat the case of caps and floors (that do not fit directly into the case of European derivatives due to the presence of multiple maturities) and we show that our approach gives, in the Gaussian case, an extension of a result of [6].

We give now an outline of the problem with a formulation that uses forward prices (see e.g. [7]). Such outline can be found also in the quoted papers, but we prefer to repeat it here for the reader's convenience. We have a riskless asset, whose value we suppose constant through time, and n risky assets whose prices $S = (S^1, \dots, S^n)$ follow the dynamic

$$dS_t = \bar{S}_t \sigma_t \, dW_t,$$

where W is a n-dimensional Brownian motion under the so-called forward neutral measure $\mathbb{Q}$, the matrix process σ is adapted and takes values in a closed bounded

2000 *Mathematics Subject Classification.* 91B28, 49L25, 93E20.
Key words and phrases. Derivatives pricing, superreplication approach, Black-Scholes-Barenblatt equation, optimal stochastic control, Hamilton-Jacobi-Bellman equation.

set $\Sigma \subset M(n,n,\mathbb{R})$, and $\bar{S}_t = \text{diag }(S_t)$. We then consider an agent who wants to price and hedge a European contingent claim whose payoff is a deterministic function $h(\cdot)$ which is globally Lipschitz continuous, calculated in S_T. Since the market could be incomplete because of the stochastic volatility σ and the agent is not able to hedge the volatility, he chooses to hedge the option by using the superhedging approach. Following e.g. [2, 14, 17], we assume that he fixes the price of the option as $C_t = C(t, S_t)$ and builds a self-financing portfolio consisting of a quantity $\Delta_t^i = \frac{\partial C}{\partial S_i}(t, S_t)$ of each asset $S^i, i = 1, \dots, n$, where $C(t, S)$ is a solution of a nonlinear PDE, similar to the Black-Scholes equation, called Black-Scholes-Barenblatt equation in analogy with [2, 17]. This equation is a Hamilton-Jacobi-Bellman equation and it is linked to a stochastic control problem that has a nice financial interpretation (see Section 2 and also [13, 17]). In fact, the agent could build a "subjective" model

$$dS_t^\gamma = \bar{S}_t^\gamma \gamma_t \, dW_t,$$

where the state variable S^γ has to be controlled by the "subjective" volatility γ in order to maximize the payoff function

$$\mathbb{E}_{\mathbb{Q}} \left[e^{-r(T-t)} h(S_T^\gamma) \mid \mathcal{F}_t \right]$$

corresponding to the no-arbitrage price of the contingent claim at time t. In this way, the agent wants to maximize in γ his payoff protecting himself against "the worst possible case", and the BSB equation is the Hamilton-Jacobi-Bellman equation corresponding to this optimal control problem.

The main known results on the relationship between the BSB equation and the superhedging of a European contingent claim are substantially Theorem 2.3 and Theorem 2.6 (recalled in Section 2 for the reader's convenience). Theorem 2.3 characterizes superstrategies in terms of the solutions of the BSB equation *when they are regular enough to apply Itô formula* (C^1 in time and C^2 in space). This result has been proved in the single asset case in [2] and in the multiasset case in [14] where a detailed discussion on the model and on its financial interpretation is given, (see also [17] for a simpler proof of it), and it is also shown (using known results on PDEs) that if the BSB equation is uniformly parabolic, then its solution is smooth enough to apply Itô formula. Uniform parabolicity of the BSB equation depends essentially on the set Σ being composed of invertible matrices. This is not verified in some financial examples (see Section 8 of [17] for a typical situation). For this reason it is interesting to see what can be said in non uniformly parabolic cases and Theorem 2.6 presented and proved in [13] furnishes a partial answer in this case. We notice that many other papers on this subject and related ones have been written. We recall e.g. [9] where the problem was considered first and also [1, 2, 6, 8, 10, 12, 18]. We refer to [13] and [17] for a wider introduction to the problem and a more complete bibliography on it.

In this paper the results above are applied to the case of interest rates derivatives. We point out that such extension is not trivial since the setup for interest rates derivatives is in fact different and some basic assumptions of the general model of [13] and [17] are not satisfied (in particular for the case of caps and floors, where we have multiple maturities).

The paper is organized as follows: Section 2 is devoted to present the problem and to collect some known material (mainly taken from [13] and [17]) needed in the sequel. In Section 3 we present the main results on superreplication of interest rates derivatives with a single maturity (e.g. swaptions); in Section 4 we consider the case of caps and we show that our approach can characterize the superstrategies and, in the case of Gaussian interest rates models can allow to calculate explicitly such superstrategies.

2. The general model

This section is devoted to collect the material (mostly taken from [13] and [17]) concerning the general problem of superreplication of European multiasset derivatives via the BSB equation. It is divided in subsection as follows:

- the first one devoted to the general setting,
- the second one devoted to the BSB equation,
- the third one devoted to characterisation theorems for superstrategies,
- the fourth one containing a result of [17] that, in some case, allows to find an explicit solution of BSB equation and an example where it can be applied (which is exploited in Section 4 to study the case of caps).

2.1. Setting of the model

We suppose that there exists a riskless asset M and n risky assets S^i, $i = 1, \dots, n$ in the market. We make the usual assumptions that there exist a probability space $(\Omega, \mathcal{F}, \mathbb{P})$, a filtration $(\mathcal{F}_t)_{t\in[0,T]}$ complete and right continuous, where $\mathcal{F}_t$ represents the information available up to time t and that M and S^i are stochastic processes adapted to $(\mathcal{F}_t)_t$. Besides, we assume that there exists a probability measure $\mathbb{Q}$, equivalent to $\mathbb{P}$, which is called **forward-neutral probability** [7], such that the value of the riskless asset M remains constantly equal to 1 through time, and the dynamics of the assets S^i under $\mathbb{Q}$ are the following:

$$dS_t^i = S_t^i \langle \sigma_t^i, dW_t \rangle,$$

where $(W_t)_t$ is a d-dimensional $\mathbb{Q}$-Brownian motion adapted to $(\mathcal{F}_t)_t$, $\langle \cdot, \cdot \rangle$ is the Euclidean scalar product in $\mathbb{R}^d$ and σ^i is a d-dimensional process such that $\sigma = (\sigma^i)_i \in \mathcal{A}(\Sigma)$, where Σ is a closed bounded set in the space of $n \times d$ real matrices $M(n, d, \mathbb{R})$ and $\mathcal{A}(\Sigma)$ (which we call set of **admissible volatilities**) is the set of Σ-valued processes progressively measurable with respect to $(\mathcal{F}_t)_t$.

We can write the dynamics of the risky assets in a more compact vectorial notation in this way:

$$dS_t = \bar{S}_t \sigma_t \, dW_t,$$

where we let $\bar{S}_t = \operatorname{diag}(S_t)$.

Now we consider an operator in the market who wants to create and sell a European derivative asset with payoff $h(S_T) = h(S_T^1, \dots, S_T^n)$, where h is a locally Lipschitz continuous function having polynomial growth.

Remark 2.1. We do not assume that the market is complete. In particular, the filtration $(\mathcal{F}_t)_t$ could be strictly larger than the one generated by W or than the one generated by S. In this case we have a genuine stochastic volatility model and the market is incomplete. For this reason, we can assume that the number of assets n is different from the number of Brownian motions d in the model. Moreover, the assumption that the interest rate is zero can be achieved by a so-called change of numeraire, that is by expressing the prices of all the assets in the market in units of a zero coupon bond with maturity T (for more details, see [7]).

In order to price and hedge the asset, the agent fixes a price C_t for the asset at time t and builds a self-financing portfolio Π, holding η_t units of the money market account M and Δ_t^i units of the asset S^i at time t. We indicate with Π_t the value of the portfolio at time t, defined by:

$$\Pi_t = \eta_t + \langle \Delta_t, S_t \rangle , \tag{1}$$

where $\Delta_t = (\Delta_t^1, \dots, \Delta_t^n)$. In order to prevent arbitrage opportunities, we say that the portfolio is **admissible** if Π is a supermartingale: in this way, if $\Pi_0 = 0$, we have $\mathbb{E}[\Pi_T] \leq 0$, so there are not arbitrage opportunities in the market. We say that the portfolio is **self-financing** if Π follows the dynamic

$$\begin{cases} d\Pi_t = \langle \Delta_t, dS_t \rangle \\ \Pi_0 = C_0 \, . \end{cases} \tag{2}$$

If the portfolio is self-financing, the process Δ is sufficient to characterize it, and $\eta_t = \Pi_t - \langle \Delta_t, S_t \rangle$.

The market is not complete because of the stochastic volatility σ, so the agent chooses to hedge the claim using the superhedging approach. To this aim, we define the **tracking error**:

$$e_t = \Pi_t - C_t$$

so $e_0 = 0$ by definition of Π. Intuitively, the tracking error gives the error made by the operator in estimation, or better, the difference between the hedging portfolio held by the agent and the option sold. If the portfolio is admissible and self-financing, we say that (C, Δ) is a

- **superhedging strategy**, or simply **superstrategy** if e is a non decreasing process
- **subhedging strategy**, or simply **substrategy**, if e is a non increasing process
- **hedging strategy** if e is identically equal to zero.

If one builds a superstrategy, one can successfully hedge a short position in the contingent claim C_t, while if one builds a substrategy, one can successfully hedge a long position in C_t. In particular with a superstrategy we have that $e_T \geq 0$, that is $\Pi_T \geq C_T = h(S_T)$, so the portfolio succeeds in superhedging the contingent claim. We notice that superstrategies and substrategies are good candidates to be arbitrages: in fact if we have (say) $\Pi_T \geq C_T$ $\mathbb{Q}$-a.s. and $\Pi_T > C_T$ with positive probability, then the agent succeeds in making a profit with no initial endowment. This means that the value C_t in a superstrategy has to be interpreted as an arbitrage upper bound for the price of the claim at time t; similarly, the value C_t in a substrategy has to be interpreted as a lower bound. In fact, if the price of the claim is C_t or less, one could build an arbitrage by buying the claim and by selling short the subhedging portfolio for the same price, thus making the final profit $C_T - \Pi_T = -e_T \geq 0$.

Remark 2.2. The substrategy case is in some sense symmetric to the superstrategy one. In fact, if C defines a superstrategy for the claim defined by h, then $-C$ defines a substrategy for the claim defined by $-h$. This allows us to analyse only one of the two cases, and to have automatically results for the other. For this reason, in the rest of the paper we will analyse only the superstrategy case, and it will be implicit that analogous results will hold also for substrategies.

As already outlined by [2, 8, 13, 14, 17], in order to have a superstrategy, a natural procedure is this: the agent fixes the price C_t of the option and the quantities Δ_t of the risky assets in the hedging portfolio Π at time t as

$$C_t = C(t, S_t), \qquad \Delta_t^i = \frac{\partial C}{\partial s_i}(t, S_t),$$

where C is the solution of the following partial differential equation:

$$\begin{cases} \dfrac{\partial C}{\partial t}(t,s) + \dfrac{1}{2} \max\limits_{\gamma \in \Sigma} \operatorname{tr}\left(D^2 C(t,s)(\bar{s}\gamma)(\bar{s}\gamma)^*\right) = 0, & t \in [0,T), s \in \mathbb{R}_+^n \\ C(T,s) = h(s), & s \in \mathbb{R}_+^n\,, \end{cases} \tag{3}$$

where $DC(t,s) = (\frac{\partial C}{\partial s_1}(t,s), \ldots, \frac{\partial C}{\partial s_n}(t,s))$, and $D^2 C(t,s) = (\frac{\partial^2 C}{\partial s_i \partial s_j}(t,s))_{ij}$. Equation (3), as in [2], will be called from now on the **Black-Scholes-Barenblatt (BSB) equation** for h. Moreover, by Remark 2.2, in order to have a substrategy it is sufficient to proceed as above by substituting the maximum operator in Equation (3) with a minimum.

The approach above has a stochastic control interpretation: we assume that the agent does not know the volatility σ, but in order to hedge the asset he can use the "subjective" model

$$S_t^\gamma = S_t^\sigma\,, \qquad S_\tau^\gamma = \bar{S}_\tau^\gamma \gamma_\tau\, dW_\tau\,, \qquad t < \tau < T. \tag{4}$$

Here the initial datum S_t^σ is the present market price of the assets at time t which depends of course on the real but unknown volatility σ up to time t. The process $\gamma \in \mathcal{A}_t(\Sigma)$ is another admissible volatility starting from time t, that we can interpret as a control corresponding to the subjective volatility of the agent used in order to decide his strategy. If the market were complete and the dynamics were given by Equation (4), then the arbitrage free price of the European contingent claim would be

$$C_t^{\mathbb{Q},\gamma} = \mathbb{E}_{\mathbb{Q}}\left[h(S_T^\gamma) \mid S_t^\sigma\right], \tag{5}$$

where $\mathbb{Q}$ would be the unique forward-neutral measure of the problem. Since the market is not complete, the agent fixes the price of the option by maximising the quantity above both in $\mathbb{Q}$ as in γ, thus protecting himself against "the worst possible case". It turns out that the BSB equation is the Hamilton-Jacobi-Bellman equation corresponding to this optimal control problem. Under the assumptions of our model, the value function of the optimal control problem above, defined by

$$V_{\mathbb{Q}}^{+}(t,s) = \sup_{\gamma\in\mathcal{A}(\Sigma)} \mathbb{E}_{\mathbb{Q}}\left[h(S_T^\gamma) \mid S_t^\gamma = s\right] \tag{6}$$

is the unique viscosity solution C of Equation (3) (see Theorem 2.4 in the following subsection) and does not depend on the particular choice of $\mathbb{Q}$. Thus we have that $C(t, S_t^\sigma)$ defines the upper bound of the admissible arbitrage free prices for the claim at time t, in the sense that if the price of the claim is greater than $C(t, S_t^\sigma)$, then it is possible to build an arbitrage in the market by selling the claim and buying the superreplicating portfolio. The same reasoning can be repeated for the substrategy.

2.2. The characterization theorems

We now present the characterization theorems for superstrategies. We indicate here with $C_p^{1,2}([0,T)\times\mathbb{R}_+^n)$ the space of functions C that are continuous and with polynomial growth on $[0,T)\times\mathbb{R}_+^n$ together with their first derivative in t and first and second derivatives in s.

Theorem 2.3. *If we restrict ourselves to the case*

$$C_t = C(t, S_t), \quad \textit{with} \quad C \in C_p^{1,2}\left([0,T)\times\mathbb{R}_+^n\right),$$

then:

i) (C,Δ) *is a superstrategy if and only if* $\Delta_t = DC(t,S_t)$ *and*

$$R_t = -\frac{\partial C}{\partial t}(t,S_t) - \frac{1}{2}\mathrm{tr}\ D^2C(t,S_t)(\bar{S}_t\sigma_t)(\bar{S}_t\sigma_t)^*$$

is non negative $\mathbb{Q}$*-a.s. for all* $t\in[0,T]$.

ii) *If* C *is the solution of Equation (3) and* $\Delta_t = DC(t,S_t)$, *then* (C,Δ) *is a superstrategy. Moreover there is not any superstrategy* (C',Δ') *such that* $C'_t < C(t,S_t^\gamma)$ *with positive probability for some* t *and for all* $\gamma\in\mathcal{A}(\Sigma)$.

This is a characterization theorem of the superstrategies. In fact, provided the value function $C \in C_p^{1,2}$, the theorem gives the minimal superstrategy, in the sense that there does not exist a cheaper strategy that succeeds in giving an increasing tracking error for all $\gamma\in\mathcal{A}(\Sigma)$. Since the agent does not know the true volatility σ, he must protect himself against all the possible volatilities $\gamma\in\mathcal{A}(\Sigma)$ and this is the best result he can expect.

An analogous of Theorem 2 holds for the case of a substrategy, provided we reverse the inequalities and substitute the max operator in Equation (3) with a min.

The final conclusion of Theorem 2.3 is that if a smooth solution of Equation (3) exists, then there exists a Markov superstrategy, and it is given by the space derivative of the solution of Equation (3). Conversely, if this does not happen we cannot apply Theorem 2.3 without proving new regularity results that are unknown at this stage. Then the characterization theorems for superstrategies strongly depend on the properties of solutions of the BSB equation, especially the regularity of them. For this reason we give then a brief look to the known results on this topic (see [13] for bibliography, statement and proofs, and [11] for a wide introduction to viscosity solutions).

The first step is to recall a general result of existence and uniqueness of viscosity solutions for the BSB equation (3) that always holds true in our case.

Theorem 2.4. *Let Σ be compact, h be locally Lipschitz continuous and h, Dh have polynomial growth. Then the value function $C = V^+$ defined by Equation* (6) *is a viscosity solution of Equation* (3) *in $[0,T] \times \mathbb{R}^n_+$. Moreover, C is the unique viscosity solution having polynomial growth that satisfies the boundary condition $C(T,s) = h(s)$ for all $s \in \mathbb{R}^n_+$.*

The second step is to prove smoothness of the viscosity solution. In order to do this we use results on uniformly parabolic equations. For sake of brevity we only give a sketch of the procedure without precise statements. The first thing is to make the change of variable $y_i = \log s_i$ in the BSB equation. Then the BSB equation transforms into a nonlinear PDE with constant coefficients which is uniformly parabolic if and only if $\det \gamma\gamma^* \neq 0$ for all $\gamma \in \Sigma$. If this condition is satisfied we can use results on uniformly parabolic equations of [4, 19] obtaining the following result.

Theorem 2.5. *Let Σ be compact, h be locally Lipschitz continuous and h, Dh have polynomial growth. If $\Sigma^2 \subseteq \mathrm{GL}(n,\mathbb{R})$ then the viscosity solution C of the BSB equation* (3) *belongs to $C^{1,2}_p([0,T) \times \mathbb{R}^n_+)$.*

If the BSB equation (3) is not uniformly parabolic (i.e. if there exists $\gamma \in \Sigma$ such that $\det \gamma\gamma^* = 0$), a version of Theorem 2.3 can be proved under stronger assumptions. More precisely it can be proved that the process (C, Δ) defined of Section 2 gives a superstrategy when the final payoff h is a convex (or semiconvex) function of the assets and for all t the law of the random variable S_t is absolutely continuous with respect to the Lebesgue measure. This case does not cover all the possible payoffs (for example, it does not cover call-spreads), but indeed many of the multiasset European options (call options on the maximum, call and put options on the mean, exchange options, etc.) are covered.

We also observe that the analogous results for a substrategy needs to assume that h is concave, not convex and so the symmetry between sub- and superstrategies results here is broken. In fact, most of the examples require h to be convex (and generally not concave). So, if we want to get a result on existence of sub-strategies when h is convex we need the additional assumption that the set Σ be convex (see [13]).

We have the following result, proved in [13].

Theorem 2.6. *If the assumptions*

(i) *The payoff h is convex (or semiconvex) and Lipschitz continuous and h, Dh have polynomial growth,*
(ii) *For every $t > 0$ the law of the random variable S_t^σ has a density with respect to the Lebesgue measure which is absolutely continuous,*

hold and C denotes the unique viscosity solution of Eq. (3), then $\Delta_t = D_s C(t, S_t)$ is a.s. well defined and $(C_t, \Delta_t)_t$ is a superstrategy, where $C_t = C(t, S_t)$.

Remark 2.7. In fact all the results above could be generalized in the case when the set Σ of expected volatilities is time-varying. It is enough to ask a Hölder continuity of it with respect to time.

2.3. Explicit solutions via reduction to the Black and Scholes case

Now we quote some results from [17] that will be useful in the sequel.

We notice that Eq. (3) contains an optimization problem. The agent has to solve this problem if he wants to solve the BSB equation numerically, and if he wants to have a view of what could be the "worst case" against him. Moreover, it turns out that in some cases the problem gives a simple way to pass from the BSB equation to a Black-Scholes equation ((BS) in the following), which is simpler both from a theoretical point of view as well as from the numerical side. In any case, this problem is significant only if $C \in C^{1,2}$, so in the following we will make that assumption.

The optimization problem is

$$\max_{\gamma \in \Sigma} F_{t,s}(\gamma) , \tag{7}$$

where $F_{t,s} : \Sigma \to \mathbb{R}$ is defined by

$$F_{t,s}(\gamma) = \mathrm{tr}\ \left(\bar{s} D^2 C(t,s) \bar{s} \gamma \gamma^*\right) = \mathrm{tr}\ \left(A_{t,s} \gamma \gamma^*\right) , \tag{8}$$

where $A_{t,s} = \bar{s} D^2 C(t,s) \bar{s}$ and C is the solution of Equation (3). We will often write F and A instead of $F_{t,s}$ and $A_{t,s}$ if there is no risk of ambiguity.

An important case is when the optimization problem has a constant solution. In this case, we can reduce the non linear BSB equation to a standard Black-Scholes equation (see [3]) as the following result (contained in [17]) states.

Proposition 2.8. *If $C \in C^{1,2}([0,T) \times \mathbb{R}^n)$ is a solution of the Black-Scholes equation*

$$\begin{cases} \dfrac{\partial C}{\partial t}(t,s) + \dfrac{1}{2} \mathrm{tr}\ (D^2 C(t,s)(\bar{s}\bar{\gamma})(\bar{s}\bar{\gamma})^*) = 0, & t \in [0,T), s \in \mathbb{R}_+^n , \\ C(T,s) = h(s), & s \in \mathbb{R}_+^n , \end{cases} \tag{9}$$

with $\bar{\gamma} \in \Sigma$, *such that the problem* (7) *attains its maximum in* $\bar{\gamma}$ *for all* $(t,s) \in [0,T) \times \mathbb{R}^n_+$, *then* C *is also a solution of Eq.* (3).

This proposition is very useful for reducing the BSB equation, for which explicit solutions are very rare, to a BS equation, for which explicit solutions are common. We now show an example of this case: the Margrabe exchange option (see [15]), whose final payoff is

$$h(s) = (s_1 - \lambda s_2)^+ .$$

We try to apply the results above to this option. To this end, for a generic $\gamma \in \Sigma$ we calculate the Black-Scholes price $C(t, s_1, s_2)$ of the exchange option, which is the solution of Eq. (9) when we substitute $\bar{\gamma}$ with

$$\gamma = \begin{pmatrix} \gamma_{11} & \gamma_{12} \\ \gamma_{21} & \gamma_{22} \end{pmatrix} . \tag{10}$$

By Proposition 2.8, if the Black-Scholes price is such that the optimization problem (7) attains its maximum at γ for all $t \in [0,T)$, $s \in \mathbb{R}^2_+$, then $C(t, s_1, s_2)$ is also a solution of the BSB equation (3).

The Black-Scholes price of this option is:

$$C(t, s_1, s_2) = s_1 N(d_1) - \lambda s_2 N(d_2) , \tag{11}$$

where

$$d_1 = \frac{1}{\Gamma\sqrt{T-t}} \ln\left(\frac{s_1}{\lambda s_2}\right) + \frac{1}{2}\Gamma\sqrt{T-t} , \quad d_2 = d_1 - \Gamma\sqrt{T-t} ,$$

$$\Gamma = \sqrt{(\gamma_{11} - \gamma_{21})^2 + (\gamma_{12} - \gamma_{22})^2}$$

and N is the cumulative distribution function of a standard Gaussian random variable:

$$N(d) = \frac{1}{\sqrt{2\pi}} \int_{-\infty}^{d} e^{-\frac{x^2}{2}} \, dx$$

(for a proof, see [15]). In order to apply Proposition 2.8, we calculate the matrix $A_{t,s}$. After some calculations (see [17]), we arrive at

$$A_{t,s} = \bar{s} \cdot D^2 C(t,s) \cdot \bar{s} = \frac{s_1 N'(d_1)}{\Gamma\sqrt{T-t}} \begin{pmatrix} 1 & -1 \\ -1 & 1 \end{pmatrix} .$$

We notice that $A_{t,s}$ is proportional $\forall (t,s) \in [0,T) \times \mathbb{R}^2_+$ to the constant matrix

$$\begin{pmatrix} 1 & -1 \\ -1 & 1 \end{pmatrix} .$$

This means that the solution to the problem (7) is constant. Hence we easily obtain the following proposition (see [17] for the proof).

Proposition 2.9. *If* $\bar{\gamma}$ *is the maximum of* $\|\gamma_1 - \gamma_2\|_{\mathbb{R}^2}$ *on* Σ, *where* γ_1, γ_2 *are the rows of* γ, *then* (11) *is solution of Eq.* (3).

3. Interest rates derivatives

The general model in Section 2 finds a natural application in the superhedging of interest rates derivatives. In fact, the final payoff of interest rates derivatives can be often expressed as a deterministic function of suitable zero-coupon bonds. For example, if we define the forward LIBOR $f(t,T,\alpha)$ via the equation

$$1+\alpha f(t,T,\alpha)=\frac{B(t,T)}{B(t,T+\alpha)}$$

the final payoff $\alpha(f(T,T,\alpha)-K)^+$ of a caplet with payment date $T+\alpha$ on the forward LIBOR $f(T,T,\alpha)$ can be written also as

$$(1+\alpha K)\left(\frac{1}{1+\alpha K}-B(T,T+\alpha)\right)^+ .$$

It can be proved (see for example [16]) that by changing the numeraire the payoff of the caplet above is equivalent to the payoff

$$(1+\alpha K)\left(\frac{B(T,T)}{1+\alpha K}-B(T,T+\alpha)\right)^+$$

at time T. This means that a caplet can be seen as an exchange option between the two zero coupon bonds $B(\cdot,T)$ and $B(\cdot,T+\alpha)$. Thus a cap is simply a linear combination of exchange options between zero coupon bonds maturing at different times.

Interest rates derivatives with only one single maturity T (for example call, put and other kind of European options on zero coupon bonds, caplets and floorlets, swaps, swaptions) can be easily treated in the framework of our model; in fact, they all are characterized by a terminal payoff h and some bonds with maturity $T_1<T_2<\cdots<T_N$, with $T_1=T$. In particular, the final payoff can be usually written in the form

$$h\left(\frac{B(t,T_1)}{B(t,T)},\ldots,\frac{B(t,T_N)}{B(t,T)}\right) . \tag{12}$$

In order to apply Theorem 2, we define

$$S_t^i=\frac{B(t,T_i)}{B(t,T)} .$$

It follows that if the zero coupon bonds follow the dynamics

$$dB(t,T_i)=B(t,T_i)(r_t\,dt+\langle\Gamma(t,T_i),dW_t\rangle) \tag{13}$$

under $\mathbb{Q}$, then S^i follows the dynamics

$$dS_t^i=S_t^i\langle\Gamma(t,T_i)-\Gamma(t,T),dW_t'\rangle \tag{14}$$

under $\mathbb{Q}_T$, where W' is a $\mathbb{Q}_T$-Brownian motion. If we define $\sigma^i=\Gamma(t,T_i)-\Gamma(t,T)$ and if we assume that $\sigma=(\sigma_t^i)_t\in\mathcal{A}(\Sigma)$, then we are in the setting of Section 2 and we can try to apply Theorems 2.3 and 2.6. More precisely we have the following result.

Theorem 3.1. *Assume that* (12) *and* (13) *hold true. Then, if the process* S *satisfies* (14) *and the payoff* h *defined by* (12) *satisfy also assumptions of Theorem* 2.3 *or of Theorem* 2.6 *then* $(C_t, \Delta_t)_t$ *is the minimal superstrategy in the sense of Theorem* 2.3.

Remark 3.2. Observe that assumption (13) holds always true for all Gaussian models of interest rates and that in such cases Theorem 3.1 applies. Moreover, if we consider a nongaussian model for interest rates we can have (13) but in such cases the diffusion coefficient is stochastic and then it is not obvious if our regularity assumptions (like compactness of the set Σ) are verified. In such cases one has to check every particular problem. Here we will consider explicitely only applications to the Gaussian case.

The case when we have more than one maturity (for example in the case of a cap or a floor) is more complex. In fact, in principle, the total payoff can be stripped in single payoffs each one having different maturities and these single payoffs can be superhedged separately with the help of the BSB equation.

Example 3.3. Consider a cap with dates $T_1 < \ldots < T_N$, with $T_{i+1} - T_i = \alpha$, $i = 1, \ldots, N-1$ and with single payoffs $\alpha(f(T_i, T_i, \alpha) - K)^+$ at time T_{i+1}. In [6] it has been shown that in the case of a Gaussian model with stochastic volatility each caplet can be superhedged by the strategy (C^i, Δ^i) given by the BSB superstrategy, where

$$C_i(t, S_t^i, S_t^{i+1}) = S_t^i N(d_1^i) - \lambda S_t^{i+1} N(d_2^i), \tag{15}$$

where we called $\lambda = 1 + \alpha K$ and

$$d_1^i = \frac{1}{\bar{\Gamma}_i} \ln\left(\frac{S_t^i}{\lambda S_t^{i+1}}\right) + \frac{1}{2}\bar{\Gamma}_i, \quad d_2^i = d_1^i - \bar{\Gamma}_i$$

and

$$\bar{\Gamma}_i = \sqrt{T_i - t} \max_{\gamma \in \Sigma} ||\gamma_{i+1} - \gamma_i||_{\mathbb{R}^N},$$

where γ_k is the k-th row of a generic matrix $\gamma \in \Sigma$. This corresponds to the superhedging strategy for an exchange option.

The question is: is the superstrategy above the cheapest possible for the cap? We already know that it is the cheapest for each separate cap, but we do not know if it is the cheapest for the cap or if there exists another superstrategy cheaper than this one. In the next section we will answer to this question.

4. Superreplication of caps and floors

Now we show how we can superhedge a cap with our approach. In fact our approach could be used also in a more general setup but we avoid it for brevity concentrating on this main application. We take the dates $T_1 < \ldots < T_N$, with $T_{i+1} - T_i = \alpha$, $i = 1, \ldots, N-1$. A cap is then a claim corresponding to the sum of the payoff

$\alpha(f(T_i,T_i,\alpha)-K)^+$ at time T_{i+1}. In a complete market, the price of a cap at time $t<T_1$ is

$$\sum_{i=1}^{N-1} B(t,T_{i+1})(1+\alpha K)\mathbb{E}_{\mathbb{Q}_{T_{i+1}}}\left[\left(\frac{B(T_i,T_i)}{1+\alpha K}-B(T_i,T_{i+1})\right)^+\middle|\ \mathcal{F}_t\right],$$

where $\mathbb{Q}_{T_{i+1}}$ is the so-called **forward-neutral** probability for the maturity T_{i+1}. If the prices of the zero coupon bond are functions of a Markov process (for example in a short rate model like Vasicek, CIR or in a more general multifactor model), then the usual way to price a cap is this: a PDE is written to price the last caplet with maturity T_N, with terminal condition equal to the payoff of the caplet with maturity T_N; then the solution of this PDE at time T_{N-1} is summed to the payoff of the caplet with maturity T_{N-1}, and this new payoff is taken as terminal condition of a PDE pricing the last two caplets; the procedure continues in this way until time t is reached.

The procedure with our approach is similar: in fact we build a sequence of BSB equations in a similar way to obtain the price of the entire cap. More in detail, we define $s^{(i)}=(s_i,\dots,s_N)$, $\bar{s}^{(i)}=diag(0,...,0,s_i,...,s_N)$ and we write the i-th BSB equation as

$$\begin{cases} \dfrac{\partial C_i}{\partial t}(t,s^{(i)})+\dfrac{1}{2}\max_{\gamma\in\Sigma}\operatorname{tr}\left(D^2C_i(t,s^{(i)})(\bar{s}^{(i)}\gamma)(\bar{s}^{(i)}\gamma)^*\right)=0, & \\ \qquad\qquad\qquad\qquad\qquad\qquad t\in[T_{i-1},T_i),\quad s^{(i)}\in\mathbb{R}_+^{N-i+1} & \\ C_i(T_i,s^{(i)})=C_{i+1}(T_i,s^{(i+1)})+(s_i-\lambda s_{i+1})^+, \qquad\qquad s^{(i)}\in\mathbb{R}_+^{N-i+1} & \end{cases} \tag{16}$$

with $i=1,\dots,N-1$, where we put $C_N(T_{N-1},s_N)\equiv 0$, so that the terminal condition of the $(N-1)$-th equation is $C_{N-1}(T_{N-1},s_{N-1},s_N)=(s_{N-1}-\lambda s_N)^+$.

Theorem 4.1. *If for all $i=1,\dots,N$, C_i is the solution of Equation (16) and for $t\in[T_{i-1},T_i)$ we define $C_t=C_i(t,S_t^{(i)})$ and $\Delta_t=DC_i(t,S_t^{(i)})$, then (C,Δ) is the cheapest strategy (in the sense of part ii) of Theorem 2.3) strategy that superreplicates the cap.*

Proof. We proceed by backward induction starting from $i=N$. For this case, Equation (16) reduces to Equation (3), so the claim follows from Theorem 2.3. Now we assume that the thesis is true for $i+1$ and we prove that it is true for i. At time $t=T_i$, the superreplication "price" of the caplets with maturity $T_{i+1},\dots,T_N$ is given by $C_{i+1}(T_i,S_{T_i}^{(i+1)})$. Since the caplet with maturity T_i at time T_i has a deterministic payoff, the price of the cap at time T_i is given by $C_{i+1}(T_i,S_{T_i}^{(i+1)})+(s_i-\lambda s_{i+1})^+$. This means that we have to superreplicate the payoff $C_{i+1}(T_i,S_{T_i}^{(i+1)})+(s_i-\lambda s_{i+1})^+$ at time T_i. By Theorem 2.3, the cheapest superstrategy in $t\in[T_{i-1},T_i)$ is given by (C,Δ) defined by $C_t=C_i(t,S_t^{(i)})$ and $\Delta_t=DC_i(t,S_t^{(i)})$, so the claim is true also for i. The proof is complete. □

Remark 4.2. Observe that the result of the above Theorem 4.1 in fact holds every time we can apply Theorem 3.1 to every single equation. So we can state a similar result also for floors or other options with multiple maturities. This can be done also if the set Σ is different in every equation. Moreover also the case of Σ varying mildly with time along some interval (T_{N-1}, T_N) could be treated by a straightforward extension of the results of [13, 17] (see Remark 2.7).

We now use Theorem 4.1 to answer to the question set after Example 3.3. By virtue of Theorem 4.1, in order to check if a superstrategy is the cheapest, it is sufficient to check if it is solution of the BSB equation (16). To this purpose, we rewrite Proposition 2.8. We define

$$F_{t,s}^{(i)}(\gamma) = \mathrm{tr}\ \left(\bar{s}^{(i)} D^2 C_i(t, s^{(i)}) \bar{s}^{(i)} \gamma\gamma^*\right) = \mathrm{tr}\ \left(A_{t,s}^{(i)} \gamma\gamma^*\right), \tag{17}$$

where $A_{t,s}^{(i)} = \bar{s}^{(i)} D^2 C_i(t, s^{(i)}) \bar{s}^{(i)}$, $t \in [T_{i-1}, T_i)$ and C_i is the solution of Eq. (16). Notice that for all t, s, $A_{t,s}^{(i)}$ is a N-dimensional square matrix $A_{t,s}^{(i)} = (a_{jk})_{j,k=1,\ldots,N}$ such that $a_{jk} = 0$ if $j, k < i$.

Proposition 4.3. *If for all* $i = 1, \ldots, N$, $C_i \in C^{1,2}([0,T) \times \mathbb{R}_+^{N-i+1})$ *is the solution of the Black-Scholes equation*

$$\begin{cases} \dfrac{\partial C_i}{\partial t}(t, s^{(i)}) + \dfrac{1}{2}\mathrm{tr}\ (D^2 C_i(t, s^{(i)})(\bar{s}^{(i)}\bar{\gamma})(\bar{s}^{(i)}\bar{\gamma})^*) = 0, & \\ \qquad\qquad\qquad\qquad t \in [T_{i-1}, T_i), \quad s^{(i)} \in \mathbb{R}_+^{N-i+1}, & \\ C_i(T_i, s^{(i)}) = C_{i+1}(T_i, s^{(i+1)}) + (s_i - \lambda s_{i+1})^+, \qquad s \in \mathbb{R}_+^{N-i+1}, & \end{cases} \tag{18}$$

where $\bar{\gamma} = \mathrm{argmax}_{\gamma \in \Sigma} F_{t,s}^{(i)}(\gamma)$ *for all* $(t, s^{(i)}) \in [0,T) \times \mathbb{R}_+^{N-i+1}$, *then* C_i *is also a solution of Eq.* (16).

The proof is the same as that of Proposition 10 in [17].

Now we check if the strategy of Example 3.3 is the cheapest superstrategy. To this end, we calculate the quantity $A_{t,s}^{(i)}$ for the function (15). By calculations analogous to the ones of Section 6 in [17], we have that

$$A_{t,s}^{(i)} := \bar{s}^{(i)} D^2 C_i(t, s^{(i)}) \bar{s}^{(i)} = \frac{S_t^i N'(d_i)}{\Gamma_i} A_i,$$

where C_i is the price of the i-th caplet given by Equation (15), $\Gamma_i = \sqrt{T_i - t}\|\gamma_{i+1} - \gamma_i\|_{\mathbb{R}^N}$ and $A_i = (a_{jk}^i)_{j,k=1,\ldots,N}$ is such that $a_{ii}^i = a_{i+1,i+1}^i = 1$, $a_{i,i+1}^i = a_{i+1,i}^i = -1$ and $a_{jk}^i = 0$ for $j, k \neq i$, $i+1$. By making some calculations, we have that

$$\mathrm{tr}\ \left(\sum_{i=1}^{n} A_{t,s}^{(i)} \gamma\gamma^*\right) = \sum_{i=1}^{n} \frac{S_t^i N'(d_i)}{T_i - t} \Gamma_i .$$

This means that if we succeed in finding $\bar{\gamma} \in \Sigma$ such that Γ_i are all maximized *simultaneously*, then the procedure in [6] works. More in detail, we have the following theorem.

Theorem 4.4. *If there exists $\bar{\gamma} \in \Sigma$ such that the Γ_i are all maximised simultaneously, then the function $C_i(t, s^{(i)}) = \sum_{k=i}^{N} C^k(t, s_k, \ldots, s_N)$ is solution of Equation* (16), *where C^k is given by Equation* (15) *and $\bar{\Gamma}_i = \sqrt{T_i - t}\|\bar{\gamma}_{i+1} - \bar{\gamma}_i\|_{\mathbb{R}^N}$. Moreover, the superreplication strategy is given by*

$$\begin{aligned} C_t &= \sum_{k=i}^{N} C^k(t, S_t^1, \ldots, S_t^{N-k+2}) \\ \Delta_t^1 &= -\lambda N(d_2^N) \\ \Delta_t^k &= N(d_1^k) - \lambda N(d_2^k) \qquad \textit{for } k = 2, \ldots, N-i+1 \\ \Delta_t^{N-i+2} &= N(d_1^i)\,. \end{aligned}$$

Proof. The proof follows from Proposition 4.3 and Theorem 4.1. □

Remark 4.5. Observe that the assumption: "*there exists $\bar{\gamma} \in \Sigma$ such that the expressions $\|\gamma_{i+1} - \gamma_i\|_{\mathbb{R}^N}$ are all maximized simultaneously*" is not in general satisfied, in particular if the set Σ varies with time. However in some cases of interest like when Σ is a cube (i.e. all terms of the matrices belong to given confidence intervals) then such assumption holds. In fact, if

$$\Sigma = \left\{ \gamma \in \mathbb{R}^{N \times N} \,:\, \gamma_{i,j} \in \left[\bar{\gamma}_{i,j} - \varepsilon_{i,j}, \bar{\gamma}_{i,j} + \varepsilon_{i,j}\right] \right\}$$

then the supremum for every $\|\gamma_{i+1} - \gamma_i\|_{\mathbb{R}^N}$ is clearly reached by the matrix

$$\gamma_0 = \begin{pmatrix} \bar{\gamma}_1 + \varepsilon_1 \\ \bar{\gamma}_2 - \varepsilon_2 \\ \bar{\gamma}_3 + \varepsilon_3 \\ \cdots \\ \bar{\gamma}_n + (-1)^{n+1}\varepsilon_n \end{pmatrix}.$$

References

[1] H. Ahn, A. Muni, and G. Swindle, *Optimal hedging strategies for misspecified asset price models*, Applied Mathematical Finance **6** (1999), 197–208.

[2] M. Avellaneda, A. Levy and A. Parás, *Pricing and hedging derivative securities in markets with uncertain volatilities*, Applied Mathematical Finance **2** (1995), 73–88.

[3] F. Black, and M. Scholes, *The pricing of options and corporate liabilities*, Journal of Political Economy **81** (1973), 637–659.

[4] M. G. Crandall, M. Kocan, and A. Swiech, *L^p-theory for fully nonlinear parabolic equations*, Communications in Partial Differential Equations **25 (11–12)** (2000), 1997–2053.

[5] J. Cvitanic, H. Pham, and N. Touzi, *Super-replication in stochastic volatility models under portfolio constraints*, J. Appl. Prob. **36** (1999), 523–545.

[6] A. Dudenhausen, E. Schlögl, and L. Schlögl, *Robustness of Gaussian hedges and the hedging of fixed income derivatives*, preprint |http://www.finasto.uni-bonn.de/papers/bonnsfb422.pdf.

[7] N. El Karoui, H. Geman, and J. C. Rochet, *Changes of numeraire, changes of probability measures and option pricing*, J. Appl. Prob. **32** (1995), 443–458.

[8] N. El Karoui, M. Jeanblanc-Piqué, and S. E. Shreve, *Robustness of the Black and Scholes formula*, Mathematical Finance **8** (1998), no. 2, 93–126.

[9] N. El Karoui, M. Jeanblanc-Picqué, and R. Viswanathan, *Bounds for the price of options*, Applied stochastic analysis (New Brunswick, NJ, 1991), 224–237, Lecture Notes in Control and Inform. Sci., 177, Springer, Berlin, 1992.

[10] N. El Karoui and M. C. Quenez, *Dynamic programming and pricing contingent claims in incomplete markets*, SIAM Journal of Control and Optimization **33** (1995), 29–66.

[11] W. H. Fleming and H. M. Soner, *Controlled Markov processes and viscosity solutions*, Springer, **1993**.

[12] R. Frey and C. A. Sin, *Bound on european option prices under stochastic volatility*, Math. Finance **9** (1999), 97–116.

[13] F. Gozzi and T. Vargiolu, *Superreplication of European multiasset derivatives with bounded stochastic volatility*, to appear in Mathematical Methods of Operations Research **55 (1)**, (2002).

[14] T. J. Lyons, *Uncertain volatility and the risk-free synthesis of derivatives*, Applied Mathematical Finance **2** (1995), 117–133.

[15] W. Margrabe, *The price to exchange an asset for another*, The Journal of Finance **1** (1978), 177–186.

[16] M. Musiela and M. Rutkowski, *Martingale methods in financial modelling*, Springer, **1997**.

[17] S. Romagnoli and T. Vargiolu, *Robustness of the Black-Scholes formula in the case of options on several assets*, Finance and Stochastics **4** (2000), 325–341.

[18] N. Touzi, *Direct Characterization of the value of super-replication under stochastic volatility and portfolio constraints*, Stoch. Proc. Appl. **88** (2000), 305–328.

[19] L. Wang, *On the regularity theory of fully nonlinear parabolic equations: I, II*, Communications on Pure and Applied Mathematics **45** (1992), 27–76 and 141–178.

Dipartimento di Matematica per le Decisioni
Economiche Finanziarie e Assicurative
Facoltà di Economia - Università di Roma "La Sapienza"
Via del Castro Laurenziano 9
I-00146 Roma
Italy
E-mail address: gozzi@scec.eco.uniroma1.it

Dipartimento di Matematica
Pura ed Applicata
Via Belzoni
I-35131 Padova
Italy
E-mail address: vargiolu@math.unipd.it

Progress in Probability, Vol. 52, 189–204

A Complete Market Model with Poisson and Brownian Components

M. Jeanblanc and N. Privault

Abstract. We consider a complete market model with jumps, using a martingale constructed from a Brownian motion and a Poisson process that are mutually excluding each other. The chaotic calculus relative to this martingale is developed, and applications to the computation of hedging strategies are presented via a Clark formula and the method of partial differential equations.

1. Introduction

In this paper we present a complete market model that considers jumps in dynamics of asset prices. The sum of a Brownian motion and a Poisson process has been used to model discontinuous asset prices, cf. [9], however it does not have the predictable representation property, thus it does not yield market completeness. A model having market completeness with jumps has been presented in [4] using the standard Azéma martingale studied in [6].

We suggest to use a modification $(M_t)_{t\in\mathbb{R}_+}$ of another martingale also introduced in [6], which switches between a Brownian motion and a Poisson process according to a deterministic schedule. This process does have the chaos representation property because its Brownian part vanishes whenever it jumps. The martingale considered in [6] is constrained to have jump sizes linked to the intensity of the Poisson component. For this reason we use a modification of this process by regulating the intensity and jumps sizes of the Poisson part according to deterministic functions of time. As a consequence, the predictable quadratic variation of our process is no longer equal to dt. The Poisson process and Brownian motion are included as particular cases in our approach, as well as the representation formulas obtained in [8] using the Wiener chaos associated to a continuous martingale with deterministic angle bracket. Our model contains a part of determinism in the switches between different regimes. This can be used to represent certain events

---

2000 *Mathematics Subject Classification.* 91B28, 91B24, 91B26, 60H07, 60H05, 60G44, 60H30.
Key words and phrases. Market completeness, normal martingales, hedging strategies, Malliavin calculus, chaos representation property.
We thank Y. Ishikawa and the referee for a careful reading of this paper.

and announcements that occur at dates known in advance, but whose effects are not predicted with certainty.

In Section 2 we define the martingale $(M_t)_{t\in\mathbb{R}_+}$ which is a deterministic combination of a Poisson process and a Brownian motion. Section 3 presents the chaos representation property. Section 4 deals with the predictable representation property and the Clark formula with simplified proofs, using a gradient operator that plays a role in the computation of hedging strategies. Changes of probability and the Girsanov theorem for the process $(M_t)_{t\in\mathbb{R}_+}$ are considered in Section 5, with applications to the pricing of European options in Section 6.

We close this introduction with some notation. Given a martingale $(M_t)_{t\in\mathbb{R}_+}$, let $([M_t, M_t])_{t\in\mathbb{R}_+}$ denote its quadratic variation, defined by

$$[M_t, M_t] = M_t^2 - 2\int_0^t M_{t-}dM_t,$$

and let $(\langle M_t, M_t\rangle)_{t\in\mathbb{R}_+}$ denote its conditional quadratic variation, or angle bracket, such that $([M_t, M_t] - \langle M_t, M_t\rangle)_{t\in\mathbb{R}_+}$ is a martingale. If $(M_t)_{t\in\mathbb{R}_+}$ is in L^4 and has the predictable representation property then there exists a predictable process $(\phi_t)_{t\in\mathbb{R}_+}$ such that

$$[M_t, M_t] - \langle M_t, M_t\rangle = \int_0^t \phi_{s-}dM_s. \tag{1}$$

If the martingale $(M_t)_{t\in\mathbb{R}_+}$ is *normal* (cf. [3]), that is if $\langle M_t, M_t\rangle = t$, $t \in \mathbb{R}_+$, then (1) can be written as

$$d[M_t, M_t] = dt + \phi_s dM_s.$$

This type of equation has been called a structure equation in [6].

2. Solution of a deterministic structure equation

Let $\phi : \mathbb{R}_+ \longrightarrow \mathbb{R}$ and $\alpha : \mathbb{R}_+ \longrightarrow]0,\infty[$ be two deterministic functions, with $\phi \in \cap_{p\geq 1} L^p(\mathbb{R}_+, \alpha_t^2 dt)$. Let $i_t = 1_{\{\phi_t=0\}}$, and

$$\lambda_t = (1-i_t)\alpha_t^2/\phi_t^2 = \begin{cases} \alpha_t^2/\phi_t^2 & \text{if } \phi_t \neq 0, \\ 0 & \text{if } \phi_t = 0, \quad t\in\mathbb{R}_+. \end{cases}$$

Let $(B_t)_{t\in\mathbb{R}_+}$ be a standard Brownian motion, and $(N_t)_{t\in\mathbb{R}_+}$ a Poisson process with intensity $\nu_t = \int_0^t \lambda_s ds$, $t \in \mathbb{R}_+$, i.e. $(\nu_t)_{t\in\mathbb{R}_+}$ is the unique continuous deterministic function such that $(N_t - \nu_t)_{t\in\mathbb{R}_+}$ is a martingale. We assume that the processes $(B_t)_{t\in\mathbb{R}_+}$ and $(N_t)_{t\in\mathbb{R}_+}$ are independent and that $\lim_{t\to\infty}\nu_t = \infty$ and $\nu_t < \infty$, $\forall t \in \mathbb{R}_+$. Let $(X_t)_{t\in\mathbb{R}_+}$ denote the process defined as

$$dX_t = i_t dB_t + \frac{\phi_t}{\alpha_t}\left(dN_t - \lambda_t dt\right), \quad t\in\mathbb{R}_+, \qquad X_0 = 0, \tag{2}$$

which satisfies the structure equation

$$d[X, X]_t = dt + \frac{\phi_t}{\alpha_t} dX_t. \tag{3}$$

It is known from [6] that $(X_t)_{t\in\mathbb{R}_+}$ is in fact the unique solution of (3). Relation (3) implies that the process $(X_t)_{t\in\mathbb{R}_+}$ has predictable quadratic variation $d\langle X,X\rangle_t = dt$. From $(X_t)_{t\in\mathbb{R}_+}$ we construct a martingale $(M_t)_{t\in\mathbb{R}_+}$ with predictable quadratic variation $d\langle M,M\rangle_t = \alpha_t^2 dt$, as

$$dM_t = \alpha_t dX_t, \qquad t \in \mathbb{R}_+, \quad M_0 = 0, \tag{4}$$

i.e.

$$dM_t = i_t\alpha_t dB_t + \phi_t\left(dN_t - \lambda_t dt\right), \quad t \in \mathbb{R}_+, \quad M_0 = 0. \tag{5}$$

We are using $(M_t)_{t\in\mathbb{R}_+}$ instead of $(X_t)_{t\in\mathbb{R}_+}$ because the parameters $(\phi_t)_{t\in\mathbb{R}_+}$ and $(\alpha_t)_{t\in\mathbb{R}_+}$ have independent roles in the definition of $(M_t)_{t\in\mathbb{R}_+}$: $(\phi_t)_{t\in\mathbb{R}_+}$ controls the height of jumps, and $(\alpha_t)_{t\in\mathbb{R}_+}$ is relative to the continuous part of $(M_t)_{t\in\mathbb{R}_+}$.

Proposition 2.1. *The martingale $(M_t)_{t\in\mathbb{R}_+}$ satisfies the deterministic structure equation*

$$d[M,M]_t = \alpha_t^2 dt + \phi_t dM_t, \qquad t \in \mathbb{R}_+, \tag{6}$$

with $d\langle M,M\rangle_t = \alpha_t^2 dt$, $t \in \mathbb{R}_+$.

Proof. Using the relations $d[B,N]_t = 0$ and $i_t\phi_t = 0$, $t \in \mathbb{R}_+$, we have:

$$\begin{aligned} d[M,M]_t &= i_t\alpha_t^2 dt + \phi_t^2 dN_t \\ &= i_t\alpha_t^2 dt + \phi_t\left(dM_t - i_t\alpha_t dB_t + (1-i_t)\frac{\alpha_t^2}{\phi_t}dt\right) \\ &= \alpha_t^2 dt + \phi_t dM_t, \quad t \in \mathbb{R}_+. \end{aligned}$$

□

At first it could seem more general to set $dM_t = \beta_t dX_t$, however this is not the case, because the functions $(\phi_t)_{t\in\mathbb{R}_+}$ and $(\alpha_t)_{t\in\mathbb{R}_+}$ are sufficient to completely characterize $(M_t)_{t\in\mathbb{R}_+}$ as a solution of (6).

3. Chaos representation property

Let $(\mathcal{F}_t)_{t\in\mathbb{R}_+}$ be the filtration generated by $(M_t)_{t\in\mathbb{R}_+}$ which is the same as the filtration generated by $(X_t)_{t\in\mathbb{R}_+}$, since α is deterministic and does not vanish. This filtration is smaller than the filtration generated by the pair Brownian motion - Poisson process. We assume that we are working on a probability space $(\Omega,\mathcal{F},P)$ with $\mathcal{F} = \mathcal{F}_\infty$. We denote by $L^2(\mathbb{R}_+,\alpha_t^2 dt)^{\circ n}$ the space of symmetric functions on $\mathbb{R}_+^n$ that are square-integrable with respect to $\alpha_{t_1}^2\cdots\alpha_{t_n}^2 dt_1\cdots dt_n$, equipped with the $L^2(\mathbb{R}_+^n,\alpha_{t_1}^2\cdots\alpha_{t_n}^2 dt_1\cdots dt_n)$-norm. Given $f_1,\ldots,f_n \in L^2(\mathbb{R}_+,\alpha_t^2 dt)$, let also

$$f_1\circ\cdots\circ f_n : \mathbb{R}_+^n \longrightarrow \mathbb{R}$$

denote the symmetrization in n variables of the function

$$(t_1,\ldots,t_n) \mapsto f_1(t_1)\cdots f_n(t_n).$$

It is known, cf. [6], that the process $(X_t)_{t\in\mathbb{R}_+}$ has the chaos representation property, i.e. any $F \in L^2(\Omega, \mathcal{F}, P)$ has a unique decomposition

$$F = E[F] + \sum_{n=1}^{\infty} n! \int_0^{\infty} \int_0^{t_n} \cdots \int_0^{t_2} g_n(t_1, \ldots, t_n) dX_{t_1} \cdots dX_{t_n}, \qquad (7)$$

with $g_n \in L^2(\{0 \leq t_1 < \cdots < t_n\})$. Obviously a similar property holds for $(M_t)_{t\in\mathbb{R}_+}$.

Let $I_n(f_n)$ denote the multiple stochastic integral of $f_n \in L^2(\mathbb{R}_+, \alpha_t^2 dt)^{\circ n}$ with respect to $(M_t)_{t\in\mathbb{R}_+}$, defined as

$$I_n(f_n) = n! \int_0^{\infty} \int_0^{t_n} \cdots \int_0^{t_2} f_n(t_1, \ldots, t_n) dM_{t_1} \cdots dM_{t_n}, \quad n \geq 1,$$

with the convention $I_0(f_0) = f_0$, $f_0 \in \mathbb{R}$. The multiple stochastic integrals satisfy the isometry and orthogonality properties

$$E[I_n(f_n) I_m(g_m)] = \begin{cases} (f_n, g_n)_{L^2(\mathbb{R}_+, \alpha_t^2 dt)^{\circ n}} & n = m, \\ 0, & n \neq m, \end{cases} \qquad (8)$$

$f_n \in L^2(\mathbb{R}_+, \alpha_t^2 dt)^{\circ n}$, $g_m \in L^2(\mathbb{R}_+, \alpha_t^2 dt)^{\circ m}$.

Proposition 3.1. *The martingale $(M_t)_{t\in\mathbb{R}_+}$ has the chaos representation property: any $F \in L^2(\Omega, \mathcal{F}, P)$ has a unique decomposition*

$$F = E[F] + \sum_{n=1}^{\infty} I_n(f_n), \quad f_n \in L^2(\mathbb{R}_+, \alpha_t^2 dt)^{\circ n}, \quad n \geq 1.$$

Proof. It suffices to consider the expansion (7) of $F \in L^2(\Omega, \mathcal{F}, P)$ and to define $f_n \in L^2(\mathbb{R}_+, \alpha_t^2 dt)^{\circ n}$ as

$$f_n(t_1, \ldots, t_n) = (\alpha_{t_1} \cdots \alpha_{t_n})^{-1} g_n(t_1, \ldots, t_n), \quad t_1, \ldots, t_n \in \mathbb{R}_+^n, \quad n \geq 1.$$

□

The action of the conditional expectation with respect to $\mathcal{F}_t$ on multiple stochastic integrals is

$$E[I_n(f_n) \mid \mathcal{F}_t] = I_n(f_n(*) 1_{\{* \leq t\}}), \quad t \in \mathbb{R}_+, \quad f_n \in L^2(\mathbb{R}_+, \alpha_s^2 ds)^{\circ n}, \quad n \geq 1,$$

since $(M_t)_{t\in\mathbb{R}_+}$ is a martingale.

4. Gradient and divergence operators

In this section we introduce the gradient operator that will be used in the computation of hedging strategies. We denote by $\mathrm{Dom}(D)$ the set of $F \in L^2(\Omega, \mathcal{F}, P)$ whose decomposition $F = \sum_{n=0}^{\infty} I_n(f_n)$ satisfies

$$\sum_{n=1}^{\infty} n(n!) \|f_n\|^2_{L^2(\mathbb{R}_+, \alpha_t^2 dt)^{\circ n}} < \infty.$$

We let

$$D : \mathrm{Dom}(D) \longrightarrow L^2(\Omega \times \mathbb{R}_+, dP \times \alpha_t^2 dt)$$

denote the unbounded gradient operator defined as

$$D_t F = \sum_{n=1}^{\infty} n I_{n-1}(f_n(*, t)), \quad dP \times dt - a.e.,$$

if $F \in \mathrm{Dom}(D)$ is written as $F = \sum_{n=0}^{\infty} I_n(f_n)$. We have

$$\begin{aligned} \|DF\|^2_{L^2(\Omega\times\mathbb{R}_+, dP\times\alpha_t^2 dt)} &= \int_0^\infty E[(D_t F)^2] \alpha_t^2 dt \\ &= \int_0^\infty \sum_{n=1}^{\infty} n^2 (n-1)! \|f_n(*, t)\|^2_{L^2(\mathbb{R}_+, \alpha_t^2 dt)^{\circ(n-1)}} \alpha_t^2 dt \\ &= \sum_{n=1}^{\infty} n(n!) \|f_n\|^2_{L^2(\mathbb{R}_+, \alpha_t^2 dt)^{\circ n}}. \end{aligned}$$

We denote by $\mathrm{Dom}(\delta)$ the set of $u \in L^2(\Omega \times \mathbb{R}_+, dP \times \alpha_t^2 dt)$ whose decomposition $u_t = \sum_{n=0}^{\infty} I_n(u_n(*, t))$, $t \in \mathbb{R}_+$, satisfies

$$\sum_{n=0}^{\infty} (n+1)! \|\tilde{u}_n\|^2_{L^2(\mathbb{R}_+, \alpha_t^2 dt)^{\circ(n+1)}} < \infty,$$

where $\tilde{u}_n$ denotes the symmetrization in $n+1$ variables of u_n. Let

$$\delta : \mathrm{Dom}(\delta) \longrightarrow L^2(\Omega, \mathcal{F}, P)$$

be the unbounded divergence operator defined as

$$\delta(u) = \sum_{n=0}^{\infty} I_{n+1}(\tilde{u}_n), \quad dP - a.e.,$$

if $u \in \mathrm{Dom}(\delta)$ has the expression $u_t = \sum_{n=0}^{\infty} I_n(u_n(*, t))$, $t \in \mathbb{R}_+$. Let $\mathcal{S}$ denote the vector space generated by multiple stochastic integrals of the form $I_n(f_1 \circ \cdots \circ f_n)$, $f_1, \ldots, f_n \in \cap_{p\geq 1} L^p(\mathbb{R}_+, \alpha_t^2 dt)$, $n \in \mathbb{N}$.

Proposition 4.1. *The operators D and δ have the following properties:*

1. *D and δ are mutually adjoint in the following sense:*

$$E[(DF, u)_{L^2(\mathbb{R}_+, \alpha_t^2 dt)}] = E[F\delta(u)], \quad u \in \mathrm{Dom}(\delta), \quad F \in \mathrm{Dom}(D). \tag{9}$$

2. *δ coincides with the stochastic integral with respect to $(M_t)_{t\in\mathbb{R}_+}$ on the square-integrable adapted processes $u \in L^2(\Omega \times \mathbb{R}_+, dP \times \alpha_t^2 dt)$:*

$$\delta(u) = \int_0^\infty u(t) dM_t.$$

Proof. We first note that for F and u of the form $F = I_n(f_n)$ and $u_t = h_t I_m(g_m)$, (9) follows from the isometry relation (8). Hence by linearity, (9) holds on dense vector subspaces of $L^2(\Omega, \mathcal{F}, P)$ and $L^2(\Omega \times \mathbb{R}_+, dP \times \alpha_t^2 dt)$. As a consequence, D and δ are closable hence (9) holds on $\mathrm{Dom}(D)$ and $\mathrm{Dom}(\delta)$ by density. Concerning (ii) we write

$$u(t) = \sum_{n=0}^{\infty} I_n(u_n(*,t)) = \sum_{n=0}^{\infty} n! \int_0^{\infty} \int_0^{t_n} \cdots \int_0^{t_2} u_n(t_1, \ldots, t_n, t) dM_{t_1} \cdots dM_{t_n},$$

with $u_n(*,t)$ symmetric, $t \in \mathbb{R}_+$. We have $u_n(*,t) = u_n(*,t)1_{\{*\leq t\}}$, $t \in \mathbb{R}_+$, since u is $\mathcal{F}_t$-adapted, hence the symmetrization $\tilde{u}_n$ of u_n in $n+1$ variables coincides with $\frac{1}{n+1}u_n$ on $\{0 \leq t_1 < \cdots < t_n\}$. Consequently,

$$\begin{aligned}
\delta(u) &= \sum_{n=0}^{\infty} I_{n+1}(\tilde{u}_n) \\
&= \sum_{n=0}^{\infty} \frac{(n+1)!}{n+1} \int_0^{\infty} \int_0^{t_{n+1}} \cdots \int_0^{t_2} u_n(t_1, \ldots, t_n, t_{n+1}) dM_{t_1} \cdots dM_{t_{n+1}} \\
&= \int_0^{\infty} \sum_{n=0}^{\infty} n! \int_0^{t} \int_0^{t_n} \cdots \int_0^{t_2} u_n(t_1, \ldots, t_n, t) dM_{t_1} \cdots dM_{t_n} dM_t \\
&= \int_0^{\infty} \sum_{n=0}^{\infty} I_n(u_n(*,t)) dM_t = \int_0^{\infty} u(t) dM_t.
\end{aligned}$$

□

The next result states that $(M_t)_{t\in\mathbb{R}_+}$ has the predictable representation property, as a consequence of the chaos representation property.

Proposition 4.2. *Any $F \in \mathrm{Dom}(D) \subset L^2(\Omega, \mathcal{F}, P)$ has a representation*

$$F = E[F] + \int_0^{\infty} E[D_t F \mid \mathcal{F}_t] dM_t. \tag{10}$$

Proof. We write the chaos expansion of F:

$$\begin{aligned}
F &= E[F] + \sum_{n=1}^{\infty} n! \int_0^{\infty} \int_0^{t_n} \cdots \int_0^{t_2} f_n(t_1, \ldots, t_n) dM_{t_1} \cdots dM_{t_n} \\
&= E[F] + \sum_{n=1}^{\infty} n \int_0^{\infty} I_{n-1}(f_n(*,t)1_{\{*\leq t\}}) dM_t \\
&= E[F] + \int_0^{\infty} E[D_t F \mid \mathcal{F}_t] dM_t.
\end{aligned}$$

□

This formula is called the Clark formula in the case of Brownian motion, cf. [2, 11]. Combining Proposition 4.2 with (ii) of Proposition 4.1 we can also write

$$F = E[F] + \delta(E[D_. F \mid \mathcal{F}_.]).$$

Instead of the adapted projection $(E[D_tF \mid \mathcal{F}_t])_{t\in\mathbb{R}_+}$ we may also use the predictable projection $(E[D_tF \mid \mathcal{F}_{t-}])_{t\in\mathbb{R}_+}$ defined by

$$E[D_tI_n(f_n) \mid \mathcal{F}_{t-}] = I_n(f_n(*)1_{\{*<t\}}), \quad f_n \in L^2(\mathbb{R}_+, \alpha_s^2 ds)^{\circ n}, \quad t > 0, \ n \geq 1,$$

in fact this leads to the same representation since the adapted and predictable projections coincide in $L^2(\Omega \times \mathbb{R}_+, dP \times \alpha_t^2 dt)$. The following proposition gives the product rule for the operator D, which can be useful in practice for the computation of hedging strategies.

Proposition 4.3. *We have the product rule*

$$D_t(FG) = FD_tG + GD_tF + \phi_t D_tF D_tG, \qquad t \in \mathbb{R}_+, \tag{11}$$

for any F and G in the vector space $\mathcal{S}$.

Proof. Rewriting the multiplication formula for multiple stochastic integrals with respect to $(X_t)_{t\in\mathbb{R}_+}$, we obtain

$$\begin{aligned} & I_n(f^{\circ n})I_1(g) \qquad\qquad (12) \\ = \quad & I_{n+1}(g \circ f^{\circ n}) + n(f,g)_{L^2(\mathbb{R}_+,\alpha_t^2dt)}I_{n-1}(f^{\circ(n-1)}) + nI_n((\phi fg)\circ f^{\circ n-1}). \end{aligned}$$

Consequently, (11) can be proved by replicating the proof of Proposition 1 in [12]. □

From the duality between D and δ, relation (12) implies

$$F\delta(hF) = \delta(hF) + (h, DF)_{L^2(\mathbb{R}_+,\alpha_t^2dt)} + \delta(\phi h DF),$$

$h \in \bigcap_{p\geq 1} L^p(\mathbb{R}_+, \alpha_t^2 dt)$, $F \in \mathcal{S}$, which can be seen as a reformulation of (12).

Remark 4.4. Although $D : L^2(\Omega, \mathcal{F}, P) \longrightarrow L^2(\Omega \times \mathbb{R}_+, dP \times \alpha_t^2 dt)$ is unbounded, the representation formula (10) of Proposition 4.2 can be extended from $F \in \mathrm{Dom}(D)$ to $F \in L^2(\Omega, \mathcal{F}_\infty, P)$ using the fact that the operator $F \mapsto E[D.F \mid \mathcal{F}.]$ has a continuous extension from $L^2(\Omega, \mathcal{F}, P)$ into the space of square-integrable adapted processes, cf. [10, 14].

Proof. We prove the continuity of the composition of the adapted projection operator with D. Let $F = \sum_{n=0}^\infty I_n(f_n) \in Dom(\nabla)$ and $u(\cdot) = \sum_{n=0}^\infty I_n(u_n(*,\cdot)) \in L^2(\Omega \times \mathbb{R}_+, dP \times \alpha_t^2 dt)$. Let

$$\Delta_n = \{(t_1, \ldots, t_n) \in \mathbb{R}_+^n \ : \ 0 \leq t_1 < \cdots < t_n\}, \qquad n \geq 1.$$

We have, using the symmetry of f_{n+1}:

$$
\begin{aligned}
&\left|(E[D.F \mid \mathcal{F}.], u)_{L^2(\Omega\times\mathbb{R}_+, dP\times\alpha_t^2 dt)}\right| \\
&= \left|\sum_{n=0}^{\infty}(n+1)\int_0^\infty n!(f_{n+1}(*,t)1_{\{*<t\}}, u_n(*,t))_{L^2(\mathbb{R}_+,\alpha_t^2 dt)^{\circ n}}\alpha_t^2 dt\right| \\
&\le \sum_{n=0}^{\infty}(n+1)!\left|\int_0^\infty (f_{n+1}(*,t)1_{\{*<t\}}, u_n(*,t))_{L^2(\mathbb{R}_+,\alpha_t^2 dt)^{\circ n}}\alpha_t^2 dt\right| \\
&\le \sum_{n=0}^{\infty}(n+1)!\left|\int_0^\infty \|(f_{n+1}(*,t)1_{\{*<t\}}\|_{L^2(\mathbb{R}_+,\alpha_s^2 ds)^{\circ n}}\|u_n(*,t)\|_{L^2(\mathbb{R}_+,\alpha_s^2 ds)^{\circ n}}\alpha_t^2 dt\right| \\
&\le \sum_{n=0}^{\infty}(n+1)!\sqrt{n!}\left|\int_0^\infty \|f_{n+1}(*,t)1_{\Delta_n}(*)1_{\{*<t\}}\|_{L^2(\mathbb{R}_+,\alpha_s^2 ds)^{\circ n}}\right. \\
&\qquad\qquad \left.\times\|u_n(*,t)\|_{L^2(\mathbb{R}_+,\alpha_s^2 ds)^{\circ n}}\alpha_t^2 dt\right| \\
&\le \sum_{n=0}^{\infty}(n+1)!\sqrt{n!}\left(\int_0^\infty \|(f_{n+1}(*,t)1_{\Delta_n}(*)1_{\{*<t\}}\|^2_{L^2(\mathbb{R}_+,\alpha_s^2 ds)^{\circ n}}\alpha_t^2 dt\right)^{1/2} \\
&\qquad\qquad \times\left(\int_0^\infty \|u_n(*,t)\|^2_{L^2(\mathbb{R}_+,\alpha_s^2 ds)^{\circ n}}\alpha_t^2 dt\right)^{1/2} \\
&\le \sum_{n=0}^{\infty}(n+1)!\sqrt{n!}\|f_{n+1}1_{\Delta_{n+1}}\|_{L^2(\mathbb{R}_+,\alpha_s^2 ds)^{\circ(n+1)}}\|u_n\|_{L^2(\mathbb{R}_+,\alpha_s^2 ds)^{\circ(n+1)}} \\
&\le \sum_{n=0}^{\infty}\sqrt{(n+1)!}\sqrt{n!}\|f_{n+1}\|_{L^2(\mathbb{R}_+,\alpha_s^2 ds)^{\circ(n+1)}}\|u_n\|_{L^2(\mathbb{R}_+,\alpha_s^2 ds)^{\circ(n+1)}} \\
&\le \left(\sum_{n=0}^{\infty}(n+1)!\|f_{n+1}\|^2_{L^2(\mathbb{R}_+,\alpha_t^2 dt)^{\circ n}}\right)^{1/2}\left(\sum_{n=0}^{\infty}n!\|u_n\|^2_{L^2(\mathbb{R}_+,\alpha_t^2 dt)^{\circ(n+1)}}\right)^{1/2} \\
&\le \|F\|_{L^2(\Omega)}\|u\|_{L^2(\Omega\times\mathbb{R}_+,dP\times\alpha_t^2 dt)}.
\end{aligned}
$$

□

5. Change of probability

Let Q be a probability measure which is equivalent to P. Then

$$L_t = E\left[\frac{dQ}{dP} \mid \mathcal{F}_t\right], \quad t \in \mathbb{R}_+, \tag{13}$$

is a strictly positive $\mathcal{F}_t$-martingale under P. Due to the predictable representation property of $(X_t)_{t\in\mathbb{R}_+}$ and $(M_t)_{t\in\mathbb{R}_+}$ (Proposition 4.2), there exists a predictable process $(\psi_t)_{t\in\mathbb{R}_+}$ such that

$$dL_t = L_{t^-}\psi_t dM_t, \quad t\in\mathbb{R}_+, \quad L_0 = 1.$$

We have

$$L_t = \exp\left(\int_0^t i_s\alpha_s\psi_s dB_s - \frac{1}{2}\int_0^t i_s\alpha_s^2\psi_s^2 ds + \int_0^t \lambda_s\psi_s ds\right)\prod_{T_k\leq t}(1+\phi_{T_k}\psi_{T_k}),$$

where $(T_k)_{k\geq 1}$ denotes the sequence of jump times of $(N_t)_{t\in\mathbb{R}_+}$, hence the process $(\psi_t)_{t\in\mathbb{R}_+}$ satisfies $1+\phi_t\psi_t > 0$, $t\in\mathbb{R}_+$.

Proposition 5.1. *Under the probability Q, the process*

$$Z_t = M_t - \int_0^t \alpha_s^2\psi_s ds, \quad t\in\mathbb{R}_+, \tag{14}$$

is a martingale which satisfies the structure equation

$$d[Z,Z]_t = \alpha_t^2(1+\phi_t\psi_t)dt + \phi_t dZ_t. \tag{15}$$

In particular, if $(\psi_t)_{t\in\mathbb{R}_+}$ is deterministic, then $(Z_t)_{t\in\mathbb{R}_+}$ has the chaos representation property under Q.

Proof. From the classical Girsanov theorem, cf. [7], the canonical decomposition of $(M_t)_{t\in\mathbb{R}_+}$ under Q is

$$dM_t = \left(dM_t - \frac{1}{L_{t-}}d\langle L,M\rangle_t\right) + \frac{1}{L_{t-}}d\langle L,M\rangle_t, \quad t\in\mathbb{R}_+,$$

i.e. $dZ_t = dM_t - \frac{1}{L_{t-}}d\langle L,M\rangle_t$ is a martingale under Q, with $d\langle L,M\rangle_t = L_{t-}\alpha_t^2\psi_t dt$. (Under Q, the process $d\tilde{Z}_t = dM_t - \frac{1}{L_t}d[L,M]_t$ is also a martingale.) We also have

$$dZ_t = i_t\alpha_t d\tilde{B}_t + \phi_t dN_t - \lambda_t\phi_t(1+\phi_t\psi_t)dt.$$

Under the probability Q,

$$d\tilde{B}_t = dB_t - \frac{1}{L_{t-}}d\langle L,B\rangle_t = dB_t - i_t\psi_t\alpha_t dt$$

is a standard Brownian motion, and

$$dN_t - \lambda_t dt - \frac{1}{L_{t-}}d\langle L_t, N_t - \int_0^t \lambda_s ds\rangle = dN_t - \lambda_t(1+\phi_t\psi_t)dt,$$

is a martingale under Q, i.e. $(N_t)_{t\in\mathbb{R}_+}$ is a Poisson process with intensity $\gamma_t dt$, with

$$\gamma_t = \lambda_t(1+\phi_t\psi_t), \quad t\in\mathbb{R}_+.$$

Writing

$$dZ_t = i\alpha_t d\tilde{B}_t + \phi_t(dN_t - \lambda_t(1+\phi_t\psi_t)dt),$$

we have

$$\begin{aligned} d[Z,Z]_t &= i_t\alpha_t^2 dt + \phi_t^2 dN_t \\ &= i_t\alpha_t^2 dt + \phi_t(dZ_t - i_t\alpha_t d\tilde{B}_t + \lambda_t\phi_t(1+\phi_t\psi_t)dt) \\ &= i_t\alpha_t^2 dt + (1-i_t)\alpha_t^2(1+\phi_t\psi_t)dt + \phi_t dZ_t \\ &= \alpha_t^2(1+\phi_t\psi_t)dt + \phi_t dZ_t. \end{aligned}$$

From the above discussion, if $(\psi_t)_{t\in\mathbb{R}_+}$ is deterministic, then the process $(Z_t)_{t\in\mathbb{R}_+}$ has the chaos representation property since it is a martingale that satisfies the deterministic structure equation (15). □

Proposition 5.1 can be obtained in a different way, applying separately the Girsanov theorem on the Wiener and Poisson spaces. The process $dB_t - i_t\psi_t\alpha_t dt$ is a standard Brownian motion under the probability

$$\exp\left(\int_0^\infty i_s\psi_s\alpha_s dB_s - \frac{1}{2}\int_0^\infty i_s\psi_s^2\alpha_s^2 ds\right) dP.$$

let $(\nu_t^{-1})_{t\in\mathbb{R}_+}$ denote the right-continuous inverse of $(\nu_t)_{t\in\mathbb{R}_+}$.

The process $(N_{\nu_t^{-1}})_{t\in\mathbb{R}_+}$ is Poisson with intensity 1 under P, and has intensity $(1+\phi_t\psi_t)dt$ under the probability

$$\lim_{t\to\infty} \exp\left(\int_0^t (\phi\psi)(\nu_s^{-1})ds\right)\prod_{k=1}^{k=N_t}(1+(\phi\psi)(\nu^{-1}(\nu_{T_k})))dP,$$

where $(\nu_{T_k})_{k\geq 1}$ denotes the jumps of $(N_{\nu_t^{-1}})_{t\in\mathbb{R}_+}$. Hence $(N_t)_{t\in\mathbb{R}_+}$ is a Poisson process with intensity

$$\frac{d}{dt}\left(\nu_t + \int_0^t (1-i_s)\frac{\psi_s\alpha_s^2}{\phi_s}ds\right) = (1+\phi_t\psi_t)\lambda_t dt$$

under the probability Q, and

$$i_t\alpha_t dB_t - i_t\psi_t\alpha_t dt + \phi_t(dN_t - \lambda_t(1+\phi_t\psi_t)dt) = dM_t - \psi_t\alpha_t^2 dt$$

is a martingale under Q.

6. Application: European call

6.1. The model

Let $(\mu_t)_{t\in\mathbb{R}_+}$ be a deterministic process such that $\int_0^t |\mu_s|ds < \infty$, $\forall t\in\mathbb{R}_+$, and let $\sigma : \mathbb{R}_+ \longrightarrow \mathbb{R}$ be a deterministic bounded function satisfying $1+\sigma_t\phi_t > 0$, $t\in\mathbb{R}_+$. Let $(S_t)_{t\in\mathbb{R}_+}$ denote the solution of the equation

$$dS_t = \mu_t S_t dt + \sigma_t S_{t^-} dM_t, \quad t\in\mathbb{R}_+,$$

with deterministic initial condition S_0. We have

$$\begin{aligned} S_t &= S_0 \exp\left(\int_0^t \sigma_s\alpha_s i_s dB_s + \int_0^t (\mu_s - \phi_s\lambda_s\sigma_s)ds - \frac{1}{2}\int_0^t i_s\sigma_s^2\alpha_s^2 ds\right) \\ &\quad \times \prod_{k=1}^{k=N_t}(1+\sigma_{T_k}\phi_{T_k}), \quad t\in\mathbb{R}_+, \end{aligned}$$

where $(T_k)_{k\geq 1}$ denotes the jump times of $(N_t)_{t\in\mathbb{R}_+}$. We assume that $(S_t)_{t\in\mathbb{R}_+}$ represents the price of a risky asset, and that $(r_t)_{t\in\mathbb{R}_+}$ is a deterministic non-negative function which models the spot rate of a risk-less asset.

As it is well known, if there exists a probability Q equivalent to P such that under Q, $\left(S_t \exp\left(-\int_0^t r_s ds\right)\right)_{t\in\mathbb{R}_+}$ is a martingale, the market is arbitrage free. Such probabilities are called equivalent martingale measure (EMM). Since the market is complete (Proposition 4.2), the EMM given by Proposition 5.1 is unique, see e.g. [13]. In the following we construct explicitly the probability Q via the Girsanov theorem. Unlike in the continuous case, the law of $(S_t)_{t\in\mathbb{R}_+}$ under Q will also depend on $(\mu_t)_{t\in\mathbb{R}_+}$ by means of the intensity of the jump process.

Proposition 6.1. *Let us assume that* $1 + (r_t - \mu_t)\phi_t/(\sigma_t\alpha_t^2) > 0$, $t \in \mathbb{R}_+$, *and let* $(\psi_t)_{t\in\mathbb{R}_+}$ *be defined as*

$$\psi_t = \frac{r_t - \mu_t}{\sigma_t \alpha_t^2}, \quad t \in \mathbb{R}_+.$$

Then, the unique EMM is the probability Q *such that* $E\left[\frac{dQ}{dP} \mid \mathcal{F}_t\right] = L_t$, $t \in \mathbb{R}_+$, *where* $dL_t = L_{t^-}\psi_t dM_t$, $t \in \mathbb{R}_+$, $L_0 = 1$.

Proof. We have $dS_t = S_{t^-}\left(\mu_t dt + \sigma_t dZ_t + \sigma_t\psi_t\alpha_t^2 dt\right)$ and

$$\mu_t + \sigma_t\psi_t\alpha_t^2 = r_t, \quad t \in \mathbb{R}_+,$$

hence $dS_t = S_{t^-}(r_t dt + \sigma_t dZ_t)$ where $(Z_t)_{t\in\mathbb{R}_+}$, defined in Proposition 5.1, is a martingale under Q. □

If $(\psi_t)_{t\in\mathbb{R}_+}$ is defined as in Proposition 6.1, then

$$\begin{aligned} S_t &= S_0 \exp\left(\int_0^t \sigma_s\alpha_s i_s d\tilde{B}_s + \int_0^t (r_s - \phi_s\lambda_s\sigma_s(1+\phi_s\psi_s))ds - \frac{1}{2}\int_0^t i_s\sigma_s^2\alpha_s^2 ds\right) \\ &\quad \times \prod_{k=1}^{k=N_t} (1 + \sigma_{T_k}\phi_{T_k}), \quad t \in \mathbb{R}_+. \end{aligned}$$

where $\tilde{B}$ is a standard Q-Brownian motion. In the following, the probability Q will be the equivalent martingale measure constructed from the above proposition.

6.2. Pricing of a call

In order to price a European option we compute $E_Q\left[e^{-TR_T}(S_T - K)^+\right]$, where $R_T = \frac{1}{T}\int_0^T r_s ds$. (Given the uniqueness of the EMM, this is the only acceptable price of such an option, both for the buyer and the seller.) The process $(S_t)_{t\in\mathbb{R}_+}$ is a (not time homogeneous) Markov process. The price at time t is given by

$$C(t,x) = E_Q\left[\exp\left(-\int_t^T r_s ds\right)(S_T - K)^+ \mid S_t = x\right].$$

Let

$$\mathrm{BS}(x, T; r, \sigma^2; K) = E[e^{-rT}(xe^{rT-\sigma^2 T/2+\sigma W_T} - K)^+]$$

denote the classical Black-Scholes function, where W_T is a Gaussian centered random variable with variance T. In the case of deterministic volatility $(\sigma_s)_{s\in\mathbb{R}_+}$ and interest rate $(r_s)_{s\in\mathbb{R}_+}$, the price of a call in the Black-Scholes model is

$$\mathrm{BS}(x,T;R_T,\frac{1}{T}\Sigma_T;K), \quad \text{with} \quad \Sigma_T=\int_0^T \sigma_s^2 ds.$$

Let $\Gamma_t^\sigma = \int_0^t i_s\alpha_s^2\sigma_s^2 ds$ denote the variance of $\int_0^t i_s\alpha_s\sigma_s dB_s$, $t\in\mathbb{R}_+$, and $\Gamma_t = \int_0^t \gamma_s ds$, $t\in\mathbb{R}_+$, denote the intensity of $(N_t)_{t\in\mathbb{R}_+}$ under Q, where $\gamma_t=\lambda_t(1+\phi_t\psi_t)$, $t\in\mathbb{R}_+$.

Proposition 6.2. *The expectation* $E_Q\left[\exp\left(-\int_0^T r_s ds\right)(S_T-K)^+\right]$ *can be computed as*

$$E_Q\left[\exp\left(-\int_0^T r_s ds\right)(S_T-K)^+\right] = \exp\left(-\Gamma_T\right)\sum_{k=0}^{\infty}\frac{1}{k!}\int_0^T\cdots\int_0^T$$
$$BS\left(S_0\exp\left(-\int_0^T \phi_s\gamma_s\sigma_s ds\right)\prod_{i=1}^{i=k}\left(1+\sigma_{t_i}\phi_{t_i}\right),T;R_T,\frac{\Gamma_T^\sigma}{T};K\right)$$
$$\times\gamma_{t_1}\cdots\gamma_{t_k}dt_1\cdots dt_k.$$

Proof. We have

$$\begin{aligned} &E_Q\left[\exp(-TR_T)(S_T-K)^+\right] \\ &= \sum_{k=0}^{\infty} E_Q\left[\exp(-TR_T)(S_T-K)^+ \mid N_T=k\right]Q(N_T=k), \end{aligned}$$

with $Q(N_T=k)=\exp(-\Gamma_T)(\Gamma_T)^k/k!$, $k\in\mathbb{N}$. Conditionally to $\{N_T=k\}$, the jump times $(T_1,\ldots,T_k)$ have the law

$$\frac{k!}{(\Gamma_T)^k}1_{\{0<t_1<\cdots<t_k<T\}}\gamma_{t_1}\cdots\gamma_{t_k}dt_1\cdots dt_k,$$

since the process $(N_{\Gamma_t^{-1}})_{t\in\mathbb{R}_+}$ is a standard Poisson process. Hence, conditionally to $\{N(\Gamma^{-1}(\Gamma_T))=k\}=\{N_T=k\}$, its jump times $(\Gamma_{T_1},\ldots,\Gamma_{T_k})$ have a uniform law on $[0,\Gamma_T]^k$. We then use the fact that $(\tilde{B}_t)_{t\in\mathbb{R}_+}$ and $(N_t)_{t\in\mathbb{R}_+}$ are also independent under Q since $(\mu_t)_{t\in\mathbb{R}_+}$ is deterministic, and the identity in law

$$S_T \stackrel{law}{=} S_0X_T\exp\left(-\int_0^T\phi_s\lambda_s(1+\phi_s\psi_s)\sigma_s ds\right)\prod_{k=1}^{k=N_T}\left(1+\sigma_{T_k}\phi_{T_k}\right),$$

where

$$X_T=\exp\left(TR_T-\Gamma_T^\sigma/2+\left(\frac{\Gamma_T^\sigma}{T}\right)^{1/2}W_T\right),$$

and W_T is independent of N. □

6.3. The hedging strategy

An important problem is to determine the hedging strategy, i.e. to compute the processes (θ_t^0, θ_t^1) such that

$$C(t, S_t) = \theta_t^0 \exp\left(\int_0^t r_s ds\right) + \theta_t^1 S_t,$$

and

$$dC(t, S_t) = \theta_t^0 \exp\left(\int_0^t r_s ds\right) r_t dt + \theta_t^1 dS_t$$

(self-financing condition). It suffices to compute $(\theta_t^1)_{t\in\mathbb{R}_+}$ such that

$$d\left[\exp\left(-\int_0^t r_s ds\right) C(t, S_t)\right] = \theta_t^1 d\left[\exp\left(-\int_0^t r_s ds\right) S_t\right] \tag{16}$$

We shall note $\widetilde{S}_t = \exp\left(-\int_0^t r_s ds\right) S_t$, $t \in \mathbb{R}_+$, the discounted price process.

In the following proposition we still denote by D the gradient operator defined relatively to the multiple stochastic integrals with respect to $(Z_t)_{t\in\mathbb{R}_+}$ instead of $(M_t)_{t\in\mathbb{R}_+}$, ($(Z_t)_{t\in\mathbb{R}_+}$ also has the chaos representation property because $(\alpha_t^2(1 + \phi_t\psi_t))_{t\in\mathbb{R}_+}$ is deterministic). The hedging strategy is computed using the gradient D from the representation formula (10).

Proposition 6.3. *If $(r_s)_{s\in\mathbb{R}_+} = 0$, the hedging strategy is given by*

$$\theta_t^1 = \frac{1}{\sigma_t S_t} E[D_t(S_T - K)^+ \mid \mathcal{F}_t], \qquad t \in \mathbb{R}_+.$$

Proof. Let $F = (S_T - K)^+$. We have $dZ_t = (\sigma_t S_{t^-})^{-1} dS_t$, hence

$$\begin{aligned} F &= E[F] + \int_0^\infty E[D_t F \mid \mathcal{F}_t] dZ_t \\ &= E[F] + \int_0^T \frac{1}{\sigma_t S_{t^-}} E[D_t F \mid \mathcal{F}_t] dS_t, \end{aligned}$$

since F is $\mathcal{F}_T$-measurable. □

From Remark 4.4, the only regularity property that should be assumed on $F = (S_T - K)^+$ is its square-integrability, and this method can be applied to any square integrable pay-off F.

As in the standard Black-Scholes model, it is possible to determine the hedging strategy in terms of the "delta" of the price in the case $(r_t)_{t\in\mathbb{R}_+}$ is deterministic.

An application of Itô's lemma leads to

$$\begin{aligned} dC(t, S_t) &= \left[\frac{\partial C}{\partial t} + r_t S_t \frac{\partial C}{\partial x} + \frac{1}{2}\frac{\partial^2 C}{\partial x^2} i_t \alpha_t^2 S_t^2 \sigma_t^2 + \lambda_t(1 + \phi_t\psi_t)\Theta C\right](t, S_t) dt \\ &\quad + S_t \sigma_t \frac{\partial C}{\partial x}(t, S_t) dZ_t + \Theta C(t, S_t)\left[dN_t - \lambda_t(1 + \phi_t\psi_t) dt\right] \end{aligned} \tag{17}$$

where

$$\Theta C(t,S_t) = C(t,S_t(1+\sigma_t\phi_t)) - C(t,S_t) - \frac{\partial C}{\partial x}(t,S_t)S_t\sigma_t\phi_t.$$

(Note that here no smoothness property is required on $(\phi_t)_{t\in\mathbb{R}_+}$.) The process $\widetilde{C}_t = C(t,S_t)\exp\left(-\int_0^t r_s ds\right)$ is a Q-martingale, and $d\widetilde{C}_t = \theta_t^1 d\widetilde{S}_t$; therefore, by identification of (16) and (17),

$$\begin{aligned} r_t C(t,S_t) &= \left[\frac{\partial C}{\partial t} + r_t S_t \frac{\partial C}{\partial x} + \frac{1}{2} i_t \alpha_t^2 S_t^2 \sigma_t^2 \frac{\partial^2 C}{\partial x^2} + \lambda_t(1+\phi_t\psi_t)\Theta C\right](t,S_t), \\ \theta_t^1 S_t \sigma_t dZ_t &= S_t\sigma_t \frac{\partial C}{\partial x}(t,S_t)dZ_t + \Theta C(t,S_t)\left[dN_t - \lambda_t(1+\phi_t\psi_t)dt\right]. \end{aligned}$$

Therefore, by identification of the Brownian and Poisson parts,

$$\begin{cases} \theta_t^1 S_t\sigma_t i_t = S_t\sigma_t \frac{\partial C}{\partial x}(t,S_t) i_t \\ \theta_t^1 S_t\sigma_t\phi_t = S_t\sigma_t \frac{\partial C}{\partial x}(t,S_t)\phi_t + \Theta C(t,S_t). \end{cases} \tag{18}$$

On $\{t\in\mathbb{R}_+ \ :\ \phi_t = 0\} = \{t \ :\ i(t)=1\}$, the term $\Theta C(t,S_t)$ vanishes. Therefore, (18) reduces to

$$\theta_t^1 = \frac{\partial C}{\partial x}(t,S_t) + \frac{\Theta C(t,S_t)}{S_t\sigma_t\phi_t}.$$

The process $(\theta_t^1)_{t\in\mathbb{R}_+}$ is equal to the usual "delta" on the set $\{t\in\mathbb{R}_+ \ :\ i_t = 1\}$ and to $\frac{C(t,S_t(1+\phi_t\sigma_t)) - C(t,S_t)}{S_t\phi_t\sigma_t}$ on the set $\{t\in\mathbb{R}_+ \ :\ i_t = 0\}$.
The usual PDE for the price of a call splits into two parts. On $\{t : \phi(t) = 0\}$, we obtain the usual PDE

$$\frac{\partial C}{\partial t}(t,x) + r_t x\frac{\partial C}{\partial x}(t,x) + \frac{1}{2}\alpha_t^2 x^2\sigma_t^2\frac{\partial^2 C}{\partial x^2}(t,x) = r_t C(t,x),$$

whereas on the set $\{t\in\mathbb{R}_+ \ :\ \phi_t \neq 0\}$ we have

$$\frac{\partial C}{\partial t}(t,x) + r_t x\frac{\partial C}{\partial x}(t,x) + \lambda_t(1+\phi_t\psi_t)\Theta C(t,x) = r_t C(t,x).$$

Each equation has to be solved using a terminal condition, which is computed by backward induction from the condition at maturity $C(T,x) = (x-K)^+$.

6.4. The greeks

The price of the European call is still convex and increasing with respect to the value of the underlying asset. Here a new "greek" parameter has to be defined to make precise the dependence w.r.t. the drift of the asset, under the historical probability, i.e. $\Upsilon = \frac{\partial C}{\partial \mu}(t,x)$ in the case where this drift is constant.

However, it turns out that the sign of this term depends on all the parameters of the model.

7. Conclusion

The predictable representation property and the existence of an EMM give to this model the main properties for mathematical finance's purpose. For example, it is possible to solve a consumption/investment problem, and to use the backward stochastic differential equation tools. It remains to find a method to fit the parameters to some data. The continuous part of the driving martingale can have an arbitrary (deterministic) quadratic variation. Concerning the discontinuous part, its jumps as well as its intensity can be independently chosen as deterministic functions of time. This allows to model different types of transitions from continuous to discontinuous paths, for example starting with small jumps and great intensity. Transition dates between different regimes need to be known in advance. Nevertheless the behavior of the model is still random, conditionally to such information. A non-deterministic switching schedule is also possible, provided its randomness is independent of $(M_t)_{t\in\mathbb{R}_+}$.

Remark 7.1. Explicit formulas for the hedging strategies of Proposition 6.3 have been recently obtained in [1] in the Poisson case and in [5] for solutions of deterministic structure equations and for the Azéma martingales.

References

[1] K. Aase, B. Øksendal, N. Privault, and J. Ubøe, *White noise generalizations of the Clark-Haussmann-Ocone theorem with application to mathematical finance*, Finance and Stochastics, **4 (4)** (2000), 465–496.

[2] J. M. C. Clark, *The representation of functionals of Brownian motion by stochastic integrals*, Annals of Mathematical Statistics, **41** (1970), 1281–1295.

[3] C. Dellacherie, B. Maisonneuve, and P. A. Meyer, *Probabilités et Potentiel*, volume 5. Hermann, **1992**.

[4] M. Dritschel and Ph. Protter, *Complete markets with discontinuous security price*, Finance and Stochastics, **3 (2)**, 1999.

[5] Y. El-Khatib and N. Privault, *Computations of replicating portfolios in complete markets driven by normal martingales*, Preprint, 2000.

[6] M. Émery, *On the Azéma martingales*, In *Séminaire de Probabilités XXIII*, volume 1372 of Lecture Notes in Mathematics, Springer Verlag, 1990, 66–87.

[7] J. Jacod, *Calcul stochastique et problèmes de martingales*, volume 714 of Lecture Notes in Mathematics, Springer Verlag, **1979**.

[8] V. Lacoste, *Wiener chaos: a new approach to option hedging*, Math. Finance, **6 (2)** (1996), 197–213.

[9] R. Merton, *Option pricing when underlying stock returns are discontinuous*, J. of Financial Economics, **3** (1976).

[10] P.-A. Meyer and J. A. Yan, *À propos des distributions sur l'espace de Wiener*, in *Séminaire de Probabilités XXI*, volume 1247 of Lecture Notes in Mathematics, Springer Verlag, 1987, 8–26.

[11] D. Ocone, *A guide to the stochastic calculus of variations*, in H. Körezlioğlu and A.S. Üstünel, editors, Stochastic Analysis and Related Topics, volume 1316 of Lecture Notes in Mathematics, Springer-Verlag, 1988.

[12] N. Privault, *Independence of a class of multiple stochastic integrals*, in *Seminar on Stochastic Analysis, Random Fields and Applications (Ascona, 1996)*, Birkhäuser, Basel, 1999, 249–259.

[13] A. N. Shiryaev, *Essentials of stochastic finance*, World Scientific Publishing Co. Inc., River Edge, NJ, **1999**.

[14] A. S. Üstünel, *Representation of the distributions on Wiener space and stochastic calculus of variations.* J. Funct. Anal., **70** (1987), 126–139.

Département de Mathématiques
Université d'Evry
Boulevard F. Mitterrand
91025 Evry Cedex
France
E-mail address: `Monique.Jeanblanc@maths.univ-evry.fr`

Département de Mathématiques
Université de la Rochelle
Avenue Marillac
17042 La Rochelle Cedex 1
France
E-mail address: `Nicolas.Privault@univ-lr.fr`

Progress in Probability, Vol. 52, 205–217

Stochastic Calculus and Processes in Non-Commutative Space-Time

R. Vilela Mendes

Abstract. The algebra of relativistic quantum mechanics is unstable. Its stabilization requires the non-commutativity of the space-time coordinates and a fundamental length. The Heisenberg algebra is replaced by the algebras of ISO(2) and ISO(1,1). A quantum stochastic calculus is developed for these algebras. Also discussed is the construction of stochastic processes when time is an operator not commuting with the coordinates.

1. Introduction

A mathematical structure is said to be *stable* (or *rigid*) for a class of *deformations* if any deformation in this class leads to an equivalent (isomorphic) structure. Likewise a physical theory is *stable* if it does not change in a qualitative manner for a small change of parameters. If the mathematical structure of a theory is unstable, by deforming it until one falls into a stable structure, one has the chance of obtaining a generalization of wider validity.

The stable-model point of view had a large impact in the field of non-linear dynamics, where it led to the notion of *structural stability* [1, 2]. As emphasized by Flato [3] and Faddeev [4] the same pattern seems to occur in the fundamental theories of Nature. In fact, the two most important physical revolutions of this century, namely the passage from non-relativistic to relativistic and from classical to quantum mechanics, may be interpreted as the transition from two unstable theories to two stable ones.

The algebra of relativistic quantum mechanics $\Re_0$ contains the Lorentz algebra $\{M_{\mu\nu}\}$, the Heisenberg algebra for the momenta and space-time coordinates in Minkowski space $\{p_\mu, x_\nu\}$ and also the commutators that define the vector nature (under the Lorentz group) of p_μ and x_ν. This algebra turns out to be unstable as well [6] and there is a 2-parameter (ℓ, R)-deformation of $\Re_0$ to a simple algebra $\Re_{\ell,R}$ which is stable.

The R-deformation means simply that flat space is an isolated point in the set of arbitrarily curved spaces and, as long as the Poincaré group is used as the

---

2000 *Mathematics Subject Classification.* 60G20, 60H99.
Key words and phrases. Quantum stochastic calculus, non-commutative space-time.

kinematical group of the tangent space to the space-time manifold and not as a group of motions in the manifold itself, it is perfectly consistent to take $R \to \infty$ thus removing the R-deformation.

For the other deformation parameter (ℓ) there is no reason to imagine that it should vanish, even in tangent space, if one insists on the stability paradigm as a guiding principle for theory construction. One is therefore led to $\Re_{\ell,\infty}$ as our candidate for a *stable algebra of relativistic quantum mechanics* in the tangent space, namely

$$\begin{aligned}
[M_{\mu\nu}, M_{\rho\sigma}] &= i(M_{\mu\sigma}\eta_{\nu\rho} + M_{\nu\rho}\eta_{\mu\sigma} - M_{\nu\sigma}\eta_{\mu\rho} - M_{\mu\rho}\eta_{\nu\sigma}), \\
[M_{\mu\nu}, p_\lambda] &= i(p_\mu\eta_{\nu\lambda} - p_\nu\eta_{\mu\lambda}), \\
[M_{\mu\nu}, x_\lambda] &= i(x_\mu\eta_{\nu\lambda} - x_\nu\eta_{\mu\lambda}), \\
[p_\mu, p_\nu] &= 0, \\
[x_\mu, x_\nu] &= -i\epsilon\ell^2 M_{\mu\nu}, \\
[p_\mu, x_\nu] &= i\eta_{\mu\nu}\Im, \\
[p_\mu, \Im] &= 0, \\
[x_\mu, \Im] &= i\epsilon\ell^2 p_\mu, \\
[M_{\mu\nu}, \Im] &= 0,
\end{aligned} \tag{1}$$

ϵ being a $\pm$ sign, $\eta_{\mu\nu} = (1,-1,-1,-1)$, $c = \hbar = 1$ and $\Im$ a non-trivial operator that replaces the unit in the Heisenberg algebra. The main features are the non-commutativity of the x_μ coordinates and the fact that $\Im$, previously a trivial element in the center of the Heisenberg algebra, becomes now a non-trivial operator. Two constants define this deformation. One is ℓ, a fundamental length, the other the sign of ϵ. The fact that $\Im$ becomes a non-trivial operator changes the structure of the Heisenberg algebra which is replaced by the algebras of ISO(2) and ISO(1,1). This has some consequences on the construction of the state spaces even for nonrelativistic quantum mechanics. This was partly discussed in [7]. Here the emphasis will be on the stochastic calculus aspects of this algebraic structure.

Non-commutative space-time and its associate algebra constrain the type of stochastic analysis structures that may be developed in this setting. On the one hand is fact that the Heisenberg algebra is replaced by the algebras of ISO(2) and ISO(1,1). Therefore the usual boson calculus is modified. However this is not the only modification that arises from the non-commutative space-time structure. The new (ISO(1,1) and ISO(2)-based) boson calculus that is developed in Section 3 corresponds to stochastic processes that are sums of independent identically distributed random variables, the "time" of the process being simply a continuous parameter that labels the convolution semigroup. If however time is an operator with well-defined algebraic relations to the other observables, the construction has to be done in a different manner. This is discussed in Section 4.

The representation theory of algebras is the basic tool to extract physical consequences from a non-commutative geometry. This is discussed in Section 2.

2. Representations

Explicit representations of the subalgebras of $\Re_{\ell,\infty}$ in spaces of functions are the tools needed to compute the physical consequences of this type of non-commutative space-time. Here one studies in detail a few cases, starting from the representations of the 3-dimensional subalgebras that replace Heisenberg's algebra.

Consider the subalgebra associated to one-dimensional problems, that is

$$\begin{array}{rcl} [P,X] & = & -i\Im, \\ [X,\Im] & = & i\varepsilon P, \\ [P,\Im] & = & 0, \end{array} \tag{2}$$

where $P = p\ell$ and $X = \frac{x}{\ell}$. In these variables, the position is measured in units of ℓ and the momentum in units of $\frac{1}{\ell}$.

Let $\varepsilon = -1$. Then (2) is the algebra of the group of motions of the plane, ISO(2). Its irreducible representations T_r [8] are realized as operators on the space of smooth functions on S^1 with scalar product

$$(f_1, f_2) = \frac{1}{2\pi}\int_0^{2\pi} f_1(\theta) f_2^*(\theta) d\theta,$$

the operators being

$$\begin{array}{rcl} X & = & i\frac{\partial}{\partial\theta}, \\ P & = & r\sin\theta, \\ \Im & = & r\cos\theta. \end{array} \tag{3}$$

The irreducible representations are of two types. For $r \neq 0$ the irreducible representation T_r is infinite dimensional, a convenient basis being the set of exponentials $\exp(-in\theta)$

$$T_r = \left\{e^{-in\theta}; n \in Z\right\}$$

and for $r = 0$ the irreducible representations are one-dimensional

$$T_{0n} = \left\{e^{-in\theta}\right\}.$$

In T_r the operators $V_+ = iP + \Im$ and $V_- = -iP + \Im$ are raising and lowering operators

$$\begin{array}{rcl} V_+ e^{-in\theta} & = & r\, e^{-i(n-1)\theta}, \\ V_- e^{-in\theta} & = & r\, e^{-i(n-1)\theta}. \end{array}$$

Being $e^{-in\theta}$ the eigenstates of the position operator X, this one has a discrete spectrum ($= Z$ for X or $= \ell Z$ for x). The representation with $r = 0$ would correspond to a space with a single isolated point. Therefore, only the representations with $r \neq 0$ are physically useful. ℓ being the minimal fundamental length, the maximum momentum, in units of $\frac{1}{\ell}$, is one. Hence, for consistency with (3) we may choose $r = 1$.

For each localized state $e_n = \frac{1}{\sqrt{2\pi}} e^{-in\theta}$, P is a random variable with characteristic function

$$C(s) = \langle e_n, e^{isP} e_n\rangle = J_0(sr),$$

the corresponding probability density being

$$\begin{aligned} \mu(P) &= \tfrac{1}{\pi}\tfrac{1}{\sqrt{r^2-P^2}} && |P|<r, \\ &= 0 && |P|>r. \end{aligned}$$

Based on the operators of the Heisenberg algebra, a boson calculus has been developed by several authors [9, 10, 11, 12]. For $\ell \neq 0$ (and $\epsilon = -1$) the Heisenberg algebra is replaced by the algebra of ISO(2). For the calculus based on this algebra it is useful to represent it as a set of operators acting on a space of holomorphic functions

$$\begin{aligned} X &= z, \\ P &= \tfrac{1}{2i}\left(e^{\frac{\partial}{\partial z}} - e^{-\frac{\partial}{\partial z}}\right) &&= \tfrac{1}{i}\Delta_-, \\ \Im &= \tfrac{1}{2}\left(e^{\frac{\partial}{\partial z}} + e^{-\frac{\partial}{\partial z}}\right) &&= \Delta_+. \end{aligned}$$

Let T_α be the translation operator by α,

$$T_\alpha f(z) = f(z+\alpha).$$

Then $\Delta_- = \frac{1}{2}(T_1 - T_{-1})$ is a finite difference operator and $\Delta_+ = \frac{1}{2}(T_1 + T_{-1})$ a finite average operator. Therefore, instead of x and $\frac{d}{dx}$ for the Heisenberg algebra, the ISO(2) boson calculus is based on z, Δ_+, Δ_- and the relations

$$\begin{aligned} [\Delta_-, z] &= \Delta_+, \\ [\Delta_+, z] &= \Delta_-, \\ [\Delta_+, \Delta_-] &= 0. \end{aligned}$$

On the other hand, with the choice $\epsilon = -1$, the algebra for the pair $P^0 = \ell p^0$, $X^0 = \frac{x^0}{\ell}$ is the algebra of ISO(1,1)

$$\begin{aligned} [P^0, X^0] &= i\Im, \\ [X^0, \Im] &= -iP^0, \\ [P^0, \Im] &= 0, \end{aligned}$$

and, in this case, the representation as operators acting on differentiable functions on the hyperbola is

$$\begin{aligned} P^0 &= r\sinh\mu, \\ \Im &= r\cosh\mu, \\ X^0 &= -i\tfrac{d}{d\mu}. \end{aligned} \tag{4}$$

Generalized eigenvalues of the time operator are $e^{it\mu}$. Because there is no μ-periodicity in the hyperbola, there is no discrete quantization of time, as opposed to the discrete quantization of the space coordinate. This conclusion, of course, depends on the choice $\epsilon = -1$. The opposite situation would hold for $\epsilon = +1$.

Next the main steps of a calculus based on ISO(2) and ISO(1,1) will be described.

3. Quantum stochastic calculus based on ISO(1,1) and ISO(2)

Consider first the ISO(1,1) case. In the representation (4) replacing $-i\frac{d}{d\mu}$ by X^0 and μ by $i\frac{d}{dX^0}$ one obtains

$$\begin{aligned} P^0 &= \tfrac{1}{2}\left(e^{i\frac{d}{dX^0}} - e^{-i\frac{d}{dX^0}}\right) &= \tfrac{1}{2}\left(T_i - T_{-i}\right) &= i\sin\left(\tfrac{d}{dX^0}\right), \\ \Im &= \tfrac{1}{2}\left(e^{i\frac{d}{dX^0}} + e^{-i\frac{d}{dX^0}}\right) &= \tfrac{1}{2}\left(T_i + T_{-i}\right) &= \cos\left(\tfrac{d}{dX^0}\right). \end{aligned}$$

Operating on holomorphic functions of X^0, P and $\Im$ may be interpreted as a finite difference and as an averaging operator in the complex direction. Defining

$$\begin{aligned} \sin\left(\tfrac{d}{dX^0}\right) &\hat{=} D_-, \\ \cos\left(\tfrac{d}{dX^0}\right) &\hat{=} D_+, \end{aligned}$$

the commutation relations are

$$\begin{aligned} \left[D_-, X^0\right] &= D_+, \\ \left[D_+, X^0\right] &= -D_-, \\ \left[D_+, D_-\right] &= 0. \end{aligned}$$

The first step in the construction of a quantum stochastic calculus, based on this algebra, is the construction of the associated representation spaces. Let

$$\begin{aligned} A &= \tfrac{1}{\sqrt{2}}\left(X^0 + D_-\right), \\ A^\dagger &= \tfrac{1}{\sqrt{2}}\left(X^0 - D_-\right), \end{aligned} \tag{5}$$

with algebra

$$\begin{aligned} \left[A, A^\dagger\right] &= D_+, \\ \left[A, D_+\right] &= \tfrac{1}{2}\left(A - A^\dagger\right), \\ \left[A^\dagger, D_+\right] &= \tfrac{1}{2}\left(A - A^\dagger\right). \end{aligned}$$

The vacuum is chosen to be the vector that is annihilated by A

$$A\phi(X^0) = \frac{1}{\sqrt{2}}\left(X^0 + D_-\right)\phi(x^0) = 0,$$

which, in the Fourier transform, becomes

$$-i\left(\frac{d}{d\omega} + \sinh\omega\right)\phi(\omega) = 0,$$

yielding

$$\phi(\omega) = \frac{1}{\sqrt{N_\phi}}e^{-\cosh\omega} \tag{6}$$

with normalization factor

$$N_\phi = 2K_0(2) = 0.2277877...,$$

$K_\nu(z)$ being a modified Bessel function, and

$$\phi(X^0) = F^{-1}\left(\phi(\omega)\right) = \frac{1}{\sqrt{\pi K_0(2)}}K_{iX^0}(1).$$

A basis is obtained by acting on ϕ with powers of $A^\dagger$.

Lemma 3.1. *The set* $\{A^{\dagger n}\phi : n = 0, 1, 2, ...\}$ *is an orthogonal set.*

Proof. Assume that up to order n all $A^{\dagger n}\phi$ are known to be orthogonal to all states of the form $A^{\dagger}...A^{\dagger}D_{+}A^{\dagger}...A^{\dagger}\phi$ where $A^{\dagger}$ appears less than n times and D_{+} one time. Then for $\alpha \leq n$

$$\left(A^{\dagger\alpha}\phi, A^{\dagger n+1}\phi\right) = \left(A^{\dagger}AA^{\dagger\alpha-1}\phi, A^{\dagger n}\phi\right) = ... = \left(A^{\dagger\alpha}A\phi, A^{\dagger n}\phi\right) = 0.$$

The base for the induction is provided by $\left(\phi, A^{\dagger}\phi\right) = 0$ and $\left(D_{+}\phi, A^{\dagger}\phi\right) = \frac{1}{2}\left((A - A^{\dagger})\phi, \phi\right) = 0.$ □

The state ϕ may be used to define a probability distribution. An operator O in the enveloping algebra of $\{X^0, D_+, D_-\}$ becomes a random variable with expectation $(\phi, O\phi)$. In particular, for X^0 the characteristic function is

$$C_{X^0}(y) = \left(\phi, e^{iyX^0}\phi\right) = \frac{K_0(2\cosh\frac{y}{2})}{K_0(2)} = \int e^{iyx}p(dx). \tag{7}$$

The state ϕ is the analog of the harmonic oscillator ground state in the Heisenberg algebra case. Although this is the state that will always be used to define the probability structure, for the construction of the stochastic process it is more convenient to use a different basis. Define

$$\begin{aligned} H_+ &= P^0 + \Im, \\ H_- &= -P^0 + \Im. \end{aligned}$$

Then

$$\begin{aligned} \left[X^0, H_+\right] &= -iH_+, \\ \left[X^0, H_-\right] &= iH_-, \\ \left[H_+, H_-\right] &= 0, \end{aligned} \tag{8}$$

and, in the representation (4) with $r = 1$, one has the simple action

$$\begin{aligned} X^0\psi(\mu) &= -i\tfrac{d}{d\mu}\psi(\mu), \\ H_+\psi(\mu) &= e^{\mu}\psi(\mu), \\ H_-\psi(\mu) &= e^{-\mu}\psi(\mu), \end{aligned}$$

with the scalar product in the space V of square-integrable functions on the hyperbola defined by

$$(\phi, \psi)_{\mu} = \int d\mu\phi^*(\mu)\psi(\mu).$$

An equally simple representation is obtained by Fourier transform, namely

$$\begin{aligned} X^0F(\lambda) &= i\lambda F(\lambda), \\ H_+F(\lambda) &= F(\lambda + 1), \\ H_-F(\lambda) &= F(\lambda - 1). \end{aligned}$$

Let now $h = L^2(R_+)$ be the Hilbert space of square-integrable functions on the half-line $R_+ = [0, \infty]$. It is the variable in R_+ that will be the continuous index s labelling the convolution semigroup generated by the probability distribution $p(dx)$ in (7). It is interpreted as the time parameter of a stochastic process, sum of independent identically distributed random variables.

From (8) one constructs an infinite set of operators labelled by functions in h, with algebra

$$\begin{aligned}\left[X^0(f), H_+(g)\right] &= -iH_+(fg), \\ \left[X^0(f), H_-(g)\right] &= iH_-(fg), \\ \left[H_+(f), H_-(g)\right] &= 0,\end{aligned} \tag{9}$$

$f, g \in h$.

These operators are made to act on a space that is a direct integral H of spaces of functions on the hyperbola indexed by functions on h, namely

$$\psi(f) \stackrel{\circ}{=} \oplus \int d\tau f(\tau)\psi(\mu_\tau)$$

with scalar product

$$\langle \phi(f_1), \psi(f_2) \rangle = (f_1, f_2)_h (\phi, \psi)_\mu ,$$

where, in the right-hand side, the first scalar product is in h and the second in V. For $f, g \in h$ the action of the operators is

$$\begin{aligned}X^0(f)\psi(g) &= -i\partial_\mu \psi(fg), \\ H_+(f)\psi(g) &= \left(e^\mu \psi\right)(fg), \\ H_-(f)\psi(g) &= \left(e^{-\mu}\psi\right)(fg).\end{aligned} \tag{10}$$

This action satisfies the commutation relations (9).

Notice that other functional realizations of the commutation relations are possible. For example, the operators may be made to act on a space of square-integrable functionals over h in the following way

$$\begin{aligned}X^0(f)F(g) &= -iD_f F(g), \\ H_+(f)F(g) &= (\overline{f}, e^g)F(g), \\ H_-(f)F(g) &= (\overline{f}, e^{-g})F(g).\end{aligned}$$

However in this case the difficulty is in finding a scalar product for which the operators are symmetric. Therefore, only the direct integral construction will be used here.

To define adapted processes the usual splittings are considered

$$\begin{aligned}h &= L^2[0,\infty) = L^2[0,s] \oplus L^2(s,\infty) = h^s \oplus h^{(s}, \\ H &= H^s \otimes H^{(s},\end{aligned}$$

and an *adapted process* is a family $K = (K(s), s \geq 0)$ of operators in H such that for each t

$$K(s) = K^s \otimes 1.$$

The basic adapted processes are

$$\begin{aligned}X^0(s) &= X^0(\chi_{[0,s]}), \\ H_+(s) &= H_+(\chi_{[0,s]}), \\ H_-(s) &= H_-(\chi_{[0,s]}),\end{aligned}$$

$\chi_{[0,s]}$ being the indicator function of the interval $[0, s]$. Given an elementary process $E_{s_1 s_2}$, that is, a process that is constant in the interval $[s_1, s_2]$ and zero otherwise, the stochastic integral is

$$\int_0^s E_{s_1 s_2} dK = E_{s_1 s_2} \{K(\min(s, s_2)) - K(\min(s, s_1))\}.$$

The stochastic integral of a general adapted process is obtained by approximation by elementary processes and a limiting procedure. The following two propositions characterize the properties of the stochastic integral in the ISO(1,1) stochastic calculus.

Proposition 3.2. *For adapted process E_0, E_+, E_-, $f, g \in h$ and $\phi(f), \psi(g) \in H$,*

$$\begin{aligned}&\left\langle \phi(f), \textstyle\int_0^t \{E_0 dX^0 + E_+ dH_+ + E_- dH_-\} \psi(g) \right\rangle = \\ &= \textstyle\int_0^t \langle \phi(f), -iE_0(s)(\partial_\mu \psi)(g\chi_{ds}) + E_+(s)(e^{\mu}\psi)(g\chi_{ds}) + E_-(s)(e^{-\mu}\psi)(g\chi_{ds})\rangle ,\end{aligned} \tag{11}$$

where χ_{ds} denotes the indicator function of the interval $[s, s+ds]$. The proof follows from application of the operator action in (10). *Using the properties of the scalar product in H,* (11) *may be rewritten as*

$$-i(\phi, \partial_\mu \psi)_\mu (f, E_0 g)_h + (\phi, e^{\mu}\psi)_\mu (f, E_+ g)_h + (\phi, e^{-\mu}\psi)_\mu (f, E_- g)_h.$$

Proposition 3.3. *For adapted process E_0, E_+, E_- define*

$$N_i(t) = \int_0^t \left\{ E_0^{(i)} dX^0 + E_+^{(i)} dH_+ + E_-^{(i)} dH_- \right\}.$$

Then, for $f, g \in h$ and $\phi(f), \psi(g) \in H$

$$\begin{aligned}&\langle N_1(t)\phi(f), N_2(t)\psi(g)\rangle \\ &= \textstyle\int_0^t \Big\langle N_1(t)\phi(f), \Big\{ -iE_0^{(2)}(s)(\partial_\mu\psi)(g\chi_{ds}) + E_+^{(2)}(s)(e^{\mu}\psi)(g\chi_{ds}) \\ &\qquad\qquad + E_-^{(2)}(s)(e^{-\mu}\psi)(g\chi_{ds}) \Big\} \Big\rangle \\ &+ \textstyle\int_0^t \Big\langle \Big\{ -iE_0^{(1)}(s)(\partial_\mu\phi)(f\chi_{ds}) + E_+^{(1)}(s)(e^{\mu}\phi)(f\chi_{ds}) \\ &\qquad\qquad + E_-^{(1)}(s)(e^{-\mu}\phi)(f\chi_{ds}) \Big\}, N_2(t)\psi(g) \Big\rangle \\ &- \tfrac{i}{2} \textstyle\int_0^t (\phi, e^{\mu}\psi)_\mu \Big\{ \overline{E}_0^{(1)}(s)\overline{f}(s)E_+^{(2)}(s)g(s) - \overline{E}_+^{(1)}(s)\overline{f}(s)E_0^{(2)}(s)g(s) \Big\} ds \\ &+ \tfrac{i}{2} \textstyle\int_0^t (\phi, e^{-\mu}\psi)_\mu \Big\{ \overline{E}_0^{(1)}(s)\overline{f}(s)E_-^{(2)}(s)g(s) - \overline{E}_-^{(1)}(s)\overline{f}(s)E_0^{(2)}(s)g(s) \Big\} ds.\end{aligned} \tag{12}$$

The proof follows from splitting the integral in its upper and lower triangular regions plus the diagonal and using the commutation relations (9) to compute the diagonal terms. The diagonal terms (the last two terms in (12)) contain the Ito-type corrections for this stochastic calculus which may be written symbolically as

$$d(N_1 N_2) = dN_1 \cdot N_2 + N_1 \cdot dN_2 + dN_1 \cdot dN_2,$$

with multiplication rules

$$\begin{array}{rclcl} dX^0 \cdot dH_+ & = & -dH_+ \cdot dX^0 & = & -\frac{i}{2} dH_+, \\ dX^0 \cdot dH_- & = & -dH_- \cdot dX^0 & = & \frac{i}{2} dH_-, \end{array}$$

all others being identically zero.

The probability structure of the process is defined by the choice of the state that is used to compute expectations. For this purpose one uses a direct sum of states of type (6) which are annihilated by the operator A of (5)

$$\Omega = \oplus \int d\tau \phi_\tau = \oplus \int d\tau \frac{1}{N_\phi} e^{-\cosh \mu_\tau}.$$

This state belongs to the space H and splits as follows

$$\oplus \int d\tau \chi_{[0,s]} \phi_\tau + \oplus \int d\tau \chi_{[s,\infty)} \phi_\tau$$

in the decomposition $H = H^s \otimes H^{(s}$. The triple $(H, \{O_{\text{adap.}}\}, \Omega)$, where $\{O_{\text{adap.}}\}$ denotes the set of adapted operators over H, is the (non-commutative) probability space associated to ISO(1,1). For $X^0(f)$, for example, the characteristic functional is

$$C(f) = \left\langle \Omega, e^{iX^0(f)} \Omega \right\rangle = \exp \left\{ \int d\tau \log \frac{K_0\left(2 \cosh \frac{f(\tau)}{2}\right)}{K_0(2)} \right\}.$$

The construction developed so far refers to a stochastic calculus based on the ISO(1,1) algebraic structure. For the ISO(2) case, a similar construction may be carried out, which is now sketched. In this case the X, P and $\Im$ operators correspond to finite difference operators

$$\begin{array}{rcl} \Delta_- f(x) & = & \frac{1}{2} \left(f(x+1) - f(x-1) \right), \\ \Delta_+ f(x) & = & \frac{1}{2} \left(f(x+1) + f(x-1) \right), \end{array}$$

and to the representation

$$\begin{array}{rcl} X & = & x, \\ P & = & \frac{1}{i} \Delta_-, \\ \Im & = & \Delta_+, \end{array}$$

with associated operators

$$\begin{array}{rcl} B & = & \frac{1}{\sqrt{2}} \left(x + \Delta_- \right), \\ B^\dagger & = & \frac{1}{\sqrt{2}} \left(x - \Delta_- \right). \end{array}$$

An alternative representation as operators acting on functions on the circle is

$$\begin{array}{rcl} X & = & i \frac{d}{d\theta}, \\ P & = & \sin \theta, \\ \Im & = & \cos \theta. \end{array} \tag{13}$$

X has a discrete spectrum, with eigenvectors $e_n = \frac{1}{\sqrt{2\pi}} e^{in\theta}$, and, when this basis is used, states are in $\ell^2(Z)$. The state that is annihilated by B is

$$\phi_0 = \sum_n c_n^{(0)} e_n,$$

with $c_n^{(0)} \sim I_n(1)$, I_n being the modified Bessel function. In the representation (13)

$$\phi_0 = \frac{1}{\sqrt{N_\phi}} e^{\cos\theta},$$

N_ϕ being the normalization factor $2\pi I_0(2)$.

As in the ISO(1,1) case, the set $\{B^{\dagger n}\phi_0\}$ is an orthogonal set. However for the construction of the processes it is more convenient to use arbitrary square-integrable functions on the circle and the representation (13).

X is a random variable with characteristic function

$$C(s) = \langle \phi_0, e^{isX}\phi_0 \rangle = \frac{I_0\left(2\cos\frac{s}{2}\right)}{I_0(2)}.$$

To construct processes and a stochastic calculus, the commutation relations are lifted to an infinite set indexed by functions on the circle

$$\begin{aligned} [X(f), V_+(g)] &= -V_+(fg), \\ [X(f), V_-(g)] &= V_-(fg), \\ [V_+(f), V_-(g)] &= 0, \end{aligned}$$

where V_+ and V_- correspond to $\Delta_+ + \Delta_-$ and $\Delta_+ - \Delta_-$. These operators are made to act in a direct sum space H' of functions on the circle

$$\begin{aligned} X(f)\psi(g) &= i\partial_\theta\psi(fg), \\ V_+(f)\psi(g) &= \left(e^{i\theta}\psi\right)(fg), \\ V_-(f)\psi(g) &= \left(e^{-i\theta}\psi\right)(fg), \end{aligned}$$

and the construction follows the same steps as before.

4. Stochastic processes in non-commutative space-time

As stated in the introduction, the stochastic processes constructed in Section 3 are sums of independent identically distributed random variables associated to the operator sets $\{X^0, P^0, \Im\}$ and $\{X^i, P^i, \Im\}$. Therefore, the "time" label of the process is simply the continuous parameter of the probability convolution semi-group. If however time is an operator that satisfies well-defined algebraic relations with the other observables, as in the (1), the construction has to be done in a different manner. The notion of filtration, in particular, cannot be obtained simply by a splitting of the indexing space h. It must be replaced by a construction of the spaces of eigenstates of the time operator. Physically the treatment of time as a parameter still makes sense if the time scale of the processes is slow (remember that $x_0 = ct$ and then $[t, x_i] = \frac{i\ell^2}{c} M_{oi}$). However for processes with a fast time scale, a construction where time is treated as an operator is needed.

To describe time-dependent processes one needs at least one space and one time coordinate. Therefore, the minimal algebra is $\{x^0, x^1, M^{01}, p^0, p^1, \Im\}$ which, for $\varepsilon = -1$, is the algebra of $ISO(2,1)$, the group of motions of pseudo-Euclidean 3-space $E_{2,1}$. Representations may be realized on the space of functions on the double- or single-sheeted hyperboloid H_+^2 or H_-^2 and on the cone C^2, with coordinates, respectively

$$\left\{ \begin{array}{lcl} \xi_1 & = & \sinh \mu \sin \theta \\ \xi_2 & = & \sinh \mu \cos \theta \\ \xi_3 & = & \cosh \mu \end{array} \right\},$$

$$\left\{ \begin{array}{lcl} \xi_1 & = & \cosh \mu \sin \theta \\ \xi_2 & = & \cosh \mu \cos \theta \\ \xi_3 & = & \sinh \mu \end{array} \right\},$$

$$\left\{ \begin{array}{lcl} \xi_1 & = & r \sin \theta \\ \xi_2 & = & r \cos \theta \\ \xi_3 & = & r \end{array} \right\}.$$

Here representations on the space of functions on the upper sheet of the cone C^2 will be chosen. The reason for this choice is to have positive energy but no minimal nonzero energy. Then

$$\begin{array}{lcl} \ell p^1 & = & r \sin \theta, \\ \Im & = & r \cos \theta, \\ \ell p^0 & = & r, \\ \frac{x^0}{\ell} & = & -i \left\{ -\sin \theta \frac{\partial}{\partial \theta} + r \cos \theta \frac{\partial}{\partial r} \right\}, \\ \frac{x^1}{\ell} & = & i \frac{\partial}{\partial \theta}, \\ M^{01} & = & -i \left\{ \cos \theta \frac{\partial}{\partial \theta} + r \sin \theta \frac{\partial}{\partial r} \right\}, \end{array} \tag{14}$$

acting on functions $f(r,\theta)$ on C^2, square-integrable for the measure $dr d\theta$. Hermitean symmetry of the operators is obtained if either $r^{\frac{1}{2}} f(r,\theta) \to 0$ when $r \to \infty$ or $f(r,\theta) = r^{-\frac{1}{2}+i\rho} g(\theta)$. The last case corresponds to the principal series of $SO(2,1)$ representations.

One sees that the time and the space coordinates are noncommuting operators. Therefore, when describing a process, time cannot be simply considered a c-number parameter. Instead, to describe, for example, a stochastic process that at each fixed time may be sampled to find out the value of the space variable, what one has to do is to find the subspaces of time eigenvectors, corresponding to each fixed eigenvalue t. Then, in each such subspace, one has to find the possible values of x and their probabilities.

The eigenvectors of the time operator in (14) are obtained from

$$i \left\{ \sin \theta \frac{\partial}{\partial \theta} - r \cos \frac{\partial}{\partial r} \right\} f_t(r,\theta) = \frac{t}{\ell} f_t(r,\theta), \tag{15}$$

the solution being

$$f_t(r,\theta) = \left(\cot\frac{\theta}{2}\right)^{i\frac{t}{\ell}} g(r\sin\theta),$$

with $g(r\sin\theta)$ an arbitrary function of $r\sin\theta$. Now one considers the spectrum of possible values of the space coordinate x in each one of the subspaces spanned by the functions $f_t(r,\theta)$. The projection on the $e^{-in\theta}$ eigenstate of the position operator x (corresponding to the position $n\ell$) is

$$c_{-n}(t) = \frac{1}{2\pi}\int d\theta e^{in\theta}\left(\cot\frac{\theta}{2}\right)^{i\frac{t}{\ell}} g(r\sin\theta). \tag{16}$$

For a process that starts from $x = 0$ at time $t = 0$ one has $c_0(0) = 1$ and $c_n(0) = 0$ for $n \neq 0$. Therefore for such a process, $g(r\sin\theta) =$ constant. Strictly speaking a constant function is outside the L^2 domain of the operator and therefore one should consider the operator as acting on the generalized function space of a Gelfand triplet. With the choice $f_t(r,\theta) = \left(\cot\frac{\theta}{2}\right)^{i\frac{t}{\ell}}$ one obtains, by computing the integral (16)

$$c_0(t) = e^{-\frac{\pi}{2}\frac{t}{\ell}}.$$

For the other coefficients, they are more conveniently obtained by solving (15) in a $e^{-in\theta}$ basis,

$$f_t(r,\theta) = r^\sigma \sum_{n\in Z} c_n e^{in\theta},$$

which leads to the recurrence

$$(n-1-\sigma)c_{n-1} - (n+1+\sigma)c_{n+1} + i\frac{2t}{\ell}c_n = 0.$$

For $\sigma = 0$, using a recurrence relation for hypergeometric functions one obtains $c_n \sim \frac{1}{n}F(-n, -i\frac{t}{\ell}; 0; 2)$. Then

$$c_n(t) = \tfrac{1}{2}e^{in\frac{\pi}{2}}e^{-\frac{\pi}{2}\frac{t}{\ell}}P_n^0(\tfrac{t}{\ell};\tfrac{\pi}{2}) \qquad \text{for} \quad n \neq 0,$$

$P_n^0(\frac{t}{\ell};\frac{\pi}{2})$ being Pollaczek-Meixner polynomials with generating function

$$(1-it)^{ix}(1+it)^{-ix} = \sum_{n=0}^{\infty} P_n^0(x;\frac{\pi}{2})t^n.$$

Hence, without further restrictions on the dynamics, the process that at $t = 0$ starts from $x = 0$, has a probability to be found at $\pm n\ell$ at time t equal to

$$P(\pm n\ell, 0; \frac{t}{\ell}) = \frac{1}{4}e^{-\pi\frac{t}{\ell}}\left(P_n^0(\frac{t}{\ell};\frac{\pi}{2})\right)^2.$$

This probability distribution corresponds to the analog of Brownian motion in the non-commutative one-time one-space setting.

References

[1] A. Andronov and L. Pontryagin, *Systmes Grossiers*, Dokl. Akad. Nauk. USSR, **14** (1937), 247–251.

[2] S. Smale, *Differentiable dynamical systems*, Bulletin of the AMS, **73** (1967), 747–817.

[3] M. Flato, *Deformation view of physical theories*, Czech. J. Phys., **B32** (1982), 472–475.

[4] L. D. Faddeev, Asia-Pacific Physics News, **3** (1988), 21, and *Frontiers in Physics, High Technology and Mathematics*, (ed. Cerdeira and Lundqvist), 238–246, World Scientific, 1989.

[5] F. Bayen, M. Flato, C. Fronsdal, A. Lichnerowicz, and D. Sternheimer, *Quantum mechanics as a deformation of classical mechanics*, Lett. Math. Phys., **1** (1977), 521–530.

[6] R. Vilela Mendes, *Deformations, stable theories and fundamental constants*, J. Phys., **A27** (1994), 8091–8104.

[7] R. Vilela Mendes, *Quantum mechanics and non-commutative space time*, Phys. Lett., **A210** (1996), 232–240.

[8] N. Ja. Vilenkin and A. U. Klimyk, *Representation of Lie Groups and Special Functions*, vols. 1 and 2, Kluwer, Dordrecht, **1991–93**.

[9] L. Accardi, A. Frigerio and J. T. Lewis, *Quantum stochastic processes*, Publ. R.I.M.S. Kyoto, **18** (1982), 97–133.

[10] R. L. Hudson and K. R. Parthasarathy, *Quantum Ito's formula and stochastic evolutions*, Comm. Math. Phys., **93** (1984), 301–323.

[11] K. R. Parthasarathy, *An Introduction to Quantum Stochastic Calculus*, Birkhäuser, Basel, **1992**.

[12] P. A. Meyer, *Quantum Probability for Probabilists*, Springer Lecture Notes in Mathematics 1538, **1993**.

[13] M. Abramowitz and I. A. Stegun, *Handbook of Mathematical Functions*, Dover, New York, **1965**.

Complexo Interdisciplinar
Universidade de Lisboa
Av. Gama Pinto
2 - P1699 Lisboa Codex
Portugal
E-mail address: `vilela@cii.fc.ul.pt`

Progress in Probability, Vol. 52, 219–227

A Measure-Valued Process Related to the Parabolic Anderson Model

C. Mueller and R. Tribe

Abstract. We consider the following stochastic partial differential equation:

$$\frac{\partial u_t}{\partial t} = \Delta u_t + \kappa u_t \dot{F}(t, \cdot).$$

Here, $t \geq 0$, $x \in \mathbf{R}^d$ with $d \geq 3$, and $\kappa > 0$. $\dot{F}(t,x)$ is a generalized Gaussian process, with covariance

$$E\left[\dot{F}(t,x)\dot{F}(t,y)\right] = \frac{\delta(t-s)}{|x-y|^2}.$$

Our solution $u_t(dx)$ will exist as a nonnegative measure, provided $u_0(dx)$ is a nonnegative measure satisfying certain properties. We discuss various properties of the solutions. Useful tools include the Feynman-Kac formula, duality, and integral equations.

This paper is a survey of our work on a new kind of measure-valued process. We will not give rigorous proofs here, but concentrate on the background ideas. Statements like "we show" mean "we will show in our forthcoming paper". Here is the motivation behind our work. For stochastic partial differential equations (SPDE) with an additive white noise term $\dot{W}(t,x)$, it is well known that function solutions usually exist only in 1 spatial dimension, that is, if $x \in \mathbf{R}$. This statement is true, for example, for the heat equation

$$\frac{\partial u_t}{\partial t} = \Delta u_t + \dot{W}(t,x),$$

and the corresponding wave equation (see [9]). (By the way, throughout this survey, we use u_t to denote u evaluated at t, not the derivative of u with respect to t.) It is very difficult to make sense of such equations if they have nonlinear terms or noise that is not purely additive, and if $x \in \mathbf{R}^d$ with $d > 1$. To our knowledge, there are only 2 cases in which such nonlinear equations have been treated successfully. One is the Dawson-Watanabe process, sometimes called the super-Brownian motion, and its generalizations. In 2 or more dimensions, the Dawson-Watanabe process is a measure-valued process $X_t(dx)$. Although the density $u_t(x)$ of X_t does not exist,

2000 *Mathematics Subject Classification.* Primary, 60H15; Secondary, 35R60, 35L05.
Key words and phrases. Heat equation, white noise, stochastic partial differential equations.
The first author is supported by an NSF travel grant and an NSA grant.

we think of it as satisfying the following equation.

$$\frac{\partial u_t}{\partial t} = \Delta u_t + u_t^{1/2}\dot{W}(t,\cdot). \tag{1}$$

Actually, (1) has rigorous meaning in terms of a corresponding martingale problem. For a measure-valued process such as $X_t(dx)$ and a suitable function f, we write

$$X_t(f) = \int_{\mathbf{R}^d} f(x)X_t(dx).$$

Then the Dawson-Watanabe process satisfies the following martingale problem, which we regard as a rigorous version of (1). For $f \in \mathbf{C}_c^2$, we require

$$Z_t(f) = X_t(f) - \int_0^t X_s(\Delta f)ds$$

to be a continuous local martingale with square variation

$$\langle Z(f)\rangle_t = \int_0^t X_s(f^2)ds.$$

See [3] for the Dawson-Watanabe process.

The second example involves stochastic quantization. Albeverio and Röckner [1, 2] were the first to rigorously treat the following equation in 2 spatial dimensions. See also the work of Mikulevicius and Rozovskii [6].

$$\frac{\partial u_t}{\partial t} = \Delta u_t - : u_t^3 : + \dot{W}(t,\cdot).$$

Here, $: u_t^3 :$ involves the Wick product, and the solution u_t exists only as a Schwartz distribution. We will not enter into these matters further.

Since it is so hard to study SPDE with white noise in 2 or more spatial dimensions, we believe that it is time to consider other possibilities. One alternative model would be a system of SDE on the lattice, with differential operators in the x-variable replaced by the corresponding difference operators. One could then study large-scale properties of solutions, but not local properties. Another idea, which we pursue in this survey, is to replace white noise by a noise which is smoother. Population processes and particle systems naturally lead to SPDE with white noise, but many other models, such as models of turbulence, do not. There is already an extensive general theory of SPDE with smoother noise; see, for example, [4]. But even now, not much is known about the qualitative behavior of solutions (other than smoothness), and the special behavior of particular equations has been little studied. Our goal here is to elucidate the behavior of one special equation, which does not seem to fit into existing theory.

Our main motivation was from the Dawson-Watanabe process. For $d > 1$, there are no nontrivial solutions to (1) with $u^{1/2}$ replaced by u^γ and with $\gamma \neq 1/2$. We wondered if raising the power γ of u^γ and weakening the noise could lead to a new measure-valued process. For $\gamma = 1$, we succeeded. Surprisingly, our process only exists in 3 or more spatial dimensions.

Consider random measures $u_t(dx)$ solving the following SPDE.

$$\frac{\partial u_t}{\partial t} = \Delta u_t + \kappa u_t \dot{F}(t,\cdot). \tag{2}$$

Here, $t \geq 0$, $x \in \mathbf{R}^d$ with $d \geq 3$. $\dot{F}(t,x)$ is a generalized Gaussian process, with covariance

$$E\left[\dot{F}(t,x)\dot{F}(s,y)\right] = \frac{\delta(t-s)}{|x-y|^2}.$$

It turns out that for small κ, this covariance is enough to make measure-valued solutions exist. Later, we will give precise meaning to (2) in terms of a martingale problem. Although we could pose the equation in strong form, we do not see how to show strong uniqueness. Since we will be able to use duality to show weak uniqueness, we will stick to the formulation in terms of a martingale problem.

First, let us explain the exponent 2 appearing in the covariance of the noise. Suppose that we replace the noise $\dot{F}(t,x)$ with a noise $\dot{F}_p(t,x)$ with covariance

$$E\left[\dot{F}_p(t,x)\dot{F}_p(s,y)\right] = \frac{\delta(t-s)}{|x-y|^p},$$

and let $\bar{u}_t(dx)$ be the solution to

$$\frac{\partial \bar{u}_t}{\partial t} = \Delta \bar{u}_t + \dot{F}_p(t,\cdot). \tag{3}$$

We give a short argument to show the following.

Proposition 1. *Suppose that $p > 0$. The measure $\bar{u}_t(dx)$ has a density if and only if $p < 2$.*

Thus, we would expect that solutions to (2) would fail to have densities. But this leads to the unusual term $u_t(dx)\dot{F}(t,x)$, which is a product of Schwartz distributions. Although it is possible to define the corresponding stochastic integral using L^2 limits, this does not seem to lead to a proof of strong uniqueness.

It might be worthwhile to indicate briefly why we believe that the usual Picard iteration method will not give uniqueness. We show that for bounded functions $f(x)$ of compact support, and for $\kappa > 0$ small enough,

$$E\left[u_t(f)^2\right] < \infty. \tag{4}$$

Here we are using the notation

$$u_t(f) = \int_{\mathbf{R}^d} f(x)u_t(dx).$$

Since we cannot construct solutions for large values of κ, we cannot say for sure what happens. But our formulas for the second moment (4) give a value of ∞ for all values of t, when κ is sufficiently large. As far as we can see, this situation cannot arise from any construction by Picard iteration, and it seems to rule out the use of Picard iteration to prove uniqueness.

Now we move on to discuss (2). We begin by giving an outline of the proof of Proposition 1. But first we must give precise meaning to (3). The standard framework for such linear equations, as given in [9], is to regard $F_p(t,x)$ as a process taking values in the space of Schwartz distributions. Then, using the theory of linear PDE (see [5], for example), one can construct the solution $\bar{u}_t(dx)$ in the same space of distributions. In fact, one can write

$$\begin{aligned} \bar{u}_t(dx) &= \left[\int_{\mathbf{R}^d} G_t(x-y)\bar{u}_0(dy) + \int_0^t \int_{\mathbf{R}^d} G_{t-s}(x-y)F_p(dyds)\right] dx, \\ &= \left[G_t(\bar{u}_0)(x) + \int_0^t \int_{\mathbf{R}^d} G_{t-s}(x-y)F_p(dyds)\right] dx, \end{aligned} \tag{5}$$

where G_t is the heat kernel:

$$\begin{aligned} G_t(x) &= (4\pi t)^{-d/2} \exp\left(-\frac{|x|^2}{4t}\right), \\ G_t(f) &= \int_{\mathbf{R}^d} G_t(x) f(x) dx. \end{aligned}$$

The double integral in (5) is defined in terms of Schwartz distributions. Clearly $G_t(\bar{u}_0)(x)$ is a smooth function, since it is a solution of the heat equation. Therefore, the $\bar{u}(t,x)$ is a function if the double integral in (5) is a function. But this double integral is a generalized Gaussian process, and it is well known that it gives finite values for fixed t, x if and only if its covariance is finite:

$$\int_{\mathbf{R}^d}\int_{\mathbf{R}^d} G_t(x-y)G_t(x-z)\frac{1}{|y-z|^p}dydz = \int_{\mathbf{R}^d}\int_{\mathbf{R}^d} G_t(y)G_t(z)\frac{1}{|y-z|^p}dydz < \infty.$$

But a short calculation shows that this integral is finite if and only if $p < 2$.

Let $\mathcal{M}_{lf}$ denote the class of locally finite nonnegative measures on $\mathbf{R}^d$, and let $\mathbf{C}_c = \mathbf{C}_c(\mathbf{R}^d)$ be the space of continuous functions on $\mathbf{R}^d$ with compact support. Let $\mathcal{F}_t$ denote the σ-field generated by $\dot{F}$ up to time t. That is, let $\mathcal{F}_t$ be the σ-field generated by the random variables $\int_0^t \int_{\mathbf{R}^d} g(s,x)F(dxds)$ for all functions $g(t,x) \in \mathbf{C}_c(\mathbf{R}^d)$. We say that $u_t(dx) \in \mathcal{M}_{lf}$ solves (2) if it solves the following martingale problem. For all $f \in \mathbf{C}_c^2$, let $u_t(f) = \int_{\mathbf{R}^d} f(x)u_t(dx)$. We require that

$$z_t(f) = u_t(f) - \int_0^t u_s(\Delta f)ds \tag{6}$$

is a continuous local martingale with square variation

$$\langle z(f)\rangle_t = \int_0^t \int_{\mathbf{R}^d}\int_{\mathbf{R}^d} f(y)f(x)\frac{\kappa^2}{|x-y|^2}u_s(dy)u_s(dx)ds.$$

We construct solutions to (2) using an expansion related to the familiar Wiener chaos expansion.

In order to state our main theorem, we must give some definitions. We write the usual $\mathbf{L}^2$ inner product as

$$(f,g) = \int_{\mathbf{R}^d} f(x)g(x)dx.$$

We will eventually extend this inner product to certain measures on $\mathbf{R}^d$. For

$$0 < \lambda < \left(\frac{d-2}{2}\right)^2, \tag{7}$$

we define

$$\alpha(\lambda) = \frac{d-2}{2} - \left[\left(\frac{d-2}{2}\right)^2 - \lambda\right]^{1/2}. \tag{8}$$

Next, for $\lambda > 0$, we define a Hilbert space norm for signed measures $\mu(dx)$ on $\mathbf{R}^d$ as follows. We denote our Hilbert space $\mathcal{H}_\lambda$, and define its norm to be

$$\|\mu\|_{\mathcal{H}_\lambda} = \left|\int_{\mathbf{R}^d}\int_{\mathbf{R}^d} \frac{\mu(dx)\mu(dy)}{|x-y|^{2\alpha(\lambda)}}\right|^{1/2}.$$

It is easy to check that $|x|^{-2\alpha(\lambda)}$ is a positive definite function, and so $\|\cdot\|_{\mathcal{H}_\lambda}$ is a Hilbert space norm. We also note that

$$\int_{\mathbf{R}^d}\int_{\mathbf{R}^d} \frac{\mu(dx)\mu(dy)}{|x-y|^s} = \int_{\mathbf{R}^d} \left|(-\Delta)^{-s/2}\mu(x)\right|^2 dx.$$

In other words, $\|\mu\|_\lambda$ is equal to the $\mathbf{L}^2$ norm of $(-\Delta)^{-s/2}\mu$, with $s = 2\alpha(\lambda)$. The reader can check that $\mathcal{H}_\lambda$ is a complete separable Hilbert space. We will only consider nonnegative measures, not signed measures.

Finally, it is a standard fact that

$$\mathcal{H}_\lambda \subset \mathbf{H}^{-2\alpha(\lambda)},$$

where $\mathbf{H}^s$ is the Sobolev space of distributions μ whose Fourier transforms $\hat{\mu}$ satisfy

$$\|\mu\|^2_{\mathbf{H}^s} \equiv \int_{\mathbf{R}^d} |\hat{\mu}(\xi)|^2(1+4\pi^2|\xi|^2)^{s/2} d\xi < \infty.$$

For example, see Stein [8], Section 3.2.

Here is our first main theorem.

Theorem 2. *Suppose that $d \geq 3$ and that*

$$0 < \kappa < \frac{d-2}{\sqrt{2}}.$$

Suppose that $\lambda = \kappa^2/2$, and that $u_0(dx) \in \mathcal{H}_\lambda$ is a nonnegative measure on $\mathbf{R}^d$ with compact support. Then the martingale problem (6) has a solution $u_t(dx)$.

Remark 3. We also show that solutions which satisfy some second moment bounds are unique in law. We will not specify these conditions here.

Our second goal is to investigate properties of the solution $u_t(dx)$. Luckily, there are many tools available. For equations such as (2), the Feynman-Kac formula is a standard tool. It would give

$$u_t(dx) = E_x\left[u_0(B_t)\exp\left(\int_0^t \kappa\dot{F}(t-s, 2^{1/2}B_s)ds\right)\right] dx, \tag{9}$$

where E_x is expectation over the Brownian motion B_t started at x. The integral in (9) is hard to interpret, since F is a distribution. We are able to gain some useful information from (9), but other tools are more useful.

The second tool is duality, which allows us to evaluate moments of $u_t(dx)$. We give our duality in terms of Brownian motion, but unfortunately, a standard Brownian motion has generator $\Delta/2$. However, the process $2^{1/2}B_t$ has generator Δ. We call this process a modified Brownian motion.

For $f \in \mathbf{C}_c^\infty$, let $u_t(f) = \int_{\mathbf{R}^d} f(x)u_t(dx)$. Let $B_t^{(1)}, B_t^{(2)}$ be independent (modified) Brownian motions. We show that $E[u_t(dx)u_t(dy)]$ has a density, and that

$$\begin{aligned} &E\left[u_t(dx)u_t(dy)\right] \\ &= E_{x,y}\left[u_0(B_t^{(1)})u_0(B_t^{(2)})\exp\left(\int_0^t \frac{\kappa}{2\left|B_s^{(1)} - B_s^{(2)}\right|^2}ds\right)\right] dxdy. \end{aligned} \tag{10}$$

There is a similar formula for the nth moment. Let $\{B_t^{(i)}\}_{i=1}^n$ be independent Brownian motions. Then we have

$$\begin{aligned} &E\left[\prod_{i=1}^n u_t(dx_i)\right] \\ &= E_{x_1,\dots,x_n}\left[\prod_{i=1}^n u_0(B_t^{(i)})\exp\left(\sum_{1\le j<k\le n}\int_0^t \frac{\kappa}{2\left|B_s^{(j)} - B_s^{(k)}\right|^2}ds\right)\right] dx_1\dots dx_n. \end{aligned}$$

A second form of duality is also useful. Suppose that $u_t(x), v_t(x)$ are two solutions of (2), but with independent noises $\dot{F}_u(t,x), \dot{F}_v(t,x)$ respectively. We show that for appropriate functions $h \in \mathbf{C}^2(\mathbf{R})$, and for all $t > 0$,

$$E\left[h\left(\int_{\mathbf{R}^d} u_t(x)v_0(dx)\right)\right] = E\left[h\left(\int_{\mathbf{R}^d} u_0(dx)v_t(x)\right)\right]. \tag{11}$$

Thus, $u_t(x)$ is self-dual. A standard argument shows that (11) implies that given $u_0(x)$, solutions $u_t(x)$ to (2) are unique in law. For example, we could take $h(x) = e^x$. If $u_t^{(1)}, u_t^{(2)}$ are 2 solutions of (11), and $v_0 \in \mathbf{C}_c(\mathbf{R}^d)$, then (11) implies that

$$E\left[\exp\left(\int_{\mathbf{R}^d} u_t^{(1)}(x)v_0(dx)\right)\right] = E\left[\exp\left(\int_{\mathbf{R}^d} u_t^{(2)}(x)v_0(dx)\right)\right].$$

By the uniqueness of inverse Laplace transforms, it follows that for fixed times t, $u_t^{(1)}, u_t^{(2)}$ are equal in law. Using similar arguments, we can also show that the 2 processes are themselves equal in law.

Our third tool, integral equations, is familiar from [9]. We show that (2) is equivalent to:

$$u_t(dx) = \left[G_t(u_0)(x) + \kappa\int_0^t\int_{\mathbf{R}^d} G_{t-s}(x-y)u_s(y)F(dyds)\right]dx. \tag{12}$$

Again, we have to give meaning to (12).

As our fourth tool, we use the fact that (2) is a linear equation. That is, if $u_t^{(1)}(dx), u_t^{(2)}(dx)$ are solutions of (2), then $C_1u_t^{(1)}(dx) + C_2u_t^{(2)}(dx)$ are also solutions, for $C_1, C_2 \in \mathbf{R}$. As we show, this linearity implies that we can decompose solutions in the following way:

$$u_t(dx) = \int_{y\in\mathbf{R}^d} U_t(dy,dx)u_0(y).$$

Here, for fixed t, y, $U_t(dy,dx)$ is a solution to (2), with initial condition $U_0(dy,dx) = \delta(y-x)dydx$.

To construct solutions of our equation, we expand $u_t(dx)$ in terms of multiple stochastic integrals. Our expansion is related to the Wiener chaos expansion, in which multiple integrals over white noise are used. We find it more convenient, however, to take multiple integrals over the noise $\dot{F}(t,x)$. Wiener chaos expansions have been used several times before for linear equations; see Nualart and Zakai [7], for example. The idea is to start with (12) and iterate. In other words, we use the formula for $u_t(dx)$ given in (12) to substitute for the term $u_s(y)$ which appears on the right-hand side of (12). The reader can check that if we repeat this substitution a countable number of times, we will arrive at the following formula. We emphasize that for the moment, this formula is only heuristic. For a test function $\varphi(x) \in \mathbf{C}_c^\infty(\mathbf{R}^d)$,

$$u_t(\varphi) = \sum_{n=0}^{\infty} I_t^{(n)}(\varphi), \tag{13}$$

where the $I^{(n)}$ are defined as follows. First, let

$$G_t\mu(x) = \int_{\mathbf{R}} G_t(x-y)\mu(dy).$$

Then, define

$$I_t^{(0)}(\varphi) = \int_{\mathbf{R}^d} \varphi(x)G_tu_0(dy). \tag{14}$$

Next, suppose that $n \geq 1$. For convenience, let $s_{n+1} = t$ and $y_{n+1} = x$. Then let

$$\begin{aligned} I_t^{(n)}(\varphi) &= \kappa^n \int_0^t\int_0^{s_n}\cdots\int_0^{s_2}\int_{\mathbf{R}^{(n+1)d}} G_{s_1}u_0(y_1)\varphi(y_{n+1})dy_{n+1} \\ &\quad \cdot\prod_{i=1}^n G_{s_{i+1}-s_i}(y_{i+1}-y_i)\prod_{i=1}^n F(dy_ids_i). \end{aligned} \tag{15}$$

Using our formula (10) for second moments, and assuming that κ is small, we can verify that the sum on the right-hand side of (13) converges in $\mathbf{L}^2(dP)$. This calculation relies on some work of Yor [10], which allows us to simplify the right-hand side of (10).

Finally, (2) can be viewed as the Zakai equation for a filtering problem. This is not really a tool, but we mention it as an interesting connection which the reader might wish to explore.

Using these tools and some further qualitative arguments, we arrive at the following results. Some of these questions were motivated by the properties of the Dawson-Watanabe process. For example, if its initial measure is finite and if $t > 0$, the Dawson-Watanabe process at time t is supported on a set of Hausdorff measure 2. Again with a finite initial measure, the Dawson-Watanabe process dies out completely in finite time. If its initial measure is finite and compactly supported, the Dawson-Watanabe process has compact support for all times $t > 0$.

Theorem 4. *Assume that λ satisfies (7), and that $\alpha(\lambda)$ is defined as in (8). Then, for each $t > 0$, $u_t(dx)$ is supported on a set $\mathcal{S}_t$ whose Hausdorff dimension is at least $d - \alpha(\lambda)$.*

Theorem 5. *Suppose that u_0 is a measure which has a continuous density f, and assume that f is bounded, nonnegative and not identically 0. Then, with probability 1,*

(i) *For each $t \geq 0$, the support of $u_t(x)$ is dense in $\mathbf{R}^d$. It follows that $u_t(x)$ does not die out in finite time.*
(ii) *Suppose that $u_0(dx) = dx$. Let $t > 0$, $x \in \mathbf{R}^d$. Let $B_r(x)$ denote the ball of radius r centered at x. Then with probability 1,*

$$\lim_{\varepsilon \downarrow 0} E\left[\varepsilon^{-d/2} u_t(B_\varepsilon(x))^{1/2}\right] = 0.$$

Therefore, for each $t > 0$ and with probability 1, we have the following. If x does not belong to an exceptional set of Lebesgue measure 0 (which may depend on ω), then $u_t(dx)$ either has no density at x or has density 0.
(iii) *Suppose that u_0 satisfies the assumptions of Theorem 2. Further, suppose that for some $s > r > 0$ and for some open set $\mathcal{O} \in \mathbf{R}^d$, the following holds with positive probability: if $t \in (r, s)$, then $u_t(dx)$ has a density with respect to Lebesgue measure on $\mathcal{O}$. Then this density is 0.*

Strangely enough, our proof of Theorem 5 breaks down if we set up our equation (2) on a compact domain such as the d-dimensional torus. We do not know whether Theorem 5 still holds under these circumstances.

Finally, we show that although the total mass $u_t(1)$ of our process does not reach 0 in finite time, it dies out asymptotically.

Theorem 6. *Suppose that $u_0(dx)$ is a finite measure, which satisfies the assumptions of Theorem 1. Then*

$$\lim_{t \to \infty} u_t(1) = 0$$

almost surely. In other words, the total mass dies out asymptotically as $t \to \infty$.

References

[1] S. Albeverio and M. Röckner, *Dirichlet forms, quantum fields and stochastic quantization*, in *Stochastic analysis, path integration and dynamics (Warwick, 1987)*, Pitman Res. Notes Math. Ser., Longman Sci. Tech, **200** (1989), 1–21.

[2] S. Albeverio and M. Röckner, *Classical Dirichlet forms on topological vector spaces – closability and a Cameron-Martin formula* J. Funct. Anal., **88 (2)** (1990), 395–436.

[3] D. A. Dawson, *Measure-valued Markov processes*, in P. L. Hennequin (Ed.), *École d'été de probabilités de Saint-Flour, XXI-1991*, Lecture Notes in Mathematics, **1180** (1993), 1–260.

[4] G. Da Prato and J. Zabczyk, *Stochastic Equations in Infinite Dimensions*, volume 44 of Encyclopedia of mathematics and its applications, Cambridge University Press, Cambridge, New York, **1992**.

[5] L. Hörmander, *The Analysis of Linear Partial Differential Operators*, Springer-Verlag, Berlin, Heidelberg, New York, **1983-85**.

[6] R. Mikulevicius and B. L. Rozovskii, *Martingale problems for stochastic PDE's*, in *Stochastic partial differential equations: six perspectives*, Math. Surveys Monogr., Amer. Math. Soc., **64** (1999), 243–325.

[7] D. Nualart and M. Zakai, *Generalized Brownian functionals and the solution to a stochastic partial differential equation*, J. Funct. Anal., **84** (1989), 279–296.

[8] E. M. Stein, *Singular Integrals and Differentiability Properties of Functions*, Princeton University Press, Princeton, New Jersey, **1970**.

[9] J. B. Walsh, *An introduction to stochastic partial differential equations*, in P. L. Hennequin (Ed.), *École d'été de probabilités de Saint-Flour, XIV-1984*, Lecture Notes in Mathematics, **1180** (1986), 265–439.

[10] M. Yor, *Loi de l'indice du lancet Brownien, et distribution de Hartman-Watson*, Probab. Theory Related Fields, **53 (1)** (1980), 71–95.

Dept. of Mathematics
University of Rochester
Rochester, NY 14627
United States
E-mail address: `cmlr@troi.cc.rochester.edu`

Mathematics Institute
University of Warwick
Coventry CV4 7AL
United Kingdom
E-mail address: `tribe@maths.warwick.ac.uk`

Progress in Probability, Vol. 52, 229–242

Homogenization of PDEs with Non Linear Boundary Condition

Y. Ouknine and É. Pardoux

Abstract. We study the homogenization of semilinear PDEs with nonlinear Neumann boundary condition, periodic coefficients and highly oscillating drift and nonlinear term. Our method is entirely probabilistic, and builds upon earlier work of Tanaka [19] and Pardoux [12].

1. Introduction

Let L_ε be a uniformly elliptic second-order partial differential operator of the form (1) below indexed by a parameter $\varepsilon > 0$. The homogenization problem for an elliptic equation consists in computing the limit as $\varepsilon \downarrow 0$, of the solution $u_\varepsilon(x)$ of $L_\varepsilon u_\varepsilon = f$ in a domain D of $\mathbb{R}^d$ subject to an appropriate boundary condition under the assumption that the coefficients $a^{ij}(x), b^i(x)$, and $c^i(x)$ are periodic, almost periodic, or more generally, stationary random fields. In this paper, we shall only consider the case of periodic coefficients. In a probabilistic approach the problem becomes the following. What is the limit of the laws of the diffusion processes X_t^ε with generator L_ε, as $\varepsilon \to 0$? This problem has been studied for diffusion processes in the whole of $\mathbb{R}^d$ by Freidlin [2, 3] and Bensoussan, Lions, Papanicolaou [1], and in the case of the presence of boundary conditions by Tanaka [19].

In [12] the second author has combined the probabilistic approach of Chapter 3 in [1] with backward stochastic differential equations, in order to derive homogenization results for semilinear parabolic PDEs in the whole space with highly oscilating drift and nonlinear term. The aim of this paper is to pursue the same program for a semilinear PDE with nonlinear Neuman boundary condition.

The paper is organized as follows. In Section 2, we recall some results from Tanaka on homogenization of a reflected diffusion with periodic coefficients. In Section 3, we combine those results with BSDE arguments in order to take the limit law in a class of BSDEs whose data depend on the scaled reflected diffusion. Finally in Section 4 we apply our result to the homogenization of a class of semilinear parabolic PDEs with non linear Neumann boundary condition.

2000 *Mathematics Subject Classification.* 60H10, 60H30, 35K85.
Key words and phrases. Backward stochastic differential equations, homogenization, semilinear PDes with Neumann boundary condition.

2. Reflected diffusion with rapidly oscillating coefficients

In this section, we state some results which were proved in Tanaka [19], and which we will use in the rest of the paper.

Let $D = \{(x^1, \ldots, x^d) \in \mathbb{R}^d, x^1 > 0\}$ and let

$$L_\varepsilon = \sum_{i,j=1}^{d} a^{ij}(\frac{x}{\varepsilon})\partial_i\partial_j + \frac{1}{\varepsilon}\sum_{i=1}^{d} b^i(\frac{x}{\varepsilon})\partial_i + \sum_{i=1}^{d} c^i(\frac{x}{\varepsilon})\partial_i \tag{1}$$

$$\Gamma_\varepsilon = \sum_{i=1}^{d} \gamma^i\left(\frac{x}{\varepsilon}\right)\partial_i \tag{2}$$

be given (with $\varepsilon > 0$), where $\partial_i = \partial/\partial x^i$. We also set

$$L = \sum_{i,j=1}^{d} a^{ij}(x)\partial_i\partial_j + \sum_{i=1}^{d} b^i(x)\partial_i\,, \quad x \in \mathbb{R}^d$$

and require the following:

(H.1) L is uniformly elliptic and the matrix $a(x) = [a^{ij}(x)]$ is factored as $\sigma(x)\sigma^*(x)/2$.

(H.2) The functions $\sigma : \mathbb{R}^d \to \mathbb{R}^d \otimes \mathbb{R}^d$ $b : \mathbb{R}^d \to \mathbb{R}^d$ $c : \mathbb{R}^d \to \mathbb{R}^d$ are smooth (C^∞) and periodic (of period 1) in each variable, $\gamma : \partial D(\cong \mathbb{R}^{d-1}) \to \mathbb{R}^d$ is smooth (C^∞) and periodic (of period 1) in each variable, and $\gamma^1(x) = 1$.

The differential operator L_ε inside D together with the boundary condition $\Gamma_\varepsilon u = 0$ on ∂D determines a unique diffusion process X^ε in $\overline{D}$ which we call the $(L_\varepsilon, \Gamma_\varepsilon)$-diffusion. We now write the SDE satisfied by X^ε. Given a d-dimensional Brownian motion $\{B_t;\, t \geq 0\}$ defined on a probability space $(\Omega, \mathcal{F}, \mathbb{P})$, $(X^\varepsilon, \phi^\varepsilon)$ is the unique solution with values in $D \times \mathbb{R}_+$ of the following reflected SDE:

$$\begin{cases} dX_t^\varepsilon &= \sigma(X_t^\varepsilon/\varepsilon)dB_t + \frac{1}{\varepsilon}b(X_t^\varepsilon/\varepsilon)dt + c(X_t^\varepsilon/\varepsilon)dt + \gamma(X_t^\varepsilon/\varepsilon)d\phi_t^\varepsilon,\ t \geq 0 \\ X_t^{1,\varepsilon} &\geq 0,\ \phi^\varepsilon \text{ is continuous and increasing, } \int_0^t X_s^{1,\varepsilon} d\phi_s^\varepsilon = 0,\ t \geq 0, \\ X_0^\varepsilon &= x, \end{cases} \tag{3}$$

where $X^{1,\varepsilon}$ denotes the first component of the process X^ε. We recall that $D = \mathbb{R}_+^d$, so that X^ε lives is D, i.e. $X^{1,\varepsilon}$ remains nonnegative, and ϕ^ε increases when and only when $X^{1,\varepsilon}$ is zero, just to keep it nonnegative.

Let $\overline{X}$ denote the unique diffusion process with values in the d-dimensional torus $\mathbf{T}^d$, whose generator is the operator L. $\overline{X}$ is ergodic. We denote by m its unique invariant measure. In order for the process X^ε to have a limit in law as $\varepsilon \to 0$, we need that the following condition be in force:

(H.3) (Centering Condition) $\int_{\mathbf{T}^d} b(x)\, m(dx) = 0.$

First we notice that under assumption (H.3) there exists a unique periodic solution χ^k of $L\chi^k = -b^k$ for each $k = 1, \ldots, d$, with zero integral against the measure m. That solution is given by $\chi^k(x) = \int_0^\infty \mathbb{E}_x \left\{ b^k(\overline{X}_t) \right\} dt$ where, under $\mathbb{P}_x$, $\overline{X}$ starts from x. We set

$$\chi = \begin{bmatrix} \chi^1 \\ \vdots \\ \chi^d \end{bmatrix}, \quad \nabla\chi = \begin{bmatrix} \partial_1\chi^1 & \cdots & \partial_d\chi^1 \\ \vdots & \cdots & \vdots \\ \partial_1\chi^d & \cdots & \partial_d\chi^d \end{bmatrix},$$

$$\begin{aligned} a_0 &= \int_{\mathbf{T}^d} (I + \nabla\chi) a (I + \nabla\chi)^*(x) m(dx), \\ c_0 &= \int_{\mathbf{T}^d} (I + \nabla\chi) c(x) m(dx), \end{aligned}$$

$$L_0 = \sum_{i,j=1}^d a_0^{ij} \partial_i \partial_j + \sum_{i=1}^d c_0^i \partial_i.$$

We write $u = H\varphi$ for the solution u of

$$\begin{cases} Lu &= 0 \text{ in } D, \\ u &= \varphi \text{ on } \partial D. \end{cases}$$

Then H sends functions defined on ∂D to functions defined on $\overline{D}$, while ΓH sends functions on ∂D to functions on ∂D, where $\Gamma = \sum_{i=1}^d \gamma^i(x)\partial_i$. There exists a unique Markov process on ∂D with generator ΓH. By the periodicity assumption this Markov process induces a Markov process on the torus $\mathbf{T}^{d-1}$; let $\widetilde{m}$ be the invariant measure of the induced Markov process. We set

$$\gamma_0 = \int_{\mathbf{T}^{d-1}} (I + \nabla\chi)\gamma(x)\, \widetilde{m}(dx), \quad \Gamma_0 = \sum_{i=1}^d \gamma_0^i \partial_i.$$

We can now state one of the main results in Tanaka [19], which we will use below:

Theorem 2.1. *Under the assumptions* (H.1), (H.2), (H.3), *the* $(L_\varepsilon, \Gamma_\varepsilon)$*-diffusion process* X^ε *converges in law to a* (L_0, Γ_0)*-diffusion* X *as* $\varepsilon \downarrow 0$. *Moreover,*

$$(X^\varepsilon, \hat{M}^\varepsilon, \phi^\varepsilon) \Rightarrow (X, M^X, \phi),$$

where

$$\hat{M}_t^\varepsilon = \int_0^t (I + \nabla\chi)\sigma(X_s^\varepsilon/\varepsilon) dB_s,$$

M^X *is the martingale part of* X *and* ϕ *(resp.* $\phi^v arepsilon$*) is the local time of* X^1 *(resp.* $X^{1,\varepsilon}$*) at 0.*

It follows moreover easily from the results in Tanaka [19] that

Lemma 2.2. *Under the assumptions of Theorem* 2.1, *for any* $p \geq 1$,

$$\sup_{\varepsilon} \mathbb{E}\left(|X_t^\varepsilon|^p + \phi_t^\varepsilon\right) < \infty. \tag{4}$$

3. BSDE's and weak convergence

In this section, we assume that $h : \mathbb{R}^d \times \mathbb{R} \to \mathbb{R}$, $f : \mathbb{R}^d \times \mathbb{R} \to \mathbb{R}$ and $g : \mathbb{R}^d \to \mathbb{R}$ are continuous mappings which are such that for some $c > 0$, $p, q \geq 0$, $\mu \in \mathbb{R}$, $\beta < 0$, and all $x \in \mathbb{R}^d$, y, $y' \in \mathbb{R}$,

(H.4) $|g(x)| \leq c(1 + |x|^p)$,
(H.5) $|f(x, y)| \leq c(1 + |y|^2)$,
(H.6) $(y - y')(f(x, y) - f(x, y')) \leq \mu\, |y - y'|^2$,
(H.7) $(y - y')(h(x, y) - h(x, y')) \leq \beta |y - y'|^2$,
(H.8) $|h(x, y)| \leq c(1 + |y|^2)$.

For each $\varepsilon > 0$, the triple $(X^\varepsilon, M^\varepsilon, \phi^\varepsilon)$ is the one which appears in the statement of Theorem 2.1.

We now consider a type of BSDE which has been introduced in Pardoux-Zhang [16]. For each fixed $(t, x) \in [0, T] \times \overline{D}$, let $\{(Y_s^\varepsilon, U_s^\varepsilon);\ 0 \leq s \leq T\}$ be the unique solution of the BSDE

$$Y_s^\varepsilon = g(X_t^\varepsilon) + \int_s^t f\left(\frac{X_r^\varepsilon}{\varepsilon}, Y_r^\varepsilon\right) dr + \int_s^t \frac{1}{\varepsilon} e\left(\frac{X_r^\varepsilon}{\varepsilon}, Y_r^\varepsilon\right) dr + \int_s^t h\left(\frac{X_r^\varepsilon}{\varepsilon}, Y_r^\varepsilon\right) d\phi_r^\varepsilon - \int_s^t U_r^\varepsilon dM_r^{X^\varepsilon},$$

where we assume that $X_0^\varepsilon = x$ a.s..

We assume that e is a measurable mapping from $\mathbb{R}^d \times \mathbb{R}$ into $\mathbb{R}$, which is periodic, of period one in each direction, in the first argument, continuous in y uniformly with respect to x and that for all $y \in \mathbb{R}$,

$$\int_{\mathbf{T}^d} e\,(x, y)\; m(dx) = 0.$$

Moreover e is twice continuously differentiable in y, uniformly with respect to x, and there exists a constant K such that

$$|e\,(x, y)| + \left|\frac{\partial e}{\partial y}\,(x, y)\right| + \left|\frac{\partial^2 e}{\partial y^2}\,(x, y)\right| \leq K, \quad \forall x \in \mathbf{T}^d, y \in \mathbb{R}.$$

For each fixed $y \in \mathbb{R}$, we solve the Poisson equation on $\mathbf{T}^d$

$$L\hat{e}(x, y) + e(x, y) = 0.$$

More precisely, $\hat{e}$ will denote the solution of that equation which has zero mean with respect to the invariant measure m, and is given by the formula

$$\hat{e}(x, y) = \int_0^\infty \mathbb{E}_x e(\overline{X}_t, y)\; dt.$$

It follows from simple adaptations of the results in Pardoux, Veretennikov [15] that

$$\hat{e} \in C^{0,2}\left(\mathbf{T}^d \times \mathbb{R}\right) \quad \text{and} \quad \hat{e}\left(.,y\right), \frac{\partial \hat{e}}{\partial y}\left(.,y\right), \frac{\partial^2 \hat{e}}{\partial y^2}\left(.,y\right) \in W^{2,p}\left(\mathbf{T}^d\right),$$

for any $p \geq 1$ there exists $K' > 0$ such that

$$\|\hat{e}\left(.,y\right)\|_{W^{2,p}(\mathbf{T}^d)} + \left\|\frac{\partial \hat{e}}{\partial y}\left(.,y\right)\right\|_{W^{2,p}(\mathbf{T}^d)} + \left\|\frac{\partial^2 \hat{e}}{\partial y^2}\left(.,y\right)\right\|_{W^{2,p}(\mathbf{T}^d)} \leq K', \ \forall y \in \mathbb{R}.$$

In the same way we define $\{(Y_s, U_s); 0 \leq s \leq t\}$ as the unique solution of the BSDE

$$Y_s = g(X_t) + \int_s^t f_0(Y_r)dr + \int_s^t h_0(Y_r)d\phi_r - \int_s^t U_r dM_r^X,$$

where the coefficients f_0 and h_0 are those defined at the end of this section, and (X, M^X, ϕ) is the limiting triple which appears in the statement of Theorem 2.1.

We introduce the notations $M_s^\varepsilon := \displaystyle\int_0^s U_r^\varepsilon dM_r^{X^\varepsilon}$, $M_s := \int_0^s U_r dM_r^X$, $0 \leq s \leq t$, and we consider the quintuple (X, M^X, ϕ, Y, M) (*resp.*$(X^\varepsilon, M^{X^\varepsilon}, \phi^\varepsilon, Y^\varepsilon, M^\varepsilon)$) as a random element of the space $C([0,t]; \mathbb{R}^{2d+1}) \times D([0,t]; \mathbb{R}^2)$, where we equip the first factor with the sup-norm topology, and the second factor with the S-topology of Jakubowski (see [6]).

The aim of this section is to prove the following result:

Theorem 3.1. *Under the conditions* (H.1)–(H.8),

$$Y_0^\varepsilon \to Y_0 \ in \ \mathbb{R}.$$

Proof. The proof is in several steps.

Step 0: Transformation of the BSDE. Set

$$\widetilde{Y}_s^\varepsilon = Y_s^\varepsilon - \varepsilon \hat{e}(X_s^\varepsilon/\varepsilon, Y_s^\varepsilon), \quad Z_s^\varepsilon = U_s^\varepsilon \sigma(X_s^\varepsilon/\varepsilon), \quad 0 \leq s \leq t.$$

We deduce from Itô's formula that

$$\begin{aligned}
\widetilde{Y}_s^\varepsilon + \varepsilon \hat{e}(X_t^\varepsilon/\varepsilon, Y_t^\varepsilon) = g(X_t^\varepsilon) &+ \int_s^t \left[\left(\frac{\partial \hat{e}}{\partial x}, c\right) - \frac{\partial \hat{e}}{\partial y} e\right](X_r^\varepsilon/\varepsilon, Y_r^\varepsilon)\, dr \\
&+ \int_s^t \frac{\partial^2 \hat{e}}{\partial x \partial y}(X_r^\varepsilon/\varepsilon, Y_r^\varepsilon)\sigma(X_r^\varepsilon/\varepsilon)(Z_r^\varepsilon)^* dr \\
+ \int_s^t \left(1 - \varepsilon \frac{\partial \hat{e}}{\partial y}(X_r^\varepsilon/\varepsilon, Y_r^\varepsilon)\right) &[h(X_r^\varepsilon/\varepsilon, Y_r^\varepsilon)d\phi_r^\varepsilon + f(X_r^\varepsilon/\varepsilon, Y_r^\varepsilon)dr] \\
+ \int_s^t \left(\frac{\partial \hat{e}}{\partial x}(X_r^\varepsilon/\varepsilon, Y_r^\varepsilon)\sigma(X_r^\varepsilon/\varepsilon) - Z_r^\varepsilon\right) dB_r &+ \varepsilon \int_s^t \frac{\partial \hat{e}}{\partial y}(X_r^\varepsilon/\varepsilon, Y_r^\varepsilon) Z_r^\varepsilon dB_r \\
+ \frac{\varepsilon}{2} \int_s^t \frac{\partial^2 \hat{e}}{\partial y^2}(X_r^\varepsilon/\varepsilon, Y_r^\varepsilon)\left|Z_r^\varepsilon\right|^2 dr &+ \int_s^t \frac{\partial \hat{e}}{\partial x}\left(X_r^\varepsilon/\varepsilon, Y_r^\varepsilon\right)\gamma\left(X_r^\varepsilon/\varepsilon\right) d\phi_r^\varepsilon.
\end{aligned}$$

We now define

$$\widetilde{Z}_s^\varepsilon = Z_s^\varepsilon - \frac{\partial \hat{e}}{\partial x}(X_s^\varepsilon/\varepsilon, Y_s^\varepsilon)\sigma(X_s^\varepsilon/\varepsilon),\ 0 \le s \le t,$$

and note that the difference between $\widetilde{Z}_s^\varepsilon$ and Z_s^ε is a uniformly bounded process.

It then follows that

$$\begin{aligned} &Y_s^\varepsilon + \varepsilon\left(\hat{e}(X_t^\varepsilon/\varepsilon, Y_t^\varepsilon) - \hat{e}(X_s^\varepsilon/\varepsilon, Y_s^\varepsilon)\right) \\ &= g(X_t^\varepsilon) + \int_s^t \left[\left(\frac{\partial \hat{e}}{\partial x}, c\right) - \frac{\partial \hat{e}}{\partial y} e + \frac{\partial^2 \hat{e}}{\partial x \partial y} a \left(\frac{\partial \hat{e}}{\partial x}\right)^*\right](X_r^\varepsilon/\varepsilon, Y_r^\varepsilon)\, dr \\ &+ \int_s^t \left(1 - \varepsilon \frac{\partial \hat{e}}{\partial y}\right) h(X_r^\varepsilon/\varepsilon, Y_r^\varepsilon) d\phi_r^\varepsilon + \int_s^t \frac{\partial \hat{e}}{\partial x}(X_r^\varepsilon/\varepsilon, Y_r^\varepsilon)\,\gamma(X_r^\varepsilon/\varepsilon)\, d\phi_r^\varepsilon \\ &+ \int_s^t (1 - \varepsilon \frac{\partial \hat{e}}{\partial y}) f(X_r^\varepsilon/\varepsilon, Y_r^\varepsilon)\, dr + \frac{\varepsilon}{2}\int_s^t \frac{\partial^2 \hat{e}}{\partial y^2}(X_r^\varepsilon/\varepsilon, Y_r^\varepsilon)\,|Z_r^\varepsilon|^2\, dr \\ &- \int_s^t \widetilde{Z}_r^\varepsilon \left(dB_r - \sigma^* \frac{\partial^2 \hat{e}}{\partial x \partial y}(X_r^\varepsilon/\varepsilon, Y_r^\varepsilon) dr\right) + \varepsilon \int_s^t \frac{\partial \hat{e}}{\partial y}(X_r^\varepsilon/\varepsilon, Y_r^\varepsilon) Z_r^\varepsilon dB_r. \end{aligned}$$

We next let

$$\widetilde{B}_s = B_s - \int_0^s \left(\sigma^* \frac{\partial^2 \hat{e}}{\partial x \partial y}\right)(X_r^\varepsilon/\varepsilon, Y_r^\varepsilon) dr.$$

It follows from Girsanov's theorem that there exists a new probability measure $\widetilde{\mathbb{P}}$ equivalent to $\mathbb{P}$ under which $\widetilde{B}_t$ is a Brownian motion.

Let $\widetilde{X}_s^\varepsilon := X_s^\varepsilon + \varepsilon\chi(X_s^\varepsilon/\varepsilon)$. We have (with the notation $c_\varepsilon = c(X_s^\varepsilon/\varepsilon)$, and similarly for $(I + \nabla\chi)$, a, σ, γ)

$$\begin{aligned} d\widetilde{X}_s^\varepsilon &= (I + \nabla\chi)_\varepsilon \left(\sigma_\varepsilon\, d\widetilde{B}_s + c_\varepsilon\, ds + a_\varepsilon \frac{\partial^2 \hat{e}}{\partial x \partial y}(X_s^\varepsilon/\varepsilon, Y_s^\varepsilon)\, ds + \gamma_\varepsilon\, d\phi_s^\varepsilon\right), \\ \widetilde{X}_0^\varepsilon &= x + \varepsilon\chi(x/\varepsilon), \end{aligned}$$

and

$$\begin{aligned} &Y_s^\varepsilon + \varepsilon\left(\hat{e}(X_t^\varepsilon/\varepsilon, Y_t^\varepsilon) - \hat{e}(X_s^\varepsilon/\varepsilon, Y_s^\varepsilon)\right) \\ &= g(X_t^\varepsilon) + \int_s^t \left[\left(\frac{\partial \hat{e}}{\partial x}, c\right) - \frac{\partial \hat{e}}{\partial y} e + \frac{\partial^2 \hat{e}}{\partial x \partial y} a \left(\frac{\partial \hat{e}}{\partial x}\right)^*\right](X_r^\varepsilon/\varepsilon, Y_r^\varepsilon)\, dr \\ &+ \int_s^t \left(1 - \varepsilon \frac{\partial \hat{e}}{\partial y}\right) f(X_r^\varepsilon/\varepsilon, Y_r^\varepsilon) dr + \int_s^t \frac{\partial \hat{e}}{\partial x}\gamma(X_r^\varepsilon/\varepsilon, Y_r^\varepsilon)\, d\phi_r^\varepsilon \\ &+ \int_s^t \left(1 - \varepsilon \frac{\partial \hat{e}}{\partial y}\right) h(X_r^\varepsilon/\varepsilon, Y_r^\varepsilon)\, d\phi_r^\varepsilon + \frac{\varepsilon}{2}\int_s^t \frac{\partial^2 \hat{e}}{\partial y^2}(X_r^\varepsilon/\varepsilon, Y_r^\varepsilon)\,|Z_r^\varepsilon|^2\, dr \\ &- \int_s^t \widetilde{Z}_r^\varepsilon d\widetilde{B}_r + \varepsilon \int_s^t \frac{\partial \hat{e}}{\partial y}(X_r^\varepsilon/\varepsilon, Y_r^\varepsilon) Z_r^\varepsilon \left[d\widetilde{B}_r + \left(\frac{\partial^2 \hat{e}}{\partial x \partial y}\sigma\right)(X_r^\varepsilon/\varepsilon, Y_r^\varepsilon) dr\right]. \end{aligned}$$

Since $\frac{\partial^2 \hat{e}}{\partial x \partial y}$ is bounded, the Radon-Nikodym derivative $d\widetilde{\mathbb{P}}/d\mathbb{P}$ belongs to all spaces $L^q(\mathbb{P})$, hence from (4), for any $p > 0$, $\sup_\varepsilon \widetilde{\mathbb{E}} |X_t^\varepsilon|^p < \infty$, consequently for any $k > 0$, $\sup_\varepsilon \widetilde{\mathrm{E}} |g(X_t^\varepsilon)|^k < \infty$.

Step 1: A priori estimates for $(Y^\varepsilon, Z^\varepsilon)$. We need to bound appropriate moments of Y^ε and Z^ε under $\widetilde{\mathbb{P}}$. We first go back to our unperturbed BSDE under the new Brownian motion $\widetilde{B}$

$$Y_s^\varepsilon = g(X_t^\varepsilon) + \int_s^t \left[\frac{1}{\varepsilon} e(X_r^\varepsilon/\varepsilon, Y_r^\varepsilon) + f(X_r^\varepsilon/\varepsilon, Y_r^\varepsilon) - Z_r^\varepsilon \left(\frac{\partial^2 \hat{e}}{\partial x \partial y}\sigma\right)(X_r^\varepsilon/\varepsilon, Y_r^\varepsilon)\right] dr$$
$$+ \int_s^t h(X_r^\varepsilon/\varepsilon, Y_r^\varepsilon) d\phi_r^\varepsilon - \int_s^t Z_r^\varepsilon d\widetilde{B}_r.$$

Applying Itô's formula to develop $e^{\nu s}|Y_s^\varepsilon|^3$, with $\nu > 0$ large enough, and using similar arguments as those leading to (4.13) in [12], we deduce

$$\varepsilon \widetilde{\mathbb{E}} \int_s^t |Y_r^\varepsilon| \times |Z_r^\varepsilon|^2 dr \le c \left[\varepsilon + \widetilde{\mathbb{E}} \left(\int_s^t |Y_r^\varepsilon|^2 dr + \varepsilon(\phi_t^\varepsilon - \phi_s^\varepsilon)\right)\right]. \tag{5}$$

We deduce from Itô's formula and the last relation in Step 0 that

$$\left|\widetilde{Y}_r^\varepsilon\right|^2 + \int_s^t \left|\widetilde{Z}_r^\varepsilon - \varepsilon Z_r^\varepsilon \left(\frac{\partial \hat{e}}{\partial y}\right)(X_r^\varepsilon/\varepsilon, Y_r^\varepsilon)\right|^2 dr = |g(X_t^\varepsilon) - \varepsilon \hat{e}(X_t^\varepsilon/\varepsilon, Y_t^\varepsilon)|^2$$
$$+ 2\int_s^t \widetilde{Y}_r^\varepsilon \left[(\nabla_x \hat{e}, c) - \frac{\partial \hat{e}}{\partial y} e + \left(1 - \varepsilon \frac{\partial \hat{e}}{\partial y}\right) f + \frac{\partial^2 \hat{e}}{\partial x \partial y} a \nabla_x \hat{e}^*\right](X_r^\varepsilon/\varepsilon, Y_r^\varepsilon) dr$$
$$+ 2\int_s^t \widetilde{Y}_r^\varepsilon \left[\left(1 - \varepsilon \frac{\partial \hat{e}}{\partial y}\right) h + \frac{\partial \hat{e}}{\partial x}\gamma\right](X_r^\varepsilon/\varepsilon, Y_r^\varepsilon) d\phi_r^\varepsilon - 2\int_s^t \widetilde{Y}_r^\varepsilon \widetilde{Z}^\varepsilon(r) d\widetilde{B}_r$$
$$+ 2\varepsilon \int_s^t \widetilde{Y}_r^\varepsilon Z_r^\varepsilon \left(\frac{\partial \hat{e}}{\partial y}\right)(X_r^\varepsilon/\varepsilon, Y_r^\varepsilon)\left[d\widetilde{B}_r + \left(\frac{\partial^2 \hat{e}}{\partial x \partial y}\sigma\right)(X_r^\varepsilon/\varepsilon, Y_r^\varepsilon) dr\right]$$
$$+ \varepsilon \int_s^t \frac{\partial^2 \hat{e}}{\partial y^2}(X_r^\varepsilon/\varepsilon, Y_r^\varepsilon)\widetilde{Y}_r^\varepsilon |Z_r^\varepsilon|^2 dr.$$

Exploiting (4), (5), together with the fact that $1 - \varepsilon \frac{\partial \hat{e}}{\partial y}(X_r^\varepsilon/\varepsilon, Y_r^\varepsilon) \ge 1/2$ for ε small enough, and standard inequalities, we deduce that

$$\widetilde{\mathbb{E}}\left(|Y_s^\varepsilon|^2 + \frac{1}{2}\int_s^t \left|\widetilde{Z}_r^\varepsilon\right|^2 dr + \frac{|\beta|}{4}\int_s^t |Y_r^\varepsilon|^2 d\phi_r^\varepsilon\right) < C(1 + \widetilde{\mathbb{E}}\int_s^t |Y_r^\varepsilon|^2 dr).$$

Hence from Gronwall's theorem

$$\sup_{0 \le s \le t} \widetilde{\mathbb{E}} |Y_s^\varepsilon|^2 + \frac{1}{2}\widetilde{\mathbb{E}}\int_0^t \left|\widetilde{Z}_r^\varepsilon\right|^2 dr + \frac{|\beta|}{4}\widetilde{\mathbb{E}}\int_0^t |Y_r^\varepsilon|^2 d\phi_r^\varepsilon < C$$

and finally from this last inequality, the above identity and the Davis, Burkholder, Gundy inequality,

$$\sup_{0<\varepsilon<\varepsilon_0} \widetilde{\mathbb{E}}\left(\sup_{0\le s\le t}|Y_s^\varepsilon|^2+\int_0^t\left|\widetilde{Z}_r^\varepsilon\right|^2dr+\int_0^t|Y_r^\varepsilon|^2\,d\phi_r^\varepsilon\right)<\infty.$$

Step 2: Tightness. Let $\mathcal{F}_t^\varepsilon=\mathcal{F}_t^{X^\varepsilon}$. We write our BSDE in the form

$$Y_s^\varepsilon=g(X_t^\varepsilon)+A_t^\varepsilon-A_s^\varepsilon+\bar{M}_t^\varepsilon-\bar{M}_s^\varepsilon+N_t^\varepsilon-N_s^\varepsilon,$$

where

$$\begin{aligned}
A_s^\varepsilon &=\int_0^s\left[\left(\frac{\partial\hat{e}}{\partial x},c\right)-\frac{\partial\hat{e}}{\partial y}e+\frac{\partial^2\hat{e}}{\partial x\partial y}a\left(\frac{\partial\hat{e}}{\partial x}\right)^*\right](X_r^\varepsilon/\varepsilon,Y_r^\varepsilon)\,dr,\\
&\qquad+\int_0^s f(X_r^\varepsilon/\varepsilon,Y_r^\varepsilon)dr+\int_0^s(h+\frac{\partial\hat{e}}{\partial x}\gamma)(X_r^\varepsilon/\varepsilon,Y_r^\varepsilon)d\phi_r^\varepsilon\\
\bar{M}_s^\varepsilon &=-\int_0^s\widetilde{Z}^\varepsilon(r)d\widetilde{B}_r,\\
N_s^\varepsilon &=\varepsilon\left(\hat{e}(X_s^\varepsilon/\varepsilon,Y_s^\varepsilon)-\hat{e}(X_t^\varepsilon/\varepsilon,Y_t^\varepsilon)\right)-\varepsilon\int_0^s\frac{\partial\hat{e}}{\partial y}(X_r^\varepsilon/\varepsilon,Y_r^\varepsilon)f(X_r^\varepsilon/\varepsilon,Y_r^\varepsilon)dr\\
&\quad+\varepsilon\int_0^s\frac{\partial\hat{e}}{\partial y}(X_r^\varepsilon/\varepsilon,Y_r^\varepsilon)h(X_r^\varepsilon/\varepsilon,Y_r^\varepsilon)d\phi_r^\varepsilon+\frac{\varepsilon}{2}\int_0^t\frac{\partial^2\hat{e}}{\partial y^2}(X_r^\varepsilon/\varepsilon,Y_r^\varepsilon)\,|Z_r^\varepsilon|^2\,dr\\
&\quad+\varepsilon\int_0^s\frac{\partial\hat{e}}{\partial y}(X_r^\varepsilon/\varepsilon,Y_r^\varepsilon)Z_r^\varepsilon\left[d\widetilde{B}_r+\left(\frac{\partial^2\hat{e}}{\partial x\partial y}\sigma\right)(X_r^\varepsilon/\varepsilon,Y_r^\varepsilon)dr\right].
\end{aligned}$$

It is easy to check that

$$\widetilde{\mathbb{E}}\sup_{0\le s\le t}|N_s^\varepsilon|\to 0,$$

hence $\sup_{0\le s\le t}|N_s^\varepsilon|$ tends to zero in $\widetilde{\mathbb{P}}$ probability, or equivalently in law.

In order to treat the other terms, we adopt the point of view of the S-topology of Jakubovski [6].

We define the conditional variation of the process A^ε on the interval $[0,t]$ as the quantity

$$\mathcal{C}V_t(A^\varepsilon)=\sup\widetilde{\mathbb{E}}\left(\sum_i\left|\widetilde{\mathbb{E}}\left(A_{t_{i+1}}^\varepsilon-A_{t_i}^\varepsilon/\mathcal{F}_{t_i}^\varepsilon\right)\right|\right),$$

where the supremum is taken over all the partitions of the interval $[0,t]$.

Clearly

$$\mathcal{C}V_t(A^\varepsilon)\le\widetilde{\mathbb{E}}\left(\int_0^t|f(X_s^\varepsilon,Y_s^\varepsilon)|\,ds+\int_0^t|h(X_s^\varepsilon,Y_s^\varepsilon)|\,d\phi_r^\varepsilon\right),$$

and it follows from step 1 and the asumptions (H.5), (H.8) that

$$\sup_\varepsilon\left(\mathcal{C}V_t(A^\varepsilon)+\sup_{0\le s\le t}\widetilde{\mathbb{E}}|Y_s^\varepsilon|+\sup_{0\le s\le t}\widetilde{\mathbb{E}}\left|\int_0^s\widetilde{Z}_r^\varepsilon d\widetilde{B}_r\right|\right)<+\infty,$$

hence the sequences $\{(Y_s^\varepsilon, \int_0^t \widetilde{Z}_r^\varepsilon d\widetilde{B}_r); 0 \le r \le t\}$ satisfy Meyer-Zheng's tightness criterion for quasimartingales under $\widetilde{\mathbb{P}}$ (see [6, 7, 9]).

Step 3: Passage to the limit. After extraction of a suitable subsequence, which we omit as an abuse of notations, we have that under $\widetilde{\mathbb{P}}$

$$\left(X^\varepsilon, \widetilde{M}^\varepsilon, \phi^\varepsilon, Y^\varepsilon, \int_0^\cdot \widetilde{Z}_r^\varepsilon d\widetilde{B}_s\right) \Longrightarrow \left(X, \widetilde{M}^X, \phi, Y, \overline{M}\right)$$

weakly on $C\left([0,t]\,;\mathbb{R}^{2d+1}\right) \times D\left([0,t]\,;\mathbb{R}^2\right)$ equipped with the product of the topology of uniform convergence on the first factor, and the S-topology on the second factor, where

$$\widetilde{M}_s^\varepsilon = \int_0^s (I + \nabla\chi)\sigma(X_s^\varepsilon/\varepsilon)d\widetilde{B}_s.$$

It remains to establish the two next results:

Lemma 3.2. *Let $\varphi : \mathbb{R}^d \times \mathbb{R} \to \mathbb{R}$ be measurable and locally bounded, periodic of period one in each direction with respect to its first argument, continuous with respect to its second argument, uniformly with respect to the first. Then the sequence of processes*

$$\left\{\int_0^s \varphi\left(\frac{X_r^\varepsilon}{\varepsilon}, Y_r^\varepsilon\right) dr;\ 0 \le s \le t\right\}$$

converges in law under $\widetilde{\mathbb{P}}$ to

$$\left\{\int_0^s \varphi_0(Y_r) dr;\ 0 \le s \le t\right\},$$

where

$$\varphi_0\left(y\right) = \int_{\mathbf{T}^d} \varphi\left(x,y\right) m(dx).$$

Proof. The proof is identical to that of Lemma 4.2 in [12]. □

Lemma 3.3. *Let $h : \mathbb{R}^d \times \mathbb{R} \to \mathbb{R}$ be continuous and periodic of period one in each direction with respect to its first argument. Then*

$$\sup_{0\le s\le T}\left|\int_0^s h(\frac{X_r^\varepsilon}{\varepsilon}, Y_r^\varepsilon)d\phi_r^\varepsilon - \int_0^s h_0(Y_r)d\phi_r\right| \to 0$$

in $\widetilde{\mathbb{P}}$ probability as $\varepsilon \to 0$, where

$$h_0\left(y\right) = \int_{\mathbf{T}^{d-1}} h\left(x,y\right) \widetilde{m}(dx).$$

Proof. We can consider h as a mapping from $\mathbf{T}^d \times \mathbb{R} \to \mathbb{R}$. It is sufficient to treat the case of h continuous with compact support in $\mathbf{T}^d \times \mathbb{R}$. Since such a function can be uniformly approximated by a function of the form

$$\sum_1^n h_i(x)g_i(y),$$

it is sufficient to establish the result for h of the form:

$$h(x,y) = h(x)g(y),$$

with h and g continuous and bounded. Let

$$\bar{h} = \left(\inf_{x\in\mathbf{T}^d} h(x)\right)^-, \qquad \tilde{h} = \int_{T^{d-1}} \left(h(x) + \bar{h}\right) \tilde{m}(dx).$$

It suffices to prove that

$$\left(\int_0^t (h(\frac{X_s^\varepsilon}{\varepsilon}) + \bar{h})g(Y_s^\varepsilon)d\phi_s^\varepsilon, \int_0^t g(Y_s^\varepsilon)d\phi_s^\varepsilon)\right) \stackrel{\mathcal{L}}{\to} \left(\tilde{h}\int_0^t g(Y_s)d\phi_s, \int_0^t g(Y_s)d\phi_s\right),$$

as $\varepsilon \to 0$. But

$$\left(\int_0^t (h(\frac{X_s^\varepsilon}{\varepsilon}) + \bar{h})d\phi_s^\varepsilon\ ; 0 \le t \le T\right)_{\varepsilon\searrow 0}$$

is a collection of increasing functions which converges in law uniformly in t to the continuous function $(\tilde{h}\phi_t\ ; 0 \le t \le T)$, see Lemma 6.4 in Tanaka [19]. Since moreover $\{V_t^\varepsilon \stackrel{\Delta}{=} g(Y_t^\varepsilon);\ 0 \le t \le T\}$ converges in the same sense as $\{Y_t^\varepsilon\}$, the lemma follows clearly from the fact that the mapping

$$(z,\phi) \to \int_0^t z d\phi$$

is continuous from $D([0,T]) \times V_+^c$ into $C([0,T])$, where we equip $D([0,T])$ with the S-topology, V_+^c and $C([0,T])$ with the sup-norm topology. Here and below V_+ denotes the set of cadlag increasing functions from $[0,T]$ into $\mathbb{R}$ and V_+^c the set of continuous increasing functions from $[0,T]$ into $\mathbb{R}$.

We now prove this last assertion. Let $z_n \to z$ in the sense of the S-topology, which implies that the same is true for the sequences z_n^+ and z_n^- (which converges respectively to z^+ and z^-), hence we can as well assume that $z(t) \ge 0$, $0 \le t \le T$.

Let $z_n \stackrel{S}{\to} z$, $z_n \ge 0$, and $\phi_n \to \phi$ pointwise, $\phi_n \in V_+^c$, $\phi_n(0) = 0$. Hence from Dini's theorem $\phi_n \to \phi$ uniformly. For each $n, p \in \mathbb{N}$, there exists $v_{n,p} \in V = V_+ - V_-$ such that:

$$\|z_n - v_{n,p}\|_\infty \le 1/p, \qquad v_{n,p}(t) \to v_{\infty,p}(t),$$

as $n \to \infty$, for all $t \in [0,T]\backslash D; p \in \mathbb{N}$; where D is a countable subset of $[0,T]$. Integrating by parts, one easily concludes that for each $p \in \mathbb{N}$,

$$\int_0^t v_{n,p}(s)d\phi_n(s) \to \int_0^t v_{\infty,p}(s)d\phi(s), t \in [0,T]\backslash D.$$

Consequently

$$\int_0^t z_n(s)d\phi_n(s) \to \int_0^t z(s)d\phi(s), t \in [0,T]\backslash D.$$

But since $\int_0^\cdot z(s)d\phi(s)$ is continuous increasing, it follows again from Dini's theorem that the last convergence holds uniformly on $[0,T]$. □

It is then easy to pass to the limit in the SDE and the BSDE, and obtain that for all $0 \leq s \leq t$,

$$\begin{aligned} X_s &= x + c_0 s + \int_0^s b_0(Y_r)\,dr + \widetilde{M}_s^X + \gamma_0 \phi_s, \\ Y_s &= g(X_t) + \int_s^t f_0(Y_r)dr + \int_s^t h_0(Y_r)d\phi_r + \overline{M}_s - \overline{M}_t, \end{aligned}$$

where

$$\begin{aligned} b_0(y) &= \int_{\mathbf{T}^d} \left[(I + \nabla\chi) a \frac{\partial^2 \hat{e}}{\partial x \partial y}(\cdot, y) \right](x) m(dx), \\ \langle\langle \widetilde{M}^X \rangle\rangle_s &= a_0 s, \\ f_0(y) &= \int_{\mathbf{T}^d} f(x,y)\, m(dx), \\ &+ \int_{\mathbf{T}^d} \left[\left\langle \frac{\partial \hat{e}}{\partial x}(.,y), c \right\rangle - \frac{\partial \hat{e}}{\partial y}(.,y)\, e(.,y) \right. \\ &+ \left. \frac{\partial^2 \hat{e}}{\partial x \partial y}(.,y)\, a \left(\frac{\partial \hat{e}}{\partial x} \right)^* (.,y) \right](x)\, m(dx), \\ h_0(y) &= \int_{\mathbf{T}^{d-1}} \left(h(x,y) + \frac{\partial \hat{e}}{\partial x} \gamma(x,y) \right) \widetilde{m}(dx). \end{aligned}$$

Using a similar argument as in Pardoux [11] (see also [12, 14]), one can prove that $\overline{M}$ and M^X are $\mathcal{F}_t^{X,Y}$-martingales.

Step 4: Identification of the limit. Let $(\overline{Y}, \overline{U})$ denote the unique solution of the BSDE

$$\overline{Y}_s = g(X_t) + \int_s^t f_0(\overline{Y}_r)dr + \int_s^t h_0(\overline{Y}_r)d\phi_r - \int_s^t \overline{U}_r dM_r^X$$

satisfying

$$\mathbb{E}Tr \int_0^t \overline{U}_r d\langle\langle \widetilde{M}^X \rangle\rangle_r \overline{U}_r^* < \infty,$$

and let $\widetilde{M}_s = \int_0^s \overline{U}_r d\widetilde{M}_r^X$. Since $\overline{Y}$ and $\overline{U}$ are $\mathcal{F}_t^X$-adapted, and $\widetilde{M}_r^X$ is a $\mathcal{F}_r^{X,Y}$-martingale, so is also $\widetilde{M}$. It follows from Itô's formula for possibly discontinuous semimartingales that

$$\begin{aligned} &\mathbb{E}\left|Y_s - \overline{Y}_s\right|^2 + \mathbb{E}\left[M - \widetilde{M}\right]_t - \mathbb{E}\left[M - \widetilde{M}\right]_s \\ &= 2\mathbb{E}\int_s^t \left(f_0(Y_r) - f_0(\overline{Y}_r), Y_r - \overline{Y}_r\right) dr + 2\mathbb{E}\int_s^t \left(h_0(Y_r) - h_0(\overline{Y}_r), Y_r - \bar{Y}_r\right) d\phi_r \\ &\leq 2\mu\mathbb{E}\int_s^t \left|Y_r - \overline{Y}_r\right|^2 dr. \end{aligned}$$

(We use the fact that $\beta \leq 0$).

Hence from Gronwall's lemma $Y_s = \overline{Y}_s$, $0 \le s \le t$, $M \equiv \widetilde{M}$, Y and M are continuous and have the required properties.

Finally, for $s = 0$

$$Y_0^\varepsilon = g(X_t^\varepsilon) + A_t^\varepsilon + M_t^\varepsilon + N_t^\varepsilon.$$

Since our subsequence could have been chosen in such a way that for some $a > 0$

$$\{M_{s\wedge t}^\varepsilon; 0 \le s \le t + a\}$$

converges weakly to $\left\{ \overline{M}_{s\wedge t}; 0 \le s \le t + a \right\}$ on $D\left([0, t + a]\,; \mathbb{R}\right)$, we can assume that

$$M_t^\varepsilon \Rightarrow \overline{M}_t.$$

Consequently

$$Y_0^\varepsilon \to Y_0 = g(X_t) + \int_0^t f_0(Y_r)dr + \int_s^t h_0(Y_r)d\phi_r - \overline{M}_t$$

in probability, since the limit is deterministic.

4. Homogenization of semilinear PDEs with non linear boundary condition

The result of the above section permits to us to deduce weak convergence of a sequence (Y^ε) of solutions of BSDEs from weak convergence of the sequence (X^ε). In a sense, we deduce, from the probabilistic proof of convergence of linear PDEs with linear boundary condition due to Tanaka [19], a probabilistic proof of convergence of systems of semilinear PDEs with non linear boundary condition.

For each $x \in \overline{D}$, let $\{(X_s^{\varepsilon,x}, \phi_s^{\varepsilon,x}),\ s \ge 0\}$ denote the solution of the SDE (3). Let g, f, e, h be as in the previous section. For each $(t, x) \in \mathbb{R}_+ \times \overline{D}$, let

$$u^\varepsilon(t, x) \triangleq Y_0^\varepsilon,$$

where Y^ε denotes the solution of the BSDE considered in the previous section (which of course depends on the starting point x of $X_\cdot^\varepsilon$, and the final time t for the BSDE). The continuity of u^ε follows by standard BSDE argument (see e.g. [16]) from the continuity of the mapping $x \to X_\cdot^{\varepsilon,x}$ in probability. This fact can be shown by an argument very similar to the uniqueness argument in Remark 4.2 (ii) of Lions, Sznitman [8], page 527. One can show as in Pardoux, Zhang [16] that u^ε is a viscosity solution of the semilinear parabolic PDE

$$\begin{aligned}
\frac{\partial u^\varepsilon}{\partial t}(t,x) &= L_\varepsilon u^\varepsilon(t,x) + f\left(\frac{x}{\varepsilon}, u^\varepsilon(t,x)\right) + \frac{1}{\varepsilon} e\left(\frac{x}{\varepsilon}, u^\varepsilon(t,x)\right), x \in D,\\
\Gamma_\varepsilon u^\varepsilon(t,x) + h\left(\frac{x}{\varepsilon}, u^\varepsilon(t,x)\right) &= 0, x \in \partial D; t \ge 0,\\
u^\varepsilon(0,x) &= g(x), x \in D,
\end{aligned}$$

where L_ε and Γ_ε are the operators defined by (1) and (2) respectively.

Let u be the solution of the homogenized system

$$\begin{aligned}\frac{\partial u}{\partial t}(t,x) &= L_0 u(t,x) + b_0(u(t,x))\nabla u(t,x) + f_0\left(u(t,x)\right), \ x \in D,\\ \Gamma_0 u\,(t,x) + h_0(u(t,x)) &= 0, \ x \in \partial D,\\ u(0,x) &= g(x), x \in \partial D,\end{aligned}$$

where L_0 and Γ_0 are as defined in Section 2, b_0, f_0 and h_0 as defined in Section 3.

A corollary of the above theorem is the

Theorem 4.1. *Under the assumptions (H.1)–(H.8),* $\forall (t,x) \in \mathbb{R}_+ \times \overline{D}$, $u^\varepsilon(t,x) \to u(t,x)$, *as* $\varepsilon \to 0$.

References

[1] A. Bensoussan, J. L. Lions and G. C. Papanicolaou, *Asymptotic Analysis for Periodic Structures*, North-Holland, Amsterdam, **1978**.

[2] M. Freidlin, *The Dirichlet problem for an equation with periodic coefficients depending on a small parameter*, Teor. Verojatnost. i Primenen., **9** (1964), 133–139.

[3] M. Freidlin, *Markov processes and differential equations: asymptotic problems*, Lecture Notes in Mathematics – ETH Zürich, Birkhäuser, **1996**.

[4] G. Gaudron and É. Pardoux, *EDSR, convergence en loi et homogénéisation d'EDP paraboliques semilinéaires*, Ann. I.H.P., **37 (1)** (2001), 1–42.

[5] N. Ikeda and S. Watanabe, *Stochastic Differential Equations and Diffusion Processes*, North-Holland/Kodansha, Amsterdam, **1981**.

[6] A. Jakubowski, *A non-Skorohod topology on the Skorohod space*, Electronic J. of Prob., **2 (4)** (1997), 1–21.

[7] T. G. Kurtz, *Random time changes and convergence in distribution under Meyer-Zheng conditions*, Annals of Prob., **19** (1991), 1010–1034.

[8] P. L. Lions and A. S. Sznitman, *Stochastic differential equations with reflecting boundary conditions*, Comm. Pure & Appl. Math., **37** (1984), 511–537.

[9] P. A. Meyer and W. A. Zheng, *Tightness criteria for laws of semimartingales*, Ann. Inst. Henri Poincaré, **20** (1984), 353–372.

[10] G. C. Papanicolaou, D. W. Stroock and S. R. S. Varadhan, *Martingale approach to some limit theorems*, in *Statistical Mechanics and Dynamical Systems*, Univ. Conf. Turbulence, D. Ruelle (Ed.), Duke Univ. Math. Ser., **3** (1977).

[11] É. Pardoux, *BSDE's, weak convergence and homogenization of semilinear PDE's*, in F.H. Clarke & R.J. Stern (Eds.), *Nonlinear Analysis, Differential Equations and Control*, Kluwer Acad. Pub., 1999, 503–549.

[12] É. Pardoux, *Homogenization of linear and semilinear second order parabolic PDE's with periodic coefficients: a probabilistic approach*, J. of Funct. Anal., **167** (1999), 498–520.

[13] É. Pardoux and S. Peng, *Adapted solution of a backward stochastic differential equation*, Systems Control Letters, **14** (1990), 55–61.

[14] É. Pardoux and A. Yu. Veretennikov, *Averaging of backward SDEs, with application to semilinear PDEs*, Stochastics, **60** (1997), 255–270.

[15] É. Pardoux and A. Yu. Veretennikov, *On Poisson equation and diffusion approximation 1*, Ann. Prob., to appear.

[16] É. Pardoux and S. Zhang, *Generalized BSDEs and nonlinear boundary value problems*, Prob. Theory and Rel. Fields, **110** (1998), 535–558.

[17] A. V. Skorohod, *Stochastic equations for diffusion processes in a bounded region 1, 2*, Theory Prob. Appl., **6** (1961), 264–274, **7** (1962), 3–23.

[18] H. Tanaka, *Stochastic differential equations with reflecting boundary condition in convex regions*, Hir. Math. J., **9** (1979), 163–177.

[19] H. Tanaka, *Homogenization of Diffusion Processes with boundary Conditions*, in *Stochastic Analysis and Applications*, Adv. Probab. Related Topics, **7** (1984), 411–437.

[20] S. Watanabe, *On stochastic differential equations for multidimensional diffusion processes with boundary conditions, I,II*, J. Math. Kyoto Univ., **11** (1971), 169–180, 545–551.

Department of Mathematics
Faculty of Sciences Semlalia, B.P. S15
Marrakech
Morocco
E-mail address: `ouknine@ucam.ac.ma`

Laboratoire d'Analyse, Topologie, Probabilités, CNRS–UMR 6632
Centre de Mathématiques et d'Informatique
Université de Provence
39, rue F. Joliot–Curie
13453 Marseille cedex 13
France
E-mail address: `pardoux@cmi.univ-mrs.fr`

Progress in Probability, Vol. 52, 243–258

A Bayesian Adaptive Control Approach to Risk Management in a Binomial Model

W. Runggaldier, B. Trivellato and T. Vargiolu

Abstract. We consider the problem of shortfall risk minimization when there is uncertainty about the exact stochastic dynamics of the underlying. Starting from the general discrete time model and the approach described in Runggaldier and Zaccaria (1999), we derive explicit analytic solutions for the particular case of a binomial model when there is uncertainty about the probability of an "up-movement". The solution turns out to be a rather intuitive extension of that for the classical Cox-Ross-Rubinstein model.

1. Introduction

In incomplete markets the superhedging criterion allows one to eliminate the risk completely, but it requires in general too much initial capital; it corresponds in fact to a min-max-type criterion. One may then ask by how much one can lower the initial cost if one is willing to accept some risk or, dually, what is the risk corresponding to an initial capital less than what is required for superhedging. The shortfall risk minimization approach allows one to deal with these issues.

Given a market with a non-risky and a certain number of risky assets, let H_T be a liability to be hedged at some fixed future time T. Denote by $V_T(\varphi)$ the value at T of a portfolio corresponding to a self-financing investment strategy φ, possibly satisfying some additional constraints such as a shortselling prohibition. The problem is to find

$$J_0(S_0, V_0) := \inf_{\varphi} E^P_{S_0,V_0}\left\{\ell\left([H_T - V_T(\varphi)]^+\right)\right\} \tag{1}$$

for a given initial value S_0 of the asset(s) in the portfolio, for a given initial capital V_0 and where $\ell(\cdot)$ is a suitable increasing function such that $\ell(0) = 0$ and $\ell(x) > 0$ for all $x > 0$. For $\ell(z) = 1_{\{z>0\}}$, the right-hand side in (1) corresponds to the smallest shortfall probability. Problems of the type (1) have recently attracted

2000 *Mathematics Subject Classification.* 91B28, 93E35.
Key words and phrases. Risk management, shortfall risk minimization, restricted information, dynamic programming, Bayesian inference.
The authors wish to thank an anonymous referee for useful comments and suggestions.

considerable attention (see e.g. [2, 3, 4, 5, 7, 8, 9, 10, 16]). Let, for a given S_0,

$$V_0^*(S_0) := \inf\{V_0 \mid J_0(S_0, V_0) = 0\}. \tag{2}$$

Since $J_0(S_0, V_0) = 0$ means that $H_T \leq V_T(\varphi^*)$ P-almost surely (where φ^* is the optimal strategy in (1)), $V_0^*(S_0)$ is the minimal initial capital needed to *superhedge* the claim. It follows that if $V_0 \geq V_0^*(S_0)$ then $V_T(\varphi^*) \geq H_T$, P-a.s. For superhedging, the choice of the underlying probabilistic model for the evolution of the risky assets is thus irrelevant as long as it induces probability measures that are equivalent. However for the more general problem of risk minimization in (1) the probabilistic structure of the underlying model matters, but the true model is almost never known exactly. A possibility is then to use a min-max-type criterion as in [4] trying to find

$$\inf_{\varphi} \sup_{P\in\mathcal{P}} E^P_{S_0,V_0} \left\{ \ell\left([H_T - V_T(\varphi)]^+ \right) \right\},$$

where $\mathcal{P}$ is a family of "real world probability measures". However, such a criterion does not allow one to incorporate additional information on the underlying model as it becomes successively available.

Thus, an adaptive approach, corresponding to a Bayesian-type criterion, appears more appropriate. Such adaptive approaches have already been dealt with in the literature (see e.g. [2, 3, 4, 5] and, in the context of portfolio optimization, in [1, 11, 12, 13]). In all these papers the uncertainty is only in the stock appreciation rate. The tools are mainly probabilistic in nature (involving also measure transformation) and are based on convex duality. An explicit solution is essentially possible only in simpler cases and transaction costs are not taken into account.

In the present paper we base ourselves on [17] following an approach along the lines of discrete time stochastic adaptive control. In that work the authors give a general description of this approach and they apply it, in particular, to a multinomial model for the risky assets where the probabilities are not known; for the specific case of a binomial model some numerical results are also presented. Here we focus our attention on the binomial model. By neglecting transaction costs (the portfolio is rebalanced only at discrete dates, which limits their impact) and imposing the self-financing requirement as the only constraint on the investment strategies, we succeed in deriving explicit solutions for the optimal investment strategy and for the corresponding minimal value of the shortfall risk in the case $\ell(x) = x$. We do this both for the case when the probability p of an "up-movement" is known as well as when it is unknown and, according to the Bayesian approach, treated as a random variable with a Beta-type distribution. We obtain an analytic solution, that turns out to be an interesting variant of the Cox-Ross-Rubinstein (CRR) solution (see e.g. [15]), when the initial capital is insufficient for a perfect hedge.

The paper is organised as follows. In Section 2 we briefly recall some facts from the CRR binomial model, that are relevant for the sequel. In Section 3 we present a solution method to our problem which is based on backwards *Dynamic*

Programming (DP). In Section 4, by assuming $\ell(x) = x$, we compute the minimizing admissible strategy as well as an explicit evaluation formula for the minimal discounted shortfall risk in the case when p is known. In Section 5, again by assuming $\ell(x) = x$, we compute the minimizing strategy as well as the minimal discounted shortfall risk in the case when p is unknown.

2. The Cox-Ross-Rubinstein binomial model

We consider a discrete-time market model with the set of dates $0, 1, \ldots, N$, and with two primary traded securities: a risky asset (a *stock*) S and a risk-free investment (a *bond*) B. We assume that the value of the bond is constantly equal to 1 through time, and that the stock price process S satisfies

$$S_{n+1} = S_n \omega_n, \quad n = 0, \ldots, N, \tag{3}$$

where $S_0 > 0$ is a given constant and $\{\omega_n\}_{n=0,\ldots,N}$ is a sequence of i.i.d. random variables defined on a probability space $(\Omega, \mathcal{F}, P)$, taking only two real values d and u satisfying $0 < d < 1 < u$, with probability law

$$p := P\{\omega_n = u\} = 1 - P\{\omega_n = d\}, \quad n = 0, \ldots, N.$$

We can assume without loss of generality $B_n \equiv 1$ by letting S_n be the discounted prices of the asset. Let us then denote by $\varphi_n = (\eta_n, \psi_n)$, $n = 0, \ldots, N$, an *investment strategy*, where η_n stands for the amount of the nonrisky asset and ψ_n stands for the number of units of the risky asset that are held in the portfolio in period n. We assume that φ_n is adapted to the observation σ-algebra $\mathcal{F}_n^S := \sigma\{S_m,\ m \leq n\}$, for all $n = 0, \ldots, N$, and that $\varphi = \{\varphi_n\}_{n=0,\ldots,N}$ satisfies the self-financing property

$$\begin{aligned} V_0 &= \eta_0 + \psi_0 S_0, \\ V_{n+1} &:= \eta_{n+1} + \psi_{n+1} S_{n+1} = \eta_n + \psi_n S_{n+1}, \quad n = 0, \ldots, N-1, \end{aligned}$$

where V_0 is a given constant, representing the initial value of the portfolio. We shall denote by $\mathcal{A}_{\text{ad}}$ the set of all self-financing strategies (the *admissible investment strategies*).

It is a rather classical result that V follows the dynamics

$$V_{n+1} = V_n + \psi_n S_n(\omega_n - 1) =: V_{n+1}(V_n, S_n, \omega_n, \psi_n),$$

so that one can restrict oneself to just the decision variable ψ_n.

Consider a European contingent claim $H(S_N)$ and let P^* be the martingale measure for our model. It is well known that P^* corresponds to

$$P^*\{\omega_n = u\} = p^* := \frac{1-d}{u-d}, \quad P^*\{\omega_n = d\} = 1 - p^* = \frac{u-1}{u-d}, \quad n = 0, \ldots, N,$$

(see e.g. [15, Ch. 2]). The arbitrage free price V_n^* of $H(S_N)$ at time n, where $n = 0, \ldots, N-1$, is given by the Cox-Ross-Rubinstein (CRR) evaluation formula

$$V_n^*(S_n) = E^*[H(S_N)|S_n], \tag{4}$$

where by E^* we denote the expectation with respect to P^*. In particular, $V_0^*(S_0)$ is the minimal initial value of the portfolio needed to replicate the claim defined in (2). At time n, $n = 0, \dots, N-1$, the replicating strategy ψ_n is given by

$$\psi_n = \frac{V_{n+1}^*(S_n u) - V_{n+1}^*(S_n d)}{S_n(u-d)}. \tag{5}$$

If $V_0 < V_0^*(S_0)$, then the replication of the terminal payoff is not possible. In this situation, an investor may be interested in analyzing the *shortfall risk* defined as the expectation of the terminal deficit weighted by some loss function. Let us then introduce this problem in more detail.

Denote by $V_N(\varphi) = \eta_N + \psi_N S_N$ the value of the portfolio at time N corresponding to an admissible investment strategy φ. The minimal *shortfall risk* is defined as

$$J_0(S_0, V_0) := \min_{\varphi \in \mathcal{A}_{\text{ad}}} E^P_{S_0,V_0} \left\{ \ell\left([H(S_N) - V_N(\varphi)]^+\right) \right\}, \tag{6}$$

for a given initial value S_0 of the risky asset in the portfolio and a given initial capital $V_0 < V_0^*(S_0)$, where $\ell(\cdot) \geq 0$ is a suitable *loss function*, that is an increasing function such that $\ell(0) = 0$ and $\ell(x) > 0$ for all $x > 0$.

In this paper we consider the optimization problem (6) by assuming either that the probability p is known or that it is not. For the case when p is unknown, we adopt a Bayesian-type approach which allows us to incorporate additional information on the underlying model as it becomes successively available.

3. The dynamic programming algorithm

In this section we provide a DP algorithm to compute a solution to our problem (6) both for the case when p is known and when it is not (see [6] for an analogous algorithm for the case of superhedging with transaction costs). In the case where p is unknown, adopting the Bayesian point of view, we use the ordinary Bayes formula to successively update the initial (prior) density $h(p)$ of p on the basis of $\{\mathcal{F}_n^S\}_{n=0,\dots,N}$. This leads to what is called the Bayesian DP algorithm (see e.g. [14, 18]).

3.1. DP algorithm when p is known

The DP algorithm proceeds backwards according to the following steps:

$$\begin{aligned} J_N(s,v) &= \ell((H(s)-v)^+), \\ J_{n-1}(S_{n-1},V_{n-1}) &= \inf_{\psi_{n-1}} E_{S_{n-1},V_{n-1}}\{J_n(S_n,V_n)\} \\ &= \inf_{\psi_{n-1}} \{pJ_n(S_{n-1}u, V_{n-1}+\psi_{n-1}S_{n-1}(u-1)) \\ &\quad +(1-p)J_n(S_{n-1}d, V_{n-1}+\psi_{n-1}S_{n-1}(d-1))\}, \end{aligned} \tag{7}$$

for $n = N, \dots, 1$.

3.2. DP algorithm when p is unknown

The Bayesian DP algorithm proceeds similarly to the DP algorithm:

$$\begin{aligned} J_N(s,v) &= \ell((H(s)-v)^+), \\ J_{n-1}(S_{n-1},V_{n-1}) &= \inf_{\psi_{n-1}} E^{P_{n-1}}_{S_{n-1},V_{n-1}}\{J_n(S_n,V_n)\}. \end{aligned} \tag{8}$$

Now p too is a random variable, and its distribution depends on the information $\mathcal{F}^S_n$ up to time n. We incorporate this information in the probability measure P_n. Since $P\{\omega_n|p\}$ is the binomial distribution, a conjugate family of distributions of p is that of the Beta distributions. With a prior density

$$h_0(p) \propto p^{\alpha_0}(1-p)^{\beta_0},$$

with $\alpha_0, \beta_0 \geq 0$, the posterior density in period n becomes

$$h_n(p) \propto p^{\alpha_n}(1-p)^{\beta_n},$$

where, denoting by u_n the total number of "up-movements" ($u_0 := 0$) cumulated up to time n,

$$\alpha_n = \alpha_0 + u_n\ , \quad \beta_n = \beta_0 + n - u_n.$$

In particular, for $n = 0$ and $\alpha_0 = \beta_0 = 0$ the prior density $h_0(p)$ becomes the uniform density.

Since the values of p enter the DP recursions linearly, by the "smoothing property" of conditional expectations it is easily seen that, also in the present case, the DP recursions are given by the previous steps, except that p has to be replaced in $J_{n-1}(S_{n-1},V_{n-1})$, $n = N,\ldots,1$, by $E_{\alpha_{n-1},\beta_{n-1}}[p] := E\left[p|\mathcal{F}^S_{n-1}\right]$ and $1-p$ by $1-E_{\alpha_{n-1},\beta_{n-1}}[p]$, where

$$E_{\alpha,\beta}[p] = \frac{\alpha+1}{\alpha+\beta+2}.$$

4. Explicit solutions when p is known

In this section we are concerned with the evaluation of the minimal *discounted shortfall risk* (6) and the corresponding strategy in the case when $\ell(x) = x$. Due to the possibility of making direct calculations on the DP algorithm steps, we can derive explicit evaluation formulas. These evaluation formulas are simple and meaningful, showing explicitly what was to be expected: that is, the shortfall risk is decreasing with respect to the initial capital (we will show that such dependence is linear), and it is always equal to zero when the level of the initial capital is greater than or equal to $V_0^*(S_0)$. We first consider the case when there is complete information on the underlying market and we compare our results with the well-known results on *perfect hedging* of a European contingent claim.

In the following theorem we give some formulas for the optimal discounted shortfall risk and for the corresponding investment strategy.

Theorem 4.1. *Consider a European contingent claim H on a stock whose price S is assumed to follow the CRR binomial model (3). Let $V_n^*(S_n)$, where $n = 0, \dots, N-1$, be the arbitrage free price at time n defined by the CRR evaluation formula (4). Assume the loss function $\ell(\cdot)$ in (6) is the identity function $\ell(x) = x$. Then*

i) *if $p > p^*$, then*

$$J_n(S_n, V_n) = \left(\frac{1-p}{1-p^*}\right)^{N-n} [V_n^*(S_n) - V_n]^+ , \tag{9}$$

for $n = 0, \dots, N-1$. In particular, for $n = 0$ the minimal discounted shortfall risk is

$$J_0(S_0, V_0) = \left(\frac{1-p}{1-p^*}\right)^{N} [V_0^*(S_0) - V_0]^+ ,$$

where $V_0^*(S_0) = C_0^*(S_0)$. The minimizing investment strategy is given by

$$\psi_n^1 = \frac{V_{n+1}^*(S_n u) - V_n}{S_n(u-1)}, \tag{10}$$

for $n = 0, \dots, N-1$;

ii) *if $p < p^*$, then*

$$J_n(S_n, V_n) = \left(\frac{p}{p^*}\right)^{N-n} [V_n^*(S_n) - V_n]^+ ,$$

for $n = 0, \dots, N-1$. In particular, for $n = 0$ the minimal discounted shortfall risk is

$$J_0(S_0, V_0) = \left(\frac{p}{p^*}\right)^{N} [V_0^*(S_0) - V_0]^+ ,$$

where $V_0^*(S_0) = C_0^*(S_0)$. The minimizing investment strategy is given by

$$\psi_n^2 = \frac{V_{n+1}^*(S_n d) - V_n}{S_n(d-1)}, \tag{11}$$

for $n = 0, \dots, N-1$.

Proof. We start from $n = N-1$, considering expression (7) of $J_{N-1}(S_{N-1}, V_{N-1})$. The function to be minimized in (7) is a linear combination of the two piecewise affine functions

$$[H(S_{N-1}u) - V_{N-1} - \psi_{N-1} S_{N-1}(u-1)]^+ \tag{12}$$

and

$$[H(S_{N-1}d) - V_{N-1} - \psi_{N-1} S_{N-1}(d-1)]^+ \tag{13}$$

The function in (12) is decreasing for ψ_{N-1} less than ψ_{N-1}^1 in Equation (10), since $u > 1$ and S_{N-1} is positive, and thereafter it is equal to zero, while the function in (13) is equal to zero for ψ_{N-1} less than ψ_{N-1}^2 in Equation (11), since $d < 1$ and S_{N-1} is positive, and from there on it is increasing. Therefore, if $\psi_{N-1}^1 \le \psi_{N-1}^2$

or, equivalently, if $V_{N-1} \geq E^*[H(S_N)]$, then both strategies ψ^1_{N-1} and ψ^2_{N-1} are optimal as well as any admissible strategy between them. If $V_{N-1} < E^*[H(S_N)]$, then in order to establish the infimum in (7) it suffices to analyze the sign of the slope of

$$\begin{aligned} &p[H(S_{N-1}u) - V_{N-1} - \psi_{N-1}S_{N-1}(u-1)] \\ &\quad +(1-p)[H(S_{N-1}d) - V_{N-1} - \psi_{N-1}S_{N-1}(d-1)], \end{aligned}$$

which is given by the expression

$$S_{N-1}[p(d-u) + 1 - d]. \tag{14}$$

If (14) is less than zero, or, equivalently, if $p > p^*$, then the infimum in (7) is achieved at ψ^1_{N-1}. If this is the case, by putting ψ^1_{N-1} in (7) we obtain

$$J_{N-1}(S_{N-1}, V_{N-1}) = \frac{1-p}{1-p^*}\left[C^*_{N-1}(S_{N-1}) - V_{N-1}\right]^+ .$$

Conversely, if $p < p^*$ then the infimum in (7) is achieved at ψ^2_{N-1}, and we obtain

$$J_{N-1}(S_{N-1}, V_{N-1}) = \frac{p}{p^*}\left[C^*_{N-1}(S_{N-1}) - V_{N-1}\right]^+ .$$

This shows that formula (9) is true for $n = N-1$.

We now proceed by backward induction with respect to n. Assume $p > p^*$ from now on. Assume that equality (9) holds for n, where $n = N-1, \ldots, 1$, with the minimizing strategy given by (10). We show that it also holds for $n-1$, with the same strategy as (10) for $n-1$. From (7) of the DP algorithm we have

$$\begin{aligned} J_{n-1}(S_{n-1}, V_{n-1}) \quad = \quad & \inf_{\psi_{n-1}} \{p\, J_n\left(S_{n-1}u, V_n(V_{n-1}, S_{n-1}, u, \psi_{n-1})\right) \\ & +(1-p)\, J_n\left(S_{n-1}d, V_n(V_{n-1}, S_{n-1}, d, \psi_{n-1})\right)\}, \end{aligned}$$

which implies, by induction,

$$\begin{aligned} J_{n-1}(S_{n-1}, V_{n-1}) = \left(\frac{1-p}{1-p^*}\right)^{N-n} & \inf_{\psi_{n-1}} \{p\, [V^*_n(S_{n-1}u) - V_{n-1} \\ -\psi_{n-1}S_{n-1}(u-1)]^+ + (1-p)\, & [V^*_n(S_{n-1}d) - V_{n-1} - \psi_{n-1}S_{n-1}(d-1)]^+\Big\}. \end{aligned} \tag{15}$$

Using the same arguments as in the first step, and taking into account that the slope of the expression

$$\begin{aligned} &p\,[V^*_n(S_{n-1}u) - V_{n-1} - \psi_{n-1}S_{n-1}(u-1)] \\ &\quad +(1-p)\,[V^*_n(S_{n-1}d) - V_{n-1} - \psi_{n-1}S_{n-1}(d-1)] \end{aligned}$$

is again given by (14) with S_{N-1} replaced by S_{n-1}, we have that, under the assumption $p > p^*$, the infimum in expression (15) is achieved at ψ^1_{n-1} given by (10). Putting ψ^1_{n-1} in (15) we easily obtain

$$J_{n-1}(S_{n-1}, V_{n-1}) = \left(\frac{1-p}{1-p^*}\right)^{N-n+1}\left[V^*_{n-1}(S_{n-1}) - V_{n-1}\right]^+ .$$

This ends the proof of *i*). The proof of *ii*) can be obtained by using arguments similar to those of *i*). We only observe that, under the assumption $p < p^*$, the infimum in the DP algorithm steps is achieved at ψ_{n-1}, for $n = 0, \ldots, N-1$, satisfying

$$V_n^*(S_{n-1}d) - V_{n-1} - \psi_{n-1}S_{n-1}(d-1) = 0.$$

□

Remark 4.2. Notice that this approach is linked to the CRR model: in fact, by calculating the expected shortfall risk under the *historical* probability P, we arrive at an expression containing the expected price of the claim under the *risk-neutral* probability P^*. Moreover, the hedging strategy in this case is similar to (5), which is the one of the CRR model: in fact in the CRR model the hedging strategy ψ is equal to the ratio between the difference of the expected prices of the claim in the two possible future outcomes and the different prices of the underlying; here ψ is equal to the ratio between the difference of the expected price of the claim in one of the possible future outcomes and the value the portfolio would have if it were invested in the bond B, and the difference between the price of the underlying in the same future outcome considered before and its present price as if it were invested in the bond B. In other words, it is as if we were hedging a claim having a payoff that in each state of nature could be the one of the original claim or the money corresponding to the present value of the portfolio; in the same way, the underlying could either assume the value corresponding to the value of the derivative, or the value corresponding to a riskless investment.

Remark 4.3. Differently from [8], here we have not imposed that $V \geq 0$. This leads to different results: in fact, if we imposed $V \geq 0$, we would have obtained (as they do) an optimal strategy equal to the replicating strategy of a modified claim that is between zero and the original claim; we have a different strategy, that in general gives an optimal expected shortfall *lower* than they have. In particular, our strategy succeeds in replicating perfectly the claim in all the states of nature except the least probable one (see [7] for an explicit proof), so that the expected shortfall comes entirely from this state of nature. However, if V_0 is near V_0^*, then V remains positive at all times prior to the maturity N, so the two strategies that we obtain by imposing or not the constraint $V \geq 0$ coincide.

5. Explicit solutions when p is unknown

The formulas given in the following theorem for the optimal discounted shortfall risk are similar to those given in Theorem 4.1, and the minimizing investment strategies are the same. However, while in the previous case only two alternatives for the possible values of p^* were considered, i.e. $p^* > p$ and $p^* < p$, here at each step n we have to consider several alternatives according to the estimates of p, each of them leading to a different formula for $J_n(S_n, V_n)$.

Theorem 5.1. *Consider the assumptions of Theorem 4.1 for the case when p is unknown with a prior $h_0(p) \propto p^{\alpha_0}(1-p)^{\beta_0}$, with $\alpha_0, \beta_0 \geq 0$. Then*

i) *if $p^* < E_{\alpha_n,\beta_n+N-n-1}[p]$, then*

$$J_n(S_n, V_n) = \left(\prod_{j=0}^{N-n-1} \frac{1-E_{\alpha_n,\beta_n+j}[p]}{1-p^*}\right) [V_n^*(S_n) - V_n]^+ . \tag{16}$$

The minimizing investment strategy is given by (10).

ii) *if $E_{\alpha_n+i,\beta_n+N-n-1-i}[p] < p^* < E_{\alpha_n+i+1,\beta_n+N-n-2-i}[p]$, where $i = 0, \ldots, N-n-2$, then*

$$J_n(S_n, V_n) = \left(\prod_{j=0}^{N-n-2-i} \frac{1-E_{\alpha_n,\beta_n+j}[p]}{1-p^*}\right) \times \left(\prod_{j=0}^{i} \frac{E_{\alpha_n+j,\beta_n+N-n-1-i}[p]}{p^*}\right) [V_n^*(S_n) - V_n]^+ \tag{17}$$

$$= \left(\prod_{j=0}^{i} \frac{E_{\alpha_n+j,\beta_n}[p]}{p^*}\right)\left(\prod_{j=0}^{N-n-2-i} \frac{1-E_{\alpha_n+i+1,\beta_n+j}[p]}{1-p^*}\right) [V_n^*(S_n) - V_n]^+ \tag{18}$$

Both the strategies (10) *and* (11) *are optimal as well as any admissible strategy between them.*

iii) *if $p^* > E_{\alpha_n+N-n-1,\beta_n}[p]$, then*

$$J_n(S_n, V_n) = \left(\prod_{j=0}^{N-n-1} \frac{E_{\alpha_n+j,\beta_n}[p]}{p^*}\right) [V_n^*(S_n) - V_n]^+. \tag{19}$$

The minimizing investment strategy is given by (11).

Proof. We start from $n = N-1$ by considering the expression

$$\begin{aligned} & J_{N-1}(S_{N-1}, V_{N-1}) \qquad\qquad (20)\\ &= \inf_{\psi_{N-1}} E_{\alpha_{N-1},\beta_{N-1}}[p]\,[H(S_{N-1}u) - V_{N-1} - \psi_{N-1}S_{N-1}(u-1)]^+ \\ &\quad + (1 - E_{\alpha_{N-1},\beta_{N-1}}[p])\,[H(S_{N-1}d) - V_{N-1} - \psi_{N-1}S_{N-1}(d-1)]^+ , \end{aligned}$$

whose infimum is achieved, by using arguments similar to those in the proof of Theorem 4.1, at ψ_{n-1}^1 in Equation (10) if $E_{\alpha_{N-1},\beta_{N-1}}[p] > p^*$ and at ψ_{n-1}^2 in Equation (11) if $E_{\alpha_{N-1},\beta_{N-1}}[p] < p^*$. Therefore, putting the minimizing strategy in (20) we have

$$J_{N-1}(S_{N-1}, V_{N-1}) = \frac{1-E_{\alpha_{N-1},\beta_{N-1}}[p]}{1-p^*}\left[C_{N-1}^*(S_{N-1}) - V_{N-1}\right]^+$$

if $E_{\alpha_{N-1},\beta_{N-1}}[p] > p^*$, and

$$J_{N-1}(S_{N-1}, V_{N-1}) = \frac{E_{\alpha_{N-1},\beta_{N-1}}[p]}{p^*}\left[C^*_{N-1}(S_{N-1}) - V_{N-1}\right]^+$$

if $E_{\alpha_{N-1},\beta_{N-1}}[p] < p^*$, showing that formulas (16)–(19) are true for $n = N-1$ (let us observe that the validity of *ii*) is trivial for $n = N-1$).

We now proceed by induction with respect to n. We assume that equalities (16)–(19) hold for n, where $n = 1, \ldots, N-1$, and we show that they also hold for $n-1$. As regards alternative *ii*), we shall only prove formula (17). Indeed, it is not difficult to check (we omit calculations) the validity of the equality

$$\left(\prod_{j=0}^{N-n-2-i} \frac{1 - E_{\alpha_n,\beta_n+j}[p]}{1-p^*}\right)\left(\prod_{j=0}^{i} \frac{E_{\alpha_n+j,\beta_n+N-n-1-i}[p]}{p^*}\right)$$
$$= \left(\prod_{j=0}^{i} \frac{E_{\alpha_n+j,\beta_n}[p]}{p^*}\right)\left(\prod_{j=0}^{N-n-2-i} \frac{1 - E_{\alpha_n+i+1,\beta_n+j}[p]}{1-p^*}\right),$$

corresponding to, respectively, (17) and (18) of *ii*). Let us remark that, as we shall see below, formula (17) (respectively, (18)) is obtained by always choosing strategy (10) (respectively, (11)) at each step n where, for some $i \in \{0, \ldots, N-n-2\}$, we have

$$E_{\alpha_n+i,\beta_n+N-n-1-i}[p] < p^* < E_{\alpha_n+i+1,\beta_n+N-n-2-i}[p].$$

This choice will be possible since, under the above condition for p^*, both the strategies (10) and (11), as well as any admissible strategy between them, will be optimal. In fact, other representation formulas for J_n, different from (17) and (18), but equivalent to them, could be possible, each of them corresponding to a different procedure for selecting a minimizing strategy between (10) and (11).

From (8), we have

$$J_{n-1}(S_{n-1}, V_{n-1}) = \inf_{\psi_{n-1}} E_{\alpha_{n-1},\beta_{n-1}}[p]\, J_n(S_{n-1}u, V_n(V_{n-1}, S_{n-1}, u, \psi_{n-1})) +$$
$$+ (1 - E_{\alpha_{n-1},\beta_{n-1}}[p]) J_n\left(S_{n-1}d, V_n(V_{n-1}, S_{n-1}, d, \psi_{n-1})\right),$$

which implies, by induction,

$$J_{n-1}(S_{n-1}, V_{n-1}) = \inf_{\psi_{n-1}} E_{\alpha_{n-1},\beta_{n-1}}[p]\, A(\alpha_{n-1}+1, \beta_{n-1})$$
$$\times \left[V^*_n(S_{n-1}u) - V_n\left(V_{n-1}, S_{n-1}, u, \psi_{n-1}\right)\right]^+ + \left(1 - E_{\alpha_{n-1},\beta_{n-1}}[p]\right)$$
$$\times A(\alpha_{n-1}, \beta_{n-1}+1)\left[V^*_n(S_{n-1}d) - V_n(V_{n-1}, S_{n-1}, d, \psi_{n-1})\right]^+,$$

where (choosing e.g. (17) when taking into account alternative *ii*))

$$A(\alpha,\beta) = \left(\prod_{j=0}^{N-n-1} \frac{1-E_{\alpha,\beta+j}[p]}{1-p^*}\right) 1_{\{p^*<E_{\alpha,\beta+N-n-1}[p]\}}$$

$$+ \sum_{k=0}^{N-n-2} \left(\prod_{j=0}^{N-n-2-k} \frac{1-E_{\alpha,\beta+j}[p]}{1-p^*}\right)\left(\prod_{j=0}^{k} \frac{E_{\alpha+j,\beta+N-n-1-k}[p]}{p^*}\right)$$

$$\times 1_{\{E_{\alpha+k,\beta+N-n-1-k}[p]<p^*<E_{\alpha+k+1,\beta+N-n-2-k}[p]\}} +$$

$$+ \left(\prod_{j=0}^{N-n-1} \frac{E_{\alpha+j,\beta}[p]}{p^*}\right) 1_{\{p^*>E_{\alpha+N-n-1,\beta}[p]\}}.$$

Proof of part i). If $p^* < E_{\alpha_{n-1},\beta_{n-1}+N-n}[p]$ then it is not difficult to check that only the first indicator functions in both $A(\alpha_{n-1}+1,\beta_{n-1})$ and $A(\alpha_{n-1},\beta_{n-1}+1)$ are equal to one, while all the others are equal to zero. Therefore

$$J_{n-1}(S_{n-1},V_{n-1}) = \inf_{\psi_{n-1}} E_{\alpha_{n-1},\beta_{n-1}}[p]\left(\prod_{j=0}^{N-n-1} \frac{1-E_{\alpha_{n-1}+1,\beta_{n-1}+j}[p]}{1-p^*}\right)$$

$$\times\left[V_n^*(S_{n-1}u) - V_{n-1} - \psi_{n-1}S_{n-1}(u-1)\right]^+$$

$$+\left(1-E_{\alpha_{n-1},\beta_{n-1}}[p]\right)\left(\prod_{j=0}^{N-n-1} \frac{1-E_{\alpha_{n-1},\beta_{n-1}+j+1}[p]}{1-p^*}\right)$$

$$\times\left[V_n^*(S_{n-1}d) - V_{n-1} - \psi_{n-1}S_{n-1}(d-1)\right]^+. \tag{21}$$

The sign of the slope of this linear expression in ψ_{n-1} is given by

$$E_{\alpha_{n-1},\beta_{n-1}}[p]\left(\prod_{j=0}^{N-n-1} \frac{1-E_{\alpha_{n-1}+1,\beta_{n-1}+j}[p]}{1-p^*}\right)(1-u) \tag{22}$$

$$+\left(1-E_{\alpha_{n-1},\beta_{n-1}}[p]\right)\left(\prod_{j=0}^{N-n-1} \frac{1-E_{\alpha_{n-1},\beta_{n-1}+j+1}[p]}{1-p^*}\right)(1-d).$$

After making elementary manipulations, one can rewrite (22) as the product of a suitable strictly positive term and $(p^* - E_{\alpha_{n-1},\beta_{n-1}+N-n}[p])$, which is less than zero by assumption. Therefore, using the same arguments as those in the proof of Theorem 4.1, we have that the infimum in (21) is achieved at ψ^1_{n-1} given by formula (10). Putting ψ^1_{n-1} in (21) we easily obtain

$$J_{n-1}(S_{n-1},V_{n-1}) = \left(\prod_{j=0}^{N-n} \frac{1-E_{\alpha_{n-1},\beta_{n-1}+j}[p]}{1-p^*}\right)\left[V_{n-1}^*(S_{n-1}) - V_{n-1}\right]^-,$$

which gives formula (16) with n replaced by $n-1$. This ends the proof of *i*).

Proof of part iii). It is not difficult to argue that the proof of *iii)* proceeds along the same lines as the proof of *i)*. We only observe that if $p^* > E_{\alpha_{n-1}+N-n,\beta_{n-1}}[p]$ (alternative *iii)* for $n-1$) then only the last indicator functions in both $A(\alpha_{n-1}+1,\beta_{n-1})$ and $A(\alpha_{n-1},\beta_{n-1}+1)$ are equal to one, while all the others are equal to zero. Moreover, the sign of the slope of the linear expression which arises from the DP algorithm is positive, so that the infimum is achieved at ψ_{n-1} given by formula (11), i.e. satisfying

$$V_n^*(S_{n-1}d) - V_{n-1} - \psi_{n-1}S_{n-1}(d-1) = 0.$$

Proof of part ii). Now we only have to prove *ii)*. As we shall see, the linear expressions arising from the DP algorithm will have angular coefficients equal to zero, giving the possibility to choose, as minimizing strategy, any admissible strategy between (10) and (11). In particular, choosing strategy (10) we shall prove the validity of formula (17). If

$$E_{\alpha_{n-1}+i,\beta_{n-1}+N-n-i}[p] < p^* < E_{\alpha_{n-1}+i+1,\beta_{n-1}+N-n-1-i}[p],$$

for some $i \in \{0,\dots,N-n-1\}$, then the last indicator function in $A(\alpha_{n-1}+1,\beta_{n-1})$ and the first indicator function in $A(\alpha_{n-1},\beta_{n-1}+1)$ are equal to zero, i.e.

$$1_{\{p^*>E_{\alpha_{n-1}+N-n,\beta_{n-1}}[p]\}} \equiv 0 \quad \text{and} \quad 1_{\{p^*<E_{\alpha_{n-1},\beta_{n-1}+N-n}[p]\}} \equiv 0$$

Moreover, we have to distinguish between the following three alternatives:

1. If $i = 0$, then we have

$$\begin{aligned} J_{n-1}(S_{n-1},V_{n-1}) = \inf_{\psi_{n-1}} E_{\alpha_{n-1},\beta_{n-1}}[p] \left(\prod_{j=0}^{N-n-1} \frac{1-E_{\alpha_{n-1}+1,\beta_{n-1}+j}[p]}{1-p^*}\right) \\ \times [V_n^*(S_{n-1}u) - V_{n-1} - \psi_{n-1}S_{n-1}(u-1)]^+ \\ + (1-E_{\alpha_{n-1},\beta_{n-1}}[p]) \left(\prod_{j=0}^{N-n-2} \frac{1-E_{\alpha_{n-1},\beta_{n-1}+1+j}[p]}{1-p^*}\right) \\ \times \frac{E_{\alpha_{n-1},\beta_{n-1}+N-n}[p]}{p^*} [V_n^*(S_{n-1}d) - V_{n-1} - \psi_{n-1}S_{n-1}(d-1)]^+ . \end{aligned}$$

Since the slope of this linear expression in ψ_{n-1} is equal to zero (we omit calculations), the infimum is achieved at both (10) and (11) as well as at any admissible strategy between them. Choosing e.g. (10) we obtain

$$\begin{aligned} J_{n-1}(S_{n-1},V_{n-1}) = \left(\prod_{j=0}^{N-n-1} \frac{1-E_{\alpha_{n-1},\beta_{n-1}+j}[p]}{1-p^*}\right) \\ \times \frac{E_{\alpha_{n-1},\beta_{n-1}+N-n}[p]}{p^*} \left[V_{n-1}^*(S_{n-1}) - V_{n-1}\right]^+, \end{aligned}$$

which gives formula (17) with n replaced by $n-1$ and $i = 0$.

2. If $i \in \{1, \dots, N-n-2\}$, then we have

$$
\begin{aligned}
&J_{n-1}(S_{n-1}, V_{n-1}) \\
&= \inf_{\psi_{n-1}} E_{\alpha_{n-1},\beta_{n-1}}[p] \left(\prod_{j=0}^{N-n-1-i} \frac{1 - E_{\alpha_{n-1}+1,\beta_{n-1}+j}[p]}{1-p^*} \right) \\
&\quad \times \left(\prod_{j=0}^{i-1} \frac{E_{\alpha_{n-1}+1+j,\beta_{n-1}+N-n-i}[p]}{p^*} \right) \\
&\quad \times [V_n^*(S_{n-1}u) - V_{n-1} - \psi_{n-1} S_{n-1}(u-1)]^+ \\
&\quad + (1 - E_{\alpha_{n-1},\beta_{n-1}}[p]) \left(\prod_{j=0}^{N-n-2-i} \frac{1 - E_{\alpha_{n-1},\beta_{n-1}+1+j}[p]}{1-p^*} \right) \\
&\quad \times \left(\prod_{j=0}^{i} \frac{E_{\alpha_{n-1}+j,\beta_{n-1}+N-n-i}[p]}{p^*} \right) [V_n^*(S_{n-1}d) - V_{n-1} - \psi_{n-1} S_{n-1}(d-1)]^+ .
\end{aligned}
$$

The slope of this linear expression in ψ_{n-1} is again equal to zero. Choosing e.g. (10) as ψ_{n-1} we obtain

$$
\begin{aligned}
J_{n-1}(S_{n-1}, V_{n-1}) = &\left(\prod_{j=0}^{N-n-1-i} \frac{1 - E_{\alpha_{n-1},\beta_{n-1}+j}[p]}{1-p^*} \right) \\
&\times \left(\prod_{j=0}^{i} \frac{E_{\alpha_{n-1}+j,\beta_{n-1}+N-n-i}[p]}{p^*} \right) [V_{n-1}^*(S_{n-1}) - V_{n-1}]^+ ,
\end{aligned}
$$

which gives formula (17) with n replaced by $n-1$ and $i \in \{1, \dots, N-n-2\}$.

3. If $i = N-n-1$, then we have

$$
\begin{aligned}
&J_{n-1}(S_{n-1}, V_{n-1}) \\
&= \inf_{\psi_{n-1}} E_{\alpha_{n-1},\beta_{n-1}}[p] \left(\frac{1 - E_{\alpha_{n-1}+1,\beta_{n-1}}[p]}{1-p^*} \right) \\
&\quad \times \left(\prod_{j=0}^{N-n-2} \frac{E_{\alpha_{n-1}+1+j,\beta_{n-1}+1}[p]}{p^*} \right) \\
&\quad \times [V_n^*(S_{n-1}u) - V_{n-1} - \psi_{n-1} S_{n-1}(u-1)]^+ + (1 - E_{\alpha_{n-1},\beta_{n-1}}[p]) \\
&\quad \times \left(\prod_{j=0}^{N-n-1} \frac{E_{\alpha_{n-1}+j,\beta_{n-1}+1}[p]}{p^*} \right) [V_n^*(S_{n-1}d) - V_{n-1} - \psi_{n-1} S_{n-1}(d-1)]^+ .
\end{aligned}
$$

The slope of this linear expression in ψ_{n-1} is again equal to zero. Choosing e.g. (10) as ψ_{n-1} we obtain

$$J_{n-1}(S_{n-1}, V_{n-1}) = \frac{1 - E_{\alpha_{n-1},\beta_{n-1}}[p]}{1-p^*} \times \left(\prod_{j=0}^{N-n-1} \frac{E_{\alpha_{n-1}+j,\beta_{n-1}+1}[p]}{p^*} \right) \left[V^*_{n-1}(S_{n-1}) - V_{n-1}\right]^+ .$$

The inductive step is complete, and so is the proof. □

Corollary 5.2. *Letting $n = 0$ in Theorem 5.1, the minimal discounted shortfall risk is*

i) *if $p^* < E_{\alpha_0,\beta_0+N-1}[p]$, then*

$$J_0(S_0, V_0) = \left(\prod_{j=0}^{N-1} \frac{1 - E_{\alpha_0,\beta_0+j}[p]}{1-p^*} \right) [V_0^*(S_0) - V_0]^+ ;$$

ii) *if $E_{\alpha_0+i,\beta_0+N-1-i}[p] < p^* < E_{\alpha_0+i+1,\beta_0+N-2-i}[p]$, where $i = 0, \dots, N-2$, then*

$$\begin{aligned} J_0(S_0, V_0) &= \left(\prod_{j=0}^{N-2-i} \frac{1 - E_{\alpha_0,\beta_0+j}[p]}{1-p^*} \right) \\ &\quad \times \left(\prod_{j=0}^{i} \frac{E_{\alpha_0+j,\beta_0+N-1-i}[p]}{p^*} \right) [V_0^*(S_0) - V_0]^+ \\ &= \left(\prod_{j=0}^{i} \frac{E_{\alpha_0+j,\beta_0}[p]}{p^*} \right) \left(\prod_{j=0}^{N-2-i} \frac{1 - E_{\alpha_0+i+1,\beta_0+j}[p]}{1-p^*} \right) [V_0^*(S_0) - V_0]^+ ; \end{aligned}$$

iii) *if $p^* > E_{\alpha_0+N-1,\beta_0}[p]$, then*

$$J_0(S_0, V_0) = \left(\prod_{j=0}^{N-1} \frac{E_{\alpha_0+j,\beta_0}[p]}{p^*} \right) [V_0^*(S_0) - V_0]^+ .$$

The minimizing investment strategies are those of Theorem 5.1.

Remark 5.3. In the case when p is unknown we obtain formulas that are similar to those of the case when p is known, with the following difference: while in the case when p is known we know immediately whether $p > p^*$ or not, and that relation either holds at all times n or does not hold, when p is unknown we cannot know whether $p > p^*$ or not, so we must use the Bayes estimators of p. Obviously these estimators change over time, so we obtain products of different factors depending on p^* and on the Bayes estimators of p, while in the case when p is known these factors are all equal either to p/p^* or to $(1-p)/(1-p^*)$. We also notice that the investment strategies are equal to those when p is known; this happens because the Bayes estimators of p enter linearly in the DP algorithm and do not modify the optimum.

Remark 5.4. Also when p is unknown, if we impose $V \geq 0$, we obtain different results: in fact in this case too our strategy succeeds in replicating perfectly the claim in all the states of nature but one (see [7]); with the constraint $V \geq 0$, it may happen that this is not possible, so that the optimal solution gives a shortfall higher than in our case. However, also in this case (as in the case when p is known), if V_0 is near V_0^*, then V remains positive at all times prior to the maturity N, so the two strategies that we obtain by imposing or not the constraint $V \geq 0$ coincide.

References

[1] S. Browne and W. Whitt, *Portfolio choice and the Bayesian Kelly criterion*, Adv. Appl. Prob., **28** (1996), 1145–1176.

[2] J. Cvitanić, *Methods of partial hedging*, Asia-Pacific Financial Markets, **6** (1999), 7–35.

[3] J. Cvitanić, *Minimizing expected loss of hedging in incomplete and constrained markets*, SIAM J. Control Optimization, **38 (4)** (2000), 1050–1066.

[4] J. Cvitanić and I. Karatzas, *On dynamic measures of risk*, Finance and Stochastics, **3 (4)** (1999), 451–482.

[5] J. Cvitanić and G. Spivak, *Maximizing the probability of a perfect hedge*, The Annals of Applied Probability, **9 (4)** (1999), 1303–1328.

[6] C. Edirisinghe, V. Naik and R. Uppal, *Optimal replication of options with transactions costs and trading restrictions*, J. of Financial and Quantitative Analysis, **28** (1993), 117–138.

[7] G. Favero, *Shortfall risk minimization under model uncertainty in the binomial case: adaptive and robust approaches*, Mathematical Methods of Operations Research **53 (3)** (2001), 493–503.

[8] H. Föllmer and P. Leukert, *Quantile hedging*, Finance and Stochastics, **3 (3)** (1999), 251–273.

[9] H. Föllmer and P. Leukert, *Efficient hedging: cost versus shortfall risk*, Finance and Stochastics, **4 (2)** (2000), 117–146.

[10] P. Guasoni, *Risk minimization under transaction costs*, forthcoming in Finance and Stochastics (2001).

[11] I. Karatzas and X. Zhao, *Bayesian adaptive portfolio optimization*, to appear in Advances in Mathematical Finance, Cambridge University Press, (1999).

[12] P. Lakner, *Utility maximization with partial information*, Stochastic Processes and their Applications, **56** (1995), 247–273.

[13] P. Lakner, *Optimal trading strategy for an investor: the case of partial information*, Stochastic Processes and their Applications, **76** (1998), 77–97.

[14] J. J. Martin, *Bayesian decision problems and Markov chains*, John Wiley and Sons, Inc., New York, **1967**.

[15] M. Musiela and M. Rutkowski, *Martingale methods in financial modelling*, Springer, Berlin, **1997**.

[16] H. Pham, *Dynamic L^p-hedging in discrete time under cone constraints*, SIAM J. Control Optimization, **38** **(3)** (2000), 665–682.

[17] W. J. Runggaldier and A. Zaccaria, *A stochastic control approach to risk management under restricted information*, Mathematical Finance, **10** (2000), 277–288.

[18] K. M. Van Hee, *Bayesian control of Markov chains*, Mathematical Centre Tracts 95, Amsterdam, **1978**.

Department of Pure and Applied Mathematics
University of Padova
via Belzoni 7
I-35131 Padova
Italy
E-mail address: {runggal,barbara,vargiolu}@math.unipd.it
URL: http://www.math.unipd.it/~probab/home/probgroup.html

Progress in Probability, Vol. 52, 259–268

Hölder Continuity for the Stochastic Heat Equation With Spatially Correlated Noise

M. Sanz-Solé and M. Sarrà

Abstract. We study the Hölder continuity in time and space of the solution of a stochastic heat equation with spatial parameter of any dimension d and spatially correlated noise. The conditions for this property to hold are given first in terms of the spectral measure of the noise and then by means of the correlation measure.

1. Introduction

The purpose of this paper is to study Hölder continuity properties of a class of stochastic heat equations on $[0,\infty)$ with d-dimensional spatial parameter in terms of properties on the correlation of the driving noise. More specifically, let

$$\begin{cases} LX(t,x) &= \sigma(X(t,x))\dot{F}(t,x) + b(X(t,x)), \\ X(0,x) &= X_0(x), \end{cases} \tag{1}$$

$t \geq 0$, $x \in \mathbb{R}^d$, with $L = \frac{\partial}{\partial t} - \frac{1}{2}\Delta$, σ, $b : \mathbb{R} \to \mathbb{R}$. The noise $F = \{F(\varphi),\ \varphi \in \mathcal{D}(\mathbb{R}^{d+1})\}$ is an $L^2(\Omega, \mathcal{F}, P)$-valued Gaussian process with mean zero and covariance functional given by

$$J(\varphi, \psi) = \int_{\mathbb{R}_+} ds \int_{\mathbb{R}^d} \Gamma(dx)(\varphi(s,\cdot) * \tilde{\psi}(s,\cdot))(x),$$

where $\tilde{\psi}(s,x) = \psi(s,-x)$ and Γ is a non-negative tempered measure. Denote by μ the spectral measure of Γ, which is also a tempered measure. Then

$$J(\varphi, \psi) = \int_{\mathbb{R}_+} ds \int_{\mathbb{R}^d} \mu(d\xi)\, \mathcal{F}\varphi(s,\cdot)(\xi)\overline{\mathcal{F}\psi(s,\cdot)(\xi)}.$$

Let $S(t,x)$ be the fundamental solution of the deterministic heat equation in $\mathbb{R}^d$, that is

$$S(t,x) = (2\pi t)^{-d/2} \exp\Big(-\frac{|x|^2}{2t}\Big).$$

2000 *Mathematics Subject Classification.* Primary 60H15, 60H05; Secondary 35K90.
Key words and phrases. Stochastic partial differential equations, Hölder continuity, heat equation.
This work has been partially supported by the grant PB 960088 from the *Subdirección General de formación y Promoción del Conocimiento* and the grant ERBF MRX CT960075A of the EU.

Set $G_t(x, X_0) = \int_{\mathbb{R}^d} S(t, x-y)X_0(y)dy$. A classical way to provide a meaning to the formal equation(1) is through the mild formulation. Here we take Walsh's approach (see [11] and [1]), as follows. Let $M = \{M_t(A),\ t \geq 0, A \in \mathcal{B}_b(\mathbb{R}^d)\}$ be a worthy martingale measure obtained as an extension of the process F. Then, by a solution of (1) we mean a jointly measurable real-valued process $X = \{X(t,x),\ t \geq 0,\ x \in \mathbb{R}^d\}$, adapted to the natural filtration generated by M such that

$$X(t,x) = G_t(x, X_0) + \int_0^t \int_{\mathbb{R}^d} S(t-s, x-y)\sigma(X(s,y))M(ds,dy) \\ + \int_0^t \int_{\mathbb{R}^d} S(t-s, x-y)b(X(s,y))dsdy. \quad (2)$$

We point out that, since we are considering real-valued solutions and keep the spatial dimension arbitrary, the definition of the stochastic integral in (2) needs some restriction on the correlation measure Γ. Indeed, assume that X_0 is measurable and bounded; let σ and b Lipschitz and suppose that

$$\int_{\mathbb{R}^d} \frac{\mu(d\xi)}{1+|\xi|^2} < +\infty. \quad (3)$$

Then there exists a unique solution of (2). Moreover, the process X is L^2-continuous and satisfies

$$\sup_{0 \leq t \leq T} \sup_{x \in \mathbb{R}^d} E\left(|X(t,x)|^p\right) < +\infty,$$

for any $T < 0$, $p \in [1, \infty)$. We refer the reader to [1] for the proof of this fact (see also [4] for the linear case, and [8, 9]).

Continuity in time of the process X defined in (2) follows from Theorem 2.1 in [8] assuming that μ is either finite or possess a density in L^p, $p \in (1, \frac{d}{d-2} \vee 1)$. More recently in [5] the continuity in (t,x) for a class of stochastic Volterra equations with correlated additive noise has been established.

Our aim is to have a deeper insight in the trajectories; we are interested in Hölder continuity properties as a basis for further developments like, for instance, regularity of the laws.

Results in this direction for the stochastic wave equation in dimension two can be found in [2, 6, 7]. In our previous work [10] we have proved that a reinforcing of (3), namely

$$\int_{\mathbb{R}^d} \frac{\mu(d\xi)}{(1+|\xi|^2)^\eta} < +\infty, \quad (\mathrm{H})_\eta$$

for some $\eta \in (0,1)$, provides Hölder continuity in t and x for a class of stochastic partial differential equations driven by a correlated additive noise. The class includes semilinear stochastic wave equation in spatial dimension $d \in \{1,2,3\}$ and the semilinear stochastic heat equation in arbitrary spatial dimension d. We prove here that for the nonlinear heat equation the assumption $(\mathrm{H})_\eta$ also leads to Hölder continuity. We notice that $(\mathrm{H})_\eta$ can also be formulated in terms of the correlation measure Γ (see [10], Proposition 5.3).

The proof, given in Section 2, combines the stochastic factorization method (see, e.g., [3]) and Kolmogorov's continuity criterium. The expression of the quadratic variation of the stochastic integral in terms of the Fourier transform of $S(t,x)$ and the precise expression for it allows to take advantadge of $(\mathrm{H})_\eta$ and reach the conclusion. The article ends with some examples which satisfy $(\mathrm{H})_\eta$.

2. Hölder continuity of the trajectories

Let $X = \{X(t,x),\ (t,x) \in \mathbb{R}_+ \times \mathbb{R}^d\}$ be the solution of (2). The purpose of this section is to prove the following result.

Theorem 2.1. *Assume that the initial condition X_0 is a bounded ρ-Hölder continuous function for some $\rho \in (0,1)$. We also suppose that the coefficients $\sigma,\ b : \mathbb{R} \to \mathbb{R}$ are Lipschitz functions and that conditions $(H)_\eta$ holds for some $\eta \in (0,1)$. Then the trajectories of X are a.s. β_1-Hölder continuous in t and β_2-Hölder continuous in x for any $\beta_1 \in (0, \frac{1}{2}(\rho \wedge (1-\eta)))$, $\beta_2 \in (0, \rho \wedge (1-\eta))$.*

The proof of this theorem uses the factorization method and, as usually, the Kolmogorov's continuity criterium.

For any $\alpha \in (0,1)$ set

$$Y_\alpha(r,z) = \int_0^r \int_{\mathbb{R}^d} S(r-s, z-y)\sigma(X(s,y))(r-s)^{-\alpha} M(ds,dy). \tag{4}$$

Then, the semigroup property of S and a stochastic Fubini's theorem (see, e.g., [11], Theorem 2.6) yield

$$\int_0^t \int_{\mathbb{R}^d} S(t-s, x-y)\sigma(X(s,y)) M(ds,dy)$$
$$= \frac{\sin(\pi\alpha)}{\pi} \int_0^t dr \int_{\mathbb{R}^d} dz\, S(t-r, x-z)(t-r)^{\alpha-1} Y_\alpha(r,z). \tag{5}$$

Our first purpose is to check that the process $Y_\alpha(r,z)$ is bounded in L^p. Set

$$Z(s,y) = \sigma(X(s,y)), \qquad (s,y) \in \mathbb{R}_+ \times \mathbb{R}^d.$$

Theorem 13 in [1] yields

$$\sup_{0\le s\le T}\ \sup_{y\in\mathbb{R}^d} E(|X(s,y)|^p) < \infty, \tag{6}$$

for any $T \ge 0$, $p \in [1,\infty)$. Thus,

$$\sup_{0\le s\le T}\ \sup_{y\in\mathbb{R}^d} E(|Z(s,y)|^p) < +\infty,$$

for any $T \ge 0$, $p \in [1,\infty)$. Moreover $(Z(s,y),\ s \in [0,T],\ y \in \mathbb{R}^d)$ defines a predictable spatially homogeneous stochastic process (see [1], Lemma 18). Therefore

$M_t^Z(A) = \int_0^t \int_A Z(s,y)M(ds,dy)$ defines a worthy martingale measure and $Y_\alpha(r,z)$ is a stochastic integral with respect to this worthy martingale measure, that is

$$Y_\alpha(r,z) = \Big(S(r-\cdot, z-\cdot)(r-\cdot)^{-\alpha} \cdot M^Z\Big)_r.$$

Lemma 2.2. *Assume the hypotheses of Theorem 2.1. Then, for any $T \geq 0$, $p \in [1,\infty)$, $\alpha \in (0, \frac{1-\eta}{2})$,*

$$\sup_{0\leq r\leq T} \sup_{z\in\mathbb{R}^d} E(|Y_\alpha(r,z)|^p) < +\infty.$$

Proof. Fix $p \in [1,\infty)$; Theorem 5 in [1] yields

$$E(|Y_\alpha(r,z)|^p) \leq C(p) \sup_{0\leq s\leq T} \sup_{y\in\mathbb{R}^d} E(|Z(s,y)|^p)\nu_{r,z}^{p/2}, \tag{7}$$

with

$$\nu_{r,z} = \int_0^r ds \int_{\mathbb{R}^d} \mu(d\xi) \big|\mathcal{F}\big[S(r-s, z-\cdot)(r-s)^{-\alpha}\big](\xi)\big|^2.$$

Here $\mathcal{F}S(r-s, z-\cdot)(\xi) = e^{-2\pi i\xi\cdot z}\exp(-2\pi^2(r-s)|\xi|^2)$. Thus we obtain

$$\begin{aligned} \sup_{z\in\mathbb{R}^d} \nu_{r,z} &= \int_{\mathbb{R}^d} \mu(d\xi) \int_0^r (r-s)^{-2\alpha} \exp(-4\pi^2(r-s)|\xi|^2)ds \\ &= (I_1(r) + I_2(r)), \end{aligned} \tag{8}$$

where

$$\begin{aligned} I_1(r) &= \int_{|\xi|\leq 1} \mu(d\xi) \int_0^r (r-s)^{-2\alpha} \exp(-4\pi^2(r-s)|\xi|^2)ds, \\ I_2(r) &= \int_{|\xi|>1} \mu(d\xi) \int_0^r (r-s)^{-2\alpha} \exp(-4\pi^2(r-s)|\xi|^2)ds. \end{aligned}$$

Since μ is a tempered measure,

$$I_1(r) \leq C(\alpha)r^{-2\alpha+1}. \tag{9}$$

On the other hand, for any $0 < \alpha < \frac{1}{2}$

$$I_2(r) \leq C\Gamma(1-2\alpha) \int_{|\xi|>1} \frac{\mu(d\xi)}{|\xi|^{2(1-2\alpha)}},$$

where Γ denotes the Euler gamma function. Therefore,

$$I_2(r) \leq C \int_{\mathbb{R}^d} \frac{\mu(d\xi)}{(1+|\xi|^2)^{1-2\alpha}}.$$

The assumption $(\mathrm{H})_\eta$ implies $\sup_r I_2(r) < +\infty$ whenever $\eta < 1-2\alpha$. Then we conclude using (7)–(9). □

Proof of Theorem 2.1. Fix $T,\ h>0$ and $p\in[2,\infty)$. We first show that

$$\sup_{0\le t\le T}\sup_{x\in\mathbb{R}^d} E\big(\big|X(t+h,x)-X(t,x)\big|^p\big)\le C(p,T)h^{\gamma p}, \qquad (10)$$

for any $\gamma\in(0,\frac{1}{2}(\rho\wedge(1-\eta)))$.

Indeed, let $Y_\alpha(r,z)$ be as defined in (4) with $\alpha\in(0,\frac{1-\eta}{2})$; we have

$$E\big(\big|X(t+h,x)-X(t,x)\big|^p\big)\le C(p,\alpha)\sum_{i=1}^4 I_i(t,h,x), \qquad (11)$$

where

$$\begin{aligned}
I_1(t,h,x) &= \big|G_{t+h}(x,X_0)-G_t(x,X_0)\big|^p\\
I_2(t,h,x) &= E\Bigg(\Bigg|\int_0^t dr\int_{\mathbb{R}^d}dz\,\big[S(t+h-r,x-z)(t+h-r)^{\alpha-1}\\
&\qquad -S(t-r,x-z)(t-r)^{\alpha-1}\big]Y_\alpha(r,z)\Bigg|^p\Bigg),\\
I_3(t,h,x) &= E\Bigg(\Bigg|\int_t^{t+h} dr\int_{\mathbb{R}^d}dz\,S(t+h-r,x-z)(t+h-r)^{\alpha-1}\\
&\qquad \times Y_\alpha(r,z)\Bigg|^p\Bigg),\\
I_4(t,h,x) &= E\Bigg(\Bigg|\int_0^{t+h} dr\int_{\mathbb{R}^d}dz\,S(t+h-r,x-z)b(X(r,z))\\
&\qquad -\int_0^t dr\int_{\mathbb{R}^d}dz\,S(t-r,x-z)b(X(r,z))\Bigg|^p\Bigg).
\end{aligned}$$

In [10], Theorem 4.3 we have proved

$$I_1(t,h,x)\le Ch^{\rho p/2}. \qquad (12)$$

For any $\alpha\in(0,\frac{1-\eta}{2})$ set

$$\psi^\alpha(t,x)=S(t,x)t^{\alpha-1}=(2\pi)^{-d/2}\exp\left(-\frac{|x|^2}{2t}\right)t^{\alpha-1-d/2}. \qquad (13)$$

Notice that $\alpha-1-\frac{d}{2}<0$.

Hölder's inequality with respect the measure on $[0,t]\times\mathbb{R}^d$ given by $|\psi^\alpha(t+h-r,x-z)-\psi^\alpha(t-r,x-z)|drdz$ and Lemma 2.2 yield

$$I_2(t,h,x)\le C(p)\left(\int_0^t dr\int_{\mathbb{R}^d}dz\,\big|\psi^\alpha(t+h-r,x-z)-\psi^\alpha(t-r,x-z)\big|\right)^p.$$

Set

$$
\begin{aligned}
I_{2,1}(t,h,x) &= \int_0^t dr \int_{\mathbb{R}^d} dz \exp\left(-\frac{|x-z|^2}{2(t-r)}\right)\Big|(t+h-r)^{\alpha-1-d/2} \\
&\qquad -(t-r)^{\alpha-1-d/2}\Big|,
\end{aligned}
$$

$$
\begin{aligned}
I_{2,2}(t,h,x) &= \int_0^t dr \int_{\mathbb{R}^d} dz\, (t+h-r)^{\alpha-1-d/2}\left|\exp\left(-\frac{|x-z|^2}{2(t+h-r)}\right)\right. \\
&\qquad \left. -\exp\left(-\frac{|x-z|^2}{2(t-r)}\right)\right|.
\end{aligned}
$$

Then, $I_2(t,h,x) \le C(p)\Big(I_{2,1}(t,h,x)^p + I_{2,2}(t,h,x)^p\Big)$.

Fix $\gamma \in (0,\alpha)$. The mean value theorem applied to the function $s \to (s-r)^{\alpha-1-d/2}$ yields

$$
\begin{aligned}
&\left|(t+h-r)^{\alpha-1-d/2} - (t-r)^{\alpha-1-d/2}\right| \\
&\qquad \le 2(t-r)^{(\alpha-1-d/2)(1-\gamma)}\left|(t+h-r)^{\alpha-1-d/2} - (t-r)^{\alpha-1-d/2}\right|^\gamma \\
&\qquad \le C(t-r)^{(\alpha-1-d/2)(1-\gamma)+(\alpha-2-d/2)\gamma} h^\gamma = C(t-r)^{\alpha-1-\gamma-d/2} h^\gamma.
\end{aligned}
$$

Therefore,

$$
\begin{aligned}
I_{2,1}(t,h,x) &\le Ch^\gamma \int_0^t dr\, (t-r)^{\alpha-1-\gamma} \int_{\mathbb{R}^d} dz\, (t-r)^{-d/2} \exp\left(-\frac{|x-z|^2}{2(t-r)}\right) \\
&\le Ch^\gamma.
\end{aligned}
$$

A Taylor expansion applied to the function $s \to \exp\left(-\frac{|x-z|^2}{2(s-r)}\right)$ and the inequality $x < \exp(x)$ for every $x \ge 0$ imply

$$
\begin{aligned}
&\left|\exp\left(-\frac{|x-z|^2}{2(t+h-r)}\right) - \exp\left(-\frac{|x-z|^2}{2(t-r)}\right)\right| \\
&\qquad \le 2\left|\exp\left(-\frac{|x-z|^2}{2(t+h-r)}\right) - \exp\left(-\frac{|x-z|^2}{2(t-r)}\right)\right|^\gamma \\
&\qquad \le 2^{\gamma+1} h^\gamma (t-r)^{-\gamma} \exp\left(-\frac{\gamma|x-z|^2}{4(t+h-r)}\right).
\end{aligned}
$$

Hence, if $\gamma \in (0,\alpha)$

$$\begin{aligned} I_{2,2}(t,h,x) &\le Ch^{\gamma} \int_0^t dr \int_{\mathbb{R}^d} dz\, (t+h-r)^{\alpha-1-\frac{d}{2}} (t-r)^{-\gamma} \exp\left(-\frac{\gamma|x-z|^2}{4(t+h-r)}\right) \\ &\le Ch^{\gamma}. \end{aligned}$$

Consequently,

$$I_2(t,h,x) \le C(p) h^{\gamma p}. \tag{14}$$

By Hölder's inequality and Lemma 2.2

$$\begin{aligned} I_3(t,h,x) &\le C\Big(\int_t^{t+h} dr \int_{\mathbb{R}^d} dz\, S(t+h-r,x-z)(t+h-r)^{\alpha-1}\Big)^p \\ &\le Ch^{\alpha p}. \end{aligned} \tag{15}$$

A change of variable yields

$$I_4(t,h,x) \le C(p)\big(I_{4,1}(t,h,x) + I_{4,2}(t,h,x)\big),$$

with

$$\begin{aligned} I_{4,1}(t,h,x) &= E\Big(\Big|\int_0^h dr \int_{\mathbb{R}^d} dz\, S(t+h-r,x-z)\, b(X(r,z))\Big|^p\Big), \\ I_{4,2}(t,h,x) &= E\Big(\Big|\int_0^t dr \int_{\mathbb{R}^d} dz\, S(t-r,x-z) \\ &\qquad \times \big[b(X(r+h,z)) - b(X(r,z))\big]\Big|^p\Big). \end{aligned} \tag{16}$$

We clearly have

$$I_{4,1}(t,h,x) \le Ch^p, \tag{17}$$

because of (6).

Moreover, the Lipschitz property of b implies

$$I_{4,2}(t,h,x) \le \int_0^t dr \sup_{y\in\mathbb{R}^d} E\big(\big|X(r+h,y) - X(r,y)\big|^p\big). \tag{18}$$

Then, putting together (11)–(18) we obtain

$$\begin{aligned} \sup_{x\in\mathbb{R}^d} E\big(\big|X(t+h,x) - X(t,h)\big|^p\big) &\le C_1 h^{p\min(\frac{\rho}{2},\gamma,\alpha)} \\ &\quad + C_2 \int_0^t \sup_{x\in\mathbb{R}^d} E\big(\big|X(r+t,x) - X(r,x)\big|^p\big) dr. \end{aligned}$$

Since $0<\gamma<\alpha<\frac{1-\eta}{2}$, the estimate (10) follows by Gronwall's Lemma.

Consider now the increments in the space variable. We want to check that for any $T>0$, $p\in[2,\infty)$, $x,\, z\in\mathbb{R}^d$ and $\gamma\in(0,\rho\wedge(1-\eta))$,

$$\sup_{0\le t\le T} \sup_{x\in\mathbb{R}^d} E\big(\big|X(t,x+z) - X(t,x)\big|^p\big) \le C|z|^{\gamma p}, \tag{19}$$

where C is a positive constant.

Fix $\alpha \in (0, \frac{1-\eta}{2})$. By (4), (5) we have

$$E\big(|X(t,x+z) - X(t,x)|^p\big) \le C \sum_{i=1}^{3} J_i(t,x,z),$$

with

$$\begin{aligned}
J_1(t,x,z) &= \big|G_t(x+z,X_0) - G_t(x,X_0)\big|^p,\\
J_2(t,x,z) &= E\bigg(\bigg|\int_0^t dr \int_{\mathbb{R}^d} dy \left[\psi^\alpha(t-r,x+z-y) - \psi^\alpha(t-r,x-y)\right]\\
&\qquad \times Y_\alpha(r,y)\bigg|^p\bigg),\\
J_3(t,x,z) &= E\bigg(\bigg|\int_0^t dr \int_{\mathbb{R}^d} dy \left[S(t-r,x+z-y) - S(t-r,x-y)\right]\\
&\qquad \times b(X(r,y))\bigg|^p\bigg),
\end{aligned}$$

where ψ^α is defined in (13).

In [10] we have proved

$$J_1(t,x,z) \le C|z|^{\rho p}. \tag{20}$$

A Taylor expansion to the function $x \to \exp(-\frac{|x-y|^2}{2(t-r)})$ and the Fubini's Theorem imply

$$\begin{aligned}
&\int_{\mathbb{R}^d} dy \left|\psi^\alpha(t-r,x+z-y) - \psi^\alpha(t-r,x-y)\right| \le C(t-r)^{\alpha-1-\gamma\frac{d}{2}}\\
&\qquad \times \left(\int_{\mathbb{R}^d} dy \left|\exp\left(-\frac{|x+z-y|^2}{2(t-r)}\right) - \exp\left(-\frac{|x-y|^2}{2(t-r)}\right)\right|\right)^\gamma\\
&\le C(t-r)^{\alpha-1-\gamma\frac{d}{2}} \left(\int_{\mathbb{R}^d} dy \int_0^1 d\lambda \exp\left(-\frac{|x+\lambda z-y|^2}{4(t-r)}\right)\frac{2|z|}{(t-r)^{1/2}}\right)^\gamma\\
&\le C(t-r)^{\alpha-1-\gamma/2}|z|^\gamma,
\end{aligned}$$

where $\gamma \in (0,1)$.

Therefore Hölder's inequality and Lemma 2.2 imply, for $\alpha \in (0, \frac{1-\eta}{2})$, $\gamma \in (0, 2\alpha)$,

$$\begin{aligned}
J_2(t,x,z) &\le \left(\int_0^t dr \int_{\mathbb{R}^d} dy |\psi^\alpha(t-r,x+z-y) - \psi^\alpha(t-r,x-y)|\right)^p\\
&\le C|z|^{\gamma p}.
\end{aligned} \tag{21}$$

Finally, by a change of variable, the Lipschitz property of b, and Hölder's inequality,

$$\begin{aligned} J_3(t,x,z) &\leq E\Big(\Big|\int_0^t dr \int_{\mathbb{R}^d} dy\, S(t-r,x-y)\big[b(X(r,y+z)-b(X(r,y))\big]\Big|^p\Big) \\ &\leq C\int_0^t dr \sup_{y\in\mathbb{R}^d} E\big(\big|X(r,y+z)-X(r,y)\big|^p\big). \end{aligned} \tag{22}$$

Then (19) follows from (20)–(22) and Gronwall's lemma. □

Condition $(\mathrm{H})_\eta$ can be formulated in terms of integrability conditions on the correlation measure Γ which make explicit the role of the dimension. We refer the reader to Proposition 5.3 in [10]. This allows to rephrase Theorem 2.1 as follows.

Theorem 2.3. *Suppose that X_0 is bounded and ρ-Hölder continuous for some $\rho \in (0,1)$. Consider the property*

> $(\mathrm{C}_{\beta_1,\beta_2})$: *The trajectories of X are a.s. β_1-Hölder continuous in t and β_2-Hölder continuous in x.*

Then,

(a) *Let $d=1$; $(\mathrm{C}_{\beta_1,\beta_2})$ holds for any $\beta_1 \in (0,\frac{1}{2}(\rho\wedge\frac{1}{2}))$, $\beta_2\in(0,\frac{1}{2}\wedge\rho)$. If there exists $\gamma_0\in(\frac{1}{4},\frac{1}{2})$ such that*

$$\int_{|x|\leq 1} |x|^{1-4\gamma_0}\,\Gamma(dx) < \infty,$$

then $(\mathrm{C}_{\beta_1,\beta_2})$ holds for every $\beta_1\in(0,\gamma_0\wedge\frac{\rho}{2})$, $\beta_2\in(0,2\gamma_0\wedge\rho)$.

(b) *Let $d\geq 2$. Assume that there exists $\gamma_0\in(0,\frac{1}{2})$ such that*

$$\int_{|x|\leq 1} |x|^{2-4\gamma_0-d}\,\Gamma(dx) < \infty,$$

then $(\mathrm{C}_{\beta_1,\beta_2})$ holds for every $\beta_1\in(0,\gamma_0\wedge\frac{\rho}{2})$, $\beta_2\in(0,2\gamma_0\wedge\rho)$.

Examples. We finish this section by giving some examples of correlation measures Γ such that the corresponding spectral measure μ satisfies $(\mathrm{H})_\eta$. These examples have also been considered in [2, 4, 5, 7].

1) Riesz kernels. Let $\Gamma(dx) = |x|^{-\alpha}dx$, $\alpha\in(0,d)$. It is well-known that the spectral measure is $\mu(dx) = C|x|^{\alpha-d}$. Then $(\mathrm{H})_\eta$ is satisfied if and only if $\alpha\in(0,2\eta\wedge d)$.
2) Brownian free field. Let $J(\varphi,\psi) = \langle(-\triangle+m^2)^{-1}\varphi,\psi\rangle$, where $\triangle$ is the Laplace operator in $\mathbb{R}^d$ and $m>0$. In this case $\mu(dx) = (2\pi)^{-d/2}(|x|^2+m^2)^{-1}dx$. Therefore $(\mathrm{H})_\eta$ holds if and only if $\eta\in((\frac{d-2}{2})^+\wedge 1, 1)$. That is, for $d=1,2$, $(\mathrm{H})_\eta$ holds for any $\eta\in(0,1)$; for $d=3$, $(\mathrm{H})_\eta$ holds for any $\eta\in(\frac{1}{2},1)$; for $d\geq 4$, $(\mathrm{H})_\eta$ does not hold.
3) The main results of [8] have been proved assuming that

(H3): The spectral measure μ is either finite or absolutely continuous with respect to Lebesgue measure and $\gamma(x) = \frac{d\mu}{dx} \in L^p$ for some $p \in [1, \infty)$.

Clearly for finite μ, $(\mathrm{H})_\eta$ holds for any $\eta \in (0,1)$. In the second case, $(\mathrm{H})_\eta$ holds if $\eta > \frac{d(p-1)}{2p}$. That is, if $p \in [1, \frac{d}{d-2} \vee 1)$, $(\mathrm{H})_\eta$ holds for any $\eta \in ((\frac{d(p-1)}{2p})^+, 1)$.

References

[1] R. C. Dalang, *Extending the martingale measure stochastic integral with applications to spatially homogeneous s.p.d.e's,* Electronic J. of Probab., **4** (1999), 1–29. http://www.math.washington.edu/~ejpecp/EjpVol4/paper6.abs.html.

[2] R. C. Dalang and N. E. Frangos, *The stochastic wave equation in two spatial dimensions,* Annals of Probab., **26 (1)** (1998), 187–212.

[3] G. Da Prato and J. Zabczyk, *Stochastic equations in infinite dimensions,* Encyclopedia of Mathematics and its Applications, Cambridge University Press, **1992**.

[4] A. Karczeswska and J. Zabczyk, *Stochastic PDE's with function-valued solutions,* to appear in Ph. Clément, F. den Hollander, J. van Neerven and B. de Pagter (Eds.), Infinite-Dimensional Stochastic Analysis, Proceedings of the Colloquium of the Royal Netherlands Academy of Arts and Sciences, Amsterdam, North Holland.

[5] A. Karczeswska and J. Zabczyk, *Regularity of solution to stochastic Volterra equations,* Preprints di Matematica, **n. 17** (1999). Scuola Normale Superiore. Pisa.

[6] A. Millet and P.-L. Morien, *On a stochastic wave equation in two space dimension: regularity of the solution and its density,* Stoch. Processes Appl., **86 (1)** (2000), 141–162.

[7] A. Millet and M. Sanz-Solé, *A stochastic wave equation in two space dimension: Smoothness of the law,* Annals of Probab., **27** (1999), 803–844.

[8] S. Peszat and J. Zabczyk, *Stochastic evolution equations with a spatially homogeneous Wiener process,* Stoch. Processes Appl., **72** (1997), 187–204.

[9] S. Peszat and J. Zabczyk, *Nonlinear stochastic wave and heat equations,* Probability Theory Rel. Fields., **116** (2000), 421–443.

[10] M. Sanz-Solé and M. Sarrà, *Path properties of a class of martingale measures with applications to spde's,* to appear in Canadian Mathematical Society, Conference Proceedings in honor of Sergio Albeverio.

[11] J. B. Walsh, *An introduction to stochastic partial differential equations,* cole d't de Prob. de St-Flour XIV, 1984, Lect. Notes in Math., **1180** (1986), Springer Verlag.

Facultat de Matemàtiques
Universitat de Barcelona
Gran Via, 585
08007 Barcelona
Spain
E-mail address: {sanz,sarra}@mat.ub.es

Progress in Probability, Vol. 52, 269–291

Regularity Conditions for Parabolic SPDEs on Lie Groups

S. Tindel and F. Viens

Abstract. We consider the stochastic heat equation on Lie groups, that is, equations of the form $\partial_t u = \Delta_x u + \dot{W}$ on $\mathbb{R}_+ \times G$, where G is a compact Lie group, Δ is the Laplace-Beltrami operator on G, and where $\dot{W}$ is a Gaussian space-correlated noise, which is white-noise in time. We find necessary and sufficient conditions on the space correlation of $\dot{W}$ such that u exists or is a Hölder-continuous function in the spatial variable x, using some basic tools of stochastic analysis and harmonic analysis on the Lie group G. We study the significance of these conditions on some basic examples. A sufficient condition for non-Hölder continuity is also given.

1. Introduction

This article analyses the weakest conditions under which the stochastic heat equation on a compact Lie group has a function-valued solution, those under which this solution is Hölder-continuous, and those under which this solution has some other prescribed modulus of continuity. We aim at clarity and simplicity of exposition. Thus we restrict ourselves to the linear additive stochastic heat equation, rather than a more general parabolic stochastic PDE, and we illustrate our conditions on some specific Lie groups: the circle group S^1, the unitary groups $SU(2)$ and $SU(3)$, and the orthogonal group $SO(4)$, for which most calculations are explicit. All of the sufficient conditions we consider for Hölder-continuity are valid in a much more general setting; the interested reader will consult our article [22] for details on what techniques are need for such generalizations.

We consider the *evolution* (or *mild*, or *semigroup*) form of the stochastic heat equation $\partial u/\partial t = \Delta u + \dot{W}$; that is, we use the solution to the stochastic evolution equation

$$u(t,x) = \int_G u_0(y) H_t(x,y)\, dy + \int_0^t \int_G H_{t-s}(x,y)\, W(ds,dy). \tag{1}$$

2000 *Mathematics Subject Classification.* 60H15.
Key words and phrases. Parabolic stochastic partial differential equations, Lie groups, harmonic analysis, regularity of Gaussian processes.

Since the equation is linear-additive, the solution is explictly given by the right-hand side, as long as it can be given a meaning. Here G is a compact Lie group, dy denotes Haar measure on G. $H_t(x,y)$ is the heat kernel for the Laplace-Beltrami operator Δ on G (see Section 2.1), W is a real-valued centered Gaussian generalized-function-valued random field on $\mathbb{R}_+ \times G$. We assume that it is white-noise in time, and spatially homogeneous, but it is not required even to be as regular as an L^2-measure in the variable y – in particular W may be more spatially irregular than space-time white noise. The notation $W(ds, dy)$ is thus generally abusive (see Section 2.3 for the precise definition of W), and the stochastic integral on the right-hand side of (1) is to be understood as: $\int_0^t \int_G W(ds, H_{t-s}(x,\cdot))$.

Our work is a non-euclidean take on the recent activity of several authors (see for instance [4, 5, 17, 18, 20]). The use of compact Lie groups actually has two advantages over Euclidean space: the lack of any boundary conditions in space, and the compactness, which implies that harmonic analysis in G takes the form of Fourier series. We have tried to make the scope of our conditions as wide as possible (spatial homogeneity for W rather than the isotropy used in [5] and [17]), and to prove that our conditions are necessary whenever possible. We also interpret our conditions for solution regularity in terms of almost-sure regularity of W, which we believe is the most natural interpretation. The works mentioned above, as does ours, show that the evolution form of a SPDE is better suited for studying the properties of a function-valued solution, that the so-called *weak* form (using test functions). Note however the very successful treatment of solutions in Sobolev and Hölder spaces in [13] by means of an analytic approach to weak solutions.

We have received preprints ([11, 19]) which deal with the wave and heat equation in flat space and use a general spatially homogeneous noise, exploiting harmonic analysis on $\mathbb{R}^d$ and T^d, much like we have done in non-commutative space, including the consideration of a general (positive definite) distribution covariance. We have been made aware of ongoing work by P.L. Chow ([2]) regarding the regularity of the stochastic wave equation, examining general (non-Gaussian) sufficient conditions under which Hölder continuity may be obtained.

This paper is organized as follows. A detailed review of relevant material from Lie groups, including harmonic analysis and properties of the heat kernel H, is given in Section 2. Examples of Lie groups are detailed in Section 3. Section 4 describes equivalent conditions for existence and uniqueness of a function-valued solution. Hölder continuity of the solution is established in Section 5. In Section 6, we interpret the condition for existence as the existence of a pathwise spatial "antiderivative" of W; the condition for Hölder-continuity is interpreted as the fact that W's "antiderivative" be almost-surely Hölder-continuous, and a further analysis of this condition is provided for the examples introduced in Section 3. Section 7 provides sufficient conditions for continuity that is not necessarily in the Hölder scale.

2. Preliminaries on compact Lie groups

We shall give some very basic facts about Lie groups and their harmonic analysis. A connected compact Lie group G is a group with a C^∞-manifold structure, such that the multiplication (resp. the inverse operation) is a C^∞ function from $G \times G$ to G (resp. from G to G). Let us denote by e the identity element of G as a group. The Lie algebra $\mathcal{G}$ of G can be considered as the tangent space T_eG of G at the identity element e. We set $d := \dim(G) := \dim(\mathcal{G})$. The exponential map is defined on all of $\mathcal{G}$, and is onto from $\mathcal{G}$ to G.

The scalar product $\langle . ; . \rangle$ on $\mathcal{G}$ defines a Riemannian structure on G, with an associated metric $\rho : G \times G \to \mathbb{R}_+$, and a unique volume element, called the Haar measure, denoted by dx or $|\cdot|$, satisfying the following properties:

1. For any $f \in L^1(G, dx)$ and any $g \in G$,
$$\int_G f(x)\,dx = \int_G f(gx)\,dx = \int_G f(xg)\,dx = \int_G f(x^{-1})\,dx. \tag{2}$$
$$\int_G \mathbf{1}\,dx = 1.$$
2. For any $x, y \in G$, $\rho(x, y) = \rho(e, y^{-1}x)$. If $x = \exp(X)$ for a $X \in \mathcal{G}$, and if $|X|_{\mathcal{G}}$ is small enough, then
$$\rho(e, x) = |X|_{\mathcal{G}} = \langle X; X \rangle^{1/2}.$$
3. There exist constants c_1 and c_2 such that for all $r \in (0, 1)$, the Haar measure $|B_\rho(e, r)|$ of the ρ-ball of radius r centered at the identity is estimated as:
$$c_1 r^d \le |B_\rho(e, r)| \le c_2 r^d. \tag{3}$$

In the remainder of the paper, we shall also denote $\rho(e, x)$ by $\rho(x)$ and $|.|_{\mathcal{G}}$ by $|.|$ when this is not an ambiguous notation. Let us call D_G the diameter of G, $\Re c$ the real part of a complex number c, and $\Im c$ its imaginary part. For a given function $\phi : G \to \mathbb{R}$, we shall also call $\check{\phi}$ the function defined on G by
$$\check{\phi}(g) = \phi(g^{-1}). \tag{4}$$

2.1. Representation theory and harmonic analysis

We only give here a brief survey of this subject. We refer to [8], [9] and [21] for further details.

Let G be a connected compact Lie group. Its dual $\hat{G}$ is the set of all (equivalence classes of) unitary representations of G (strongly continuous homomorphisms π from G into the group $U(H_\pi)$ of unitary operators of some Hilbert space H_π) that are irreducible (no nontrivial closed subspace of H_π is π-invariant). The irreducible unitary representations of a compact Lie group G are finite-dimensional. We denote by d_π the dimension of H_π for $\pi \in \hat{G}$. Thus $\pi := (\pi_{i,j})$ is a $d_\pi \times d_\pi$-unitary-matrix-valued homomorphism on G. The *character of* π is $\chi_\pi := \mathrm{tr}(\pi)$. Notice that χ_π is central, i.e. for all $g, h \in G$, $\chi_\pi(gh) = \chi_\pi(hg)$. We will usually omit the words "irreducible" an "unitary" when referring to a representation in $\hat{G}$.

Let $d\pi_e$ be the differential of π at e. Let T be a maximal torus of G and $\mathcal{T}$ its Lie algebra. Let $v = \exp(V)$ in T. The $d_\pi \times d_\pi$-matrix $d\pi_e(V)$ is unitarily equivalent to a diagonal matrix whose entries are the scalar products

$$\left\{ i\left\langle \mu_\pi^{(k)}; V\right\rangle : k = 1, \cdots, d_\pi \right\}$$

where the real-valued vectors $\mu_1, \ldots, \mu_{d_\pi}$ are called the *weights* of π. As a consequence, we have the following

Proposition 2.1. *For any $v = \exp(V)$ in T, the character can be expressed as:*

$$\Re\chi_\pi(v) = \sum_{k=1}^{d_\pi} \cos\left\langle \mu_\pi^{(k)}; V\right\rangle. \tag{5}$$

For any $\pi \in \hat{G}$, and any $v \in G$,

$$d_\pi - \Re\chi_\pi(v) \le \frac{d_\pi |h_\pi|^2}{2}\rho^2(v).$$

Proof. Equation (5) follows from the unitary equivalence of $d\pi_e(V)$ described above. If $V \in \mathcal{T}$ and $v = \exp V \in T$, since $1 - \cos(u) \le u^2/2$, we get

$$d_\pi - \Re\chi_\pi(v) \le \frac{1}{2}\sum_{k=1}^{d_\pi}\left\langle \mu_\pi^{(k)}; V\right\rangle^2 \le \frac{d_\pi}{2}|h_\pi|^2 |V|_{\mathcal{G}}^2 = \frac{d_\pi}{2}|h_\pi|^2\rho^2(v).$$

We get the general result for $v \in G$ noticing that v is always the conjugate of a point of the maximal torus T, since the conjugation is an isometry on G. Indeed then there exists $g \in G$, and $V \in \mathcal{T}$, such that $v = g^{-1}\exp(V)g$, so $\rho(v) = |V|_{\mathcal{G}}$ and since χ_π is central

$$d_\pi - \Re\chi_\pi(v) = d_\pi - \Re\chi_\pi(\exp V) \le \frac{d_\pi}{2}|h_\pi|^2 |V_{\mathcal{G}}| = \frac{d_\pi}{2}|h_\pi|^2\rho^2(v).$$

The lattice of all weights for all the representations $\pi \in \hat{G}$ will be denoted by P. The following fact is of fundamental importance in representation theory: there is a region delimited by hyperplanes in $\mathcal{T}^*$ called the *dominant chamber* and denoted by D such that each point of $P \cap D$ is in one-to-one correspondance with an element $\pi \in \hat{G}$ which is of maximal norm among the weights of π (see [9]). This element, denoted by h_π, is called the *highest weight* of π. Within P exist the set of roots Q and the set of positive roots Q^+ (see e.g. [7] or [9] for the exact definition of those notions) which allows us to define the half-sum of the positive roots $\xi = \frac{1}{2}\sum_{\alpha\in Q^+}\alpha$. The Weyl dimension formula then explicitly yields the dimension of an irreducible representation in terms of its highest weight, as a polynomial in h_π of degree $Card(Q^+)$:

$$d_\pi = \prod_{\alpha\in Q^+} \frac{\langle h_\pi + \xi\,;\alpha\rangle}{\langle \xi\,;\alpha\rangle}. \tag{6}$$

Here and below, $\langle\cdot;\cdot\rangle$ denotes the so-called Killing inner product on $\mathbb{R}^d$, which is generally proportional to the usual inner product, and specifically, in $SU(n)$, satisfies $\langle\cdot;\cdot\rangle = (2n)^{-1}(\cdot;\cdot)$ where $(\cdot;\cdot)$ is the usual inner product. In $SO(2n)$, we have $\langle\cdot;\cdot\rangle = (4n-4)^{-1}(\cdot;\cdot)$.

The main result of harmonic analysis is that

$$\{d_\pi^{1/2}\pi_{ij};\, i,j = 1,\ldots,d_\pi\,,\, \pi \in \hat{G}\}$$

is an orthonormal basis for $L^2(G)$. Also $\left\{\chi_\pi : \pi \in \hat{G}\right\}$ is an orthonormal basis for $ZL^2(G)$, the subspace of $L^2(G)$ formed of central functions ($f(xy) = f(yx)$). The Laplace-Beltrami operator is $\Delta = \sum_{i=1}^d X_i X_i$ where $(X_i)_{i=1}^d$ is any orthonormal basis of $\mathcal{G}$. Of crucial importance to us is its eigenvalue decomposition. □

Proposition 2.2. *With ξ the half-sum of the positive roots, for any representation π in $\hat{G}$, and any $i,j \in \{1,\ldots,d_\pi\}$, the function π_{ij} and the character χ_π are eigenfunctions of Δ, associated with the eigenvalue*

$$\lambda_\pi = \langle h_\pi; h_\pi + 2\xi\rangle.$$

This proposition shows that for large $|h_\pi|$, λ_π is of the order of $|h_\pi|^2$. Note that this asymptotics is relevant since the set $\{|h_\pi|\}_{\pi\in\hat{G}}$ is unbounded. In fact, it has no accumulation point other than infinity.

2.2. Estimates of the heat kernel

We shall give here some estimations for the heat kernel H on compact Lie groups, mainly taken from [3]. The heat kernel $(t,x,y) \mapsto H_t(x,y)$ on G is defined as the fundamental solution of the heat equation on G: $\partial_t u(t,x) = \Delta u(t,x)$. By left invariance and symmetries of the Laplacian on G, it is easily seen that H_t admits in fact a convolution kernel, called again H_t, such that $H_t \in L^2(G)$ and satisfies

$$H_t f(x) = \int_G H_t(xy^{-1}) f(y) dy.$$

$$H_t(x^{-1}) = H_t(x), \qquad H_t(xy) = H_t(yx), \tag{7}$$

$$\int_G H_t(xy^{-1}) H_s(yz^{-1})\, dy = H_{t+s}(xz^{-1}), \tag{8}$$

the last line being the semigroup property. It is shown in [3] that the following estimates for H hold:

Proposition 2.3. *For every $t \in (0,T]$, $x \in G$ and every $i \in \{1,\ldots,d\}$, we have, for some constants $c_1,\ldots,c_4 > 0$ and $d_1,\ldots,d_4 > 0$,*

$$\begin{aligned}
\frac{c_1}{t^{d/2}} \exp(\frac{-\rho^2(x)}{d_1 t}) \quad &\le \quad H_t(x) \le \frac{c_2}{t^{d/2}} \exp(\frac{-\rho^2(x)}{d_2 t}),\\
\partial_t H_t(x) \quad &\le \quad \frac{c_3}{t^{d/2+1}} \exp(\frac{-\rho^2(x)}{d_3 t}),\\
X_i H_t(x) \quad &\le \quad \frac{c_4}{t^{(d+1)/2}} \exp(\frac{-\rho^2(x)}{d_4 t}).
\end{aligned}$$

We shall need the following property in the remainder of the paper.

Proposition 2.4. *Let $V \in L^2(G)$, and $R \in ZL^2(G)$. For a $t \in (0,T]$, and any $x, a, b \in G$, set $H_t^x(y) = H_t(xy^{-1})$. Then*

$$\int_G \int_G H_t^x(y) V(\hat{y}^{-1}y) H_t^x(\hat{y})\, dy d\hat{y} = \int_G H_{2t}(v) V(v)\, dv,$$

$$\int_G \int_G H_t\left(a^{-1}zy\right) H_t\left(b^{-1}y\right) R(z)\, dydz = \int_G H_{2t}\left(va^{-1}b\right) R(v)\, dv.$$

Proof. Use the change of variable $y^{-1}x = v$ and $\hat{y}^{-1}x = \hat{v}$ and the relations (2), (7) and (8) to prove the first relation. The second is proved similarly. □

2.3. Random Fourier series

Throughout this paper, we are going to make heavy use of the Fourier representation of the Gaussian noise W. The theorem below shows that random Fourier series cover all white-noises that are spatially translation- and inverse-invariant in law. In any Abelian case, inverse-invariance is always satisfied. We refer the reader to [22] for proofs and further details.

Definition 2.5. *H_p is the set of all Fourier series*

$$f = \sum_{\pi \in \hat{G}} \sum_{i,j=1}^{d_\pi} d_\pi^{1/2} \pi_{i,j} f_{\pi,i,j},$$

whose coefficients satisfy

$$\sum_{\hat{G}} \sum_{i,j=1}^{d_\pi} |f_{\pi,i,j}|^2 (1+\lambda_\pi)^p < \infty.$$

An $\mathcal{H}_p$-valued white noise is a random field W defined on $\mathcal{B}(\mathbb{R}_+) \times H_p$ such that $W(\cdot,\varphi)$ is a white noise for any $\varphi \in H_p$, and W is linear in φ. Such a random field is called spatially homogeneous *or* translation-invariant *if for any $g \in G$, $W(\cdot,\varphi)$ and $W(\cdot,\varphi(\cdot g))$ have the same distribution. It is called* inverse-invariant *if $W(\cdot,\varphi)$ and $W(\cdot,\check{\varphi})$ have the same distribution. The* generalized covariance *function of W is defined as $\tilde{Q}(\phi,\psi) = E[W([0,1],\phi)\, W([0,1],\psi)]$.*

An $\mathcal{H}_p$-valued white noise is distribution-valued. It may not be an $L^2(\Omega)$-measure if p is large. However, it is clear that the heat kernel $t \mapsto H_t$ is in $L^2([0,T]; H_p)$ so that the evolution equation (1) is well defined if understood as follows:

$$u(t,x) = \int_G u_0(y) H_t(x,y)\, dy + \int_0^t W(ds, H_{t-s}(x,\cdot)). \tag{9}$$

The stochastic integral on the right-hand side, in the Itô sense, is the $L^2(\Omega)$-limit of its nonanticipating Riemann sums. In the case of the circle group S^1, the condition for existence and uniqueness of a solution described in Theorem 4.1 shows that indeed W need not be a measure, making it necessary to use form 9 of the equation. For example, for S^1, the homogeneous covariance defined by

$\tilde{Q}(\varphi, 1) = Q(\varphi) = \sum_{n=0}^{\infty} \sqrt{n}\chi_n(\varphi)$ is the half-derivative of the dirac delta, and thus is more irregular than any measure (and in particular than the covariance of space-time white noise), but does yield a solution.

Definition 2.6. *For a $p \geq 0$, consider $Q \in H_{-p}$, and $\phi, \psi \in H_p$. The* convolution *of ϕ and ψ is defined by $\phi * \psi(z) = \int_G \phi(zy)\psi(y^{-1})\,dy$. Q is called* of positive type *if $Q\left(\phi * \overline{(\check{\phi})}\right) \geq 0$ for all ϕ, and is* central *if $Q(\phi * \psi) = Q(\psi * \phi)$ for all $\phi, \psi \in H_p$.*

Theorem 2.7. *Let $p \geq 0$. Assume W is a white-noise-valued Gaussian field linearly indexed by H_p with possibly infinite covariance $\tilde{Q}$. The following four conditions are equivalent.*

(i) *W is a homogeneous and inverse-invariant $\mathcal{H}_p$-valued white noise.*

(ii) *there exists a sequence of non-negative numbers $(q_\pi)_{\pi\in\hat{G}}$ such that $\tilde{Q}$ is given by*

$$\tilde{Q}(\phi, \psi) = Q(\phi * \check{\psi}), \tag{10}$$

$$Q(\phi) = \sum_{\pi\in\hat{G}} q_\pi \int_G \chi_\pi(z)\phi(z)\,dz, \tag{11}$$

$$\sum_{\pi\in\hat{G}} q_\pi^2 (1+\lambda_\pi)^{-p} < \infty, \tag{12}$$

for all $\phi, \psi \in H_p$.

(iii) *There exists a sequence of non-negative numbers $(q_\pi)_{\pi\in\hat{G}}$ such that (12) holds and a family of standard real or complex Brownian motions $\left(W_{k,l}^{\pi}\right)_{k,l=1,\cdots,d_\pi}^{\pi\in\hat{G}}$ such that for all $\phi \in H_p$, we have the equality in $L^2(\Omega)$*

$$W([0,t],\phi) = \sum_{\pi\in\hat{G}} q_\pi^{1/2} \sum_{k,l=1}^{d_\pi} W_{k,l}^{\pi}(t) \int_G \phi(x)\pi_{k,l}(x)\,dx. \tag{13}$$

$W_{k,l}^{\pi}$ is real-valued iff $\pi \simeq \bar{\pi}$, i.e. iff π can be chosen real-valued. Moreover $W_{k,l}^{\pi}$ and $W_{k',l'}^{\pi'}$ are independent unless $k = k', l = l'$ and $\pi' \simeq \pi$ or $\bar{\pi}$. When $\pi' \simeq \bar{\pi}$, they are conjugate.

(iv) *$\tilde{Q}$ is given by (10) and Q is in H_{-p} and is central and of positive type.*

The fact that W is real-valued implies that the purely imaginary terms in the series (13) vanish. See [22] for a more explicitly real-valued representation.

3. Examples of Lie groups

As is apparent from the existence, uniqueness, and continuity conditions given below, a good control of the relative behavior of λ_π and d_π is desirable. We calculate these quantities in various explicit examples.

In the case of $SU(n)$, the algebra $\mathcal{T}$ of the maximal torus can be identified then with $\mathbb{R}^n$, an element $(\theta_1,\ldots,\theta_n)$ of $\mathbb{R}^n$ being in $\mathcal{T}$ if $\sum_{i=1}^n \theta_i = 0$. Let $(\theta_1^*,\ldots,\theta_n^*)$ be the canonical basis of $\mathbb{R}^n$, and set $\gamma^* = \theta_1^* - \frac{1}{n}\sum_{i=1}^n \theta_i^*$. The lattices Q, Q^+, P and D are defined then by (see e.g. [7] for more details)

$$\begin{aligned}
Q &= \left\{ \sum_{i,j=1}^{n} c_{i,j}(\theta_i^* - \theta_j^*); c_{i,j} \in \mathbb{Z} \right\}, \\
Q^+ &= \left\{ \sum_{1\le i<j\le n} c_{i,j}(\theta_i^* - \theta_j^*); c_{i,j} \in \mathbb{N} \right\}, \\
P &= \left\{ \sum_{i=1}^{n} \beta_i\theta_i^* + c\gamma^*; \beta_i, c \in \mathbb{Z}, \sum_{i=1}^{n} \beta_i = 0 \right\}, \\
D\cap P &= \left\{ h = \sum_{i=1}^{n} a_i\theta_i^*; h \in P, a_1 \ge \cdots \ge a_n \right\}.
\end{aligned}$$

The notation Q is sometimes used to refer to the basis $\{(\theta_i^* - \theta_j^*) : i,j = 1,\ldots,n\}$ rather than the entire lattice, especially when one refers to the cardinality of Q, or a set indexed by Q. The same goes for Q^+.

The case of $SO(2n)$ is similar: there P is now $\mathbb{Z}^n$, $D\cap P$ is the set of $h = \sum_{i=1}^n a_i\theta_i^*$ in P with integer a_i's such that $a_1 \ge \cdots \ge a_n \ge 0$, and in the expressions for Q and Q^+, $\theta_i^* - \theta_j^*$ are replaced by $\theta_i^* \pm \theta_j^*$.

Recall that $\hat{G}$ is in one-to-one correspondence with $D\cap P$. It will be convenient to index λ's and d's by the parameters of the corresonding $h \in D\cap P$. We will give such parametrizations in four simple examples: $S^1 = SO(2)$, $SU(2)$, $SU(3)$, and $SO(4)$. We will follow these examples further below. The case of $Spin(n)$ is also of general interest, and useful parametrizations can be found in [9, Chapter 17]. Of particular interest is the case of $Spin(6) = SU(4)$. For the sake of conciseness, we omit this example.

3.1. S^1

Note that $S^1 = SO(2)$. In this classical case, $\hat{G} = \mathbb{Z}$, $d_n = |n|$, $\lambda_n = n^2$.

3.2. $SU(2)$

$\mathcal{T} = \{(a;-a) : a \in \mathbb{R}\} \subset \mathbb{R}^2$. We calculate $\gamma^* = (1/2;-1/2)$. Since

$$P = \{(a;-a) + 1/2\,(c,-c) : a,c \in \mathbb{Z}\},$$

we have $P = \mathbb{Z}\,(1/2;-1/2)$ and $\hat{G} = D\cap P = \mathbb{N}\,(1/2;-1/2)$. Finally $Q = Q^+ = \{(1,-1)\}$. Thus we calculate $\xi = 1/2\,(1;-1)$ and for $h = 1/2\,(m;-m)$ in $D\cap P$, $d_h = 1+m$ and $\lambda_h = m(m+2)/8$ for any m in $\mathbb{N}$. Hence we conclude that $\hat{G}$ has one parameter, λ_h is of the order $m^2 \asymp \|h\|^2$ while d_h is of the order m.

3.3. $SU(3)$

$\mathcal{T} = \{(a; b; -a-b) : a, b \in \mathbb{R}\} \subset \mathbb{R}^3$. We calculate $\gamma^* = (2/3; -1/3; -1/3)$.

$$P = \{(a+b+2c/3; -a-c/3; -b-c/3) : a, b, c \in \mathbb{Z}\}.$$

To express a generic element (x, y, z) in P with a simple parametrization, then $z = -b - c/3$ while $y = -a - c/3$ so that modulo the fact that $3y = 3z$ mod 3, y and z can be any elements in $(1/3)\mathbb{Z}$. Then of course $x = -y - z$ and the conditions to belong to the dominant chamber translate as $z \leq y \leq -z/2$. Thus

$$D \cap P = \left\{h = (x, y, z) \in (1/3)\mathbb{Z}^3 : 3y = 3z \text{ mod } 3; x = -y-z; z \leq y \leq -z/2\right\}.$$

Since $Q^+ = \{(1; -1; 0); (1; 0; -1); (0; 1; -1)\}$, we calculate $\xi = (1; 0; -1)$ and

$$\begin{aligned} \lambda_h &= \left(x^2 + y^2 + z^2 + 2x - 2z\right)/6, \\ &= \left(y^2 + z^2 + (y+z)^2 - 2y - 4z\right)/6, \end{aligned}$$

while

$$\begin{aligned} d_h &= (1 + x - y)(1 + y - z)\left(1 + \frac{1}{2}(x - z)\right), \\ &= \frac{1}{2}\left(-2z^3 + 2y^3 - 3yz^2 + y^2 z + 6z^2 - 3y^2 + 6yz - 6z - 3y + 2\right). \end{aligned}$$

This shows that λ_h is clearly of order $\|h\|^2$, while a further analysis yields that d_h is of order $\|h\|^3$. Note in particular that $z \leq 0$ and since y may be as negative as z, the term $2y^3$ may cancel out the positive term $-2z^3$, in which case this will be more that made up by the positive term $-3yz^2$ since it is much larger than the negative term $y^2 z$. Also, if y is large, then so is z, but if z is large and not y, the positive term $-2z^3$ will still be dominant.

3.4. $SO(4)$

$\mathcal{T} = \mathbb{R}^2$. $P = \mathbb{Z}^2$. $Q = Q^+ = \{(1,1); (1,-1)\}$. $D = ((x, y) : x \geq y; x, y \in \mathbb{N})$. Thus we calculate $\xi = (1, 0)$ and for $h = (x, y)$, $\lambda_h = \left(x^2 + 2x + y^2\right)/4$. Also

$$d_h = (1 + (x+y))(1 + (x-y)) = 1 + 2x + x^2 - y^2.$$

This shows that although most d_h are of the order of $\|h\|^2$, as are all λ_h, those for which y is close to x are only of order $\|h\|$.

4. Existence and uniqueness

Theorem 4.1. *Let W be an $\mathcal{H}_p$-valued white-noise that is homogeneous and inverse invariant. Let $Q = \sum_{\pi \in \hat{G}} q_\pi \chi_\pi$ be its spatial covariance, as in Theorem 2.7. The following four conditions are equivalent.*

(a)
$$\sum_{\pi \in \hat{G}} \frac{q_\pi d_\pi}{1 + \lambda_\pi} < \infty. \tag{14}$$

(b) *For any bounded function u_0 in G, for any $T > 0$, there is a unique adapted solution u to equation (9) in $[0,T] \times G$ satisfying*

$$\sup_{x \in G, t \leq T} E\left|u(t,x)\right|^2 < \infty. \tag{15}$$

(c) *For all $T > 0$,*

$$\int_0^T Q(H_t)\,dt < \infty. \tag{16}$$

(d) *For any function $u_0 \in L^2(G)$, for any $T > 0$, there is a unique solution u to equation (9) in $[0,T] \times G$ satisfying*

$$\sup_{t \leq T} E\left\|u(t)\right\|^2_{L^2(G)} < \infty.$$

If Q is a non-negative measure, then condition (c) is equivalent to:

(e) *$Q(h) < \infty$, where for $d \geq 3$, $h(x) = \rho^{-d+2}(x)$, and for $d = 2$, $h(x) = \log \rho^{-1}(x)$.*

Proof. Equation (9) gives the solution u explicitly. With bounded u_0, since $H_t(x,\cdot)\,dy$ is a probability measure for any x,t, the supremum in t,x of the first term in u, which is non-random, is always finite regardless of Q. Denote by H_s^x the function $H_s(x,\cdot)$ for $x \in G$, $t > 0$, and that $H^e = H$. Using the representation (13), the isometry property of Brownian motion, as well as the identities of Subsection 2.2 (including Proposition 2.4) and the fact that χ_π is central, (b)⇔(c) follows from the calculation

$$\begin{aligned}
&E\left[\left|\int_0^t W\left(ds, H_{t-s}^x\right)\right|^2\right] \\
&= \int_0^t Q\left(H_{t-s}^x * \check{H}_{t-s}^x\right) ds \qquad (17)\\
&= \sum_{\pi \in \hat{G}} q_\pi \int_0^t \left(\int_G \int_G \chi_\pi\left(yz^{-1}\right) H_{t-s}\left(x^{-1}y\right) H_{t-s}\left(z^{-1}x\right) dydz\right) ds \\
&= \sum_{\pi \in \hat{G}} q_\pi \int_0^t \int_G \chi_\pi(v) H_{2(t-s)}(v)\, dvds \\
&= \int_0^t Q(H_{2s})\, ds.
\end{aligned}$$

An even simpler calculation yields (d)⇔(c). To establish equivalence with (a), we use the fact that, by definition, the linear operator $\phi \to \int_G H_t\left(v^{-1}x\right)\phi(v)\,dv$ can be written as $e^{s\Delta}\phi$, so that since χ_π is an eigenfunction of Δ with eigenvalue $-\lambda_\pi$, we get

$$\int_G H_t\left(v^{-1}x\right)\chi_\pi(v)\,dv = \chi_\pi(x)\exp\left(-t\lambda_\pi\right).$$

Therefore

$$\int_0^T Q(H_t)\,dt = \sum_{\pi\in\hat{G}} q_\pi \chi_\pi(e) \int_0^T \exp(-t\lambda_\pi)\,dt, \tag{18}$$
$$= q_{\mathbf{1}} d_{\mathbf{1}} T + \sum_{\pi\in\hat{G}\setminus\{\mathbf{1}\}} q_\pi d_\pi \frac{1}{\lambda_\pi}\left(1-\exp(-T\lambda_\pi)\right).$$

Since $-\Delta$ is a positive operator and has a spectral gap, the set of values $(\lambda_\pi)_{\pi\neq\mathbf{1}}$ has a strictly positive lower bound, so that the above series converges if and only if $\sum_{\pi\in\hat{G}} q_\pi d_\pi/(1+\lambda_\pi)$ converges. Lastly, using Fubini's theorem and the heat kernel bounds (2.3) on the quantity in (d), one relates it to condition (e). □

Condition (e) shows that our conditions coincide with the conditions given in [4, 5, 17, 18, 19, 20] for existence and uniqueness of the stochastic heat and wave equations in flat spaces for $d \geq 2$.

We now record some remarks regarding the equivalent conditions of Theorem 4.1. A more detailed analysis for each of the examples is given in Section 6.2. Denote by $\mathcal{H}$ the set of all formal series $Q = \sum_{\pi\in\hat{G}} q_\pi \chi_\pi$ such that $\sum_{\pi\in\hat{G}} q_\pi d_\pi/(1+\lambda_\pi)$ converges. Denote by $M(G)$ the set of all finite signed measures on G. Arguably the most important and/or basic example of a Lie group is the circle S^1. In this case, not only is $\mathcal{H}$ not included in $M(S^1)$, it is not even included in H_{-1}. However, it is included in $H_{-3/2-\varepsilon}$ for any ε, as will become apparent in Section 6.2. One can even prove that no matter what G is, $\mathcal{H} \subset H_{-2}$. Nevertheless, the case of S^1 is somewhat exotic, since it is possible to show that for many NON-commutative compact Lie groups, and for all the tori $(S^1)^d, d \geq 2$, $\mathcal{H}$ is included in $M(G)$. The results of Section 6.2 contain an illustration of how the regularity of $\mathcal{H}$ is determined by the relationship between λ_π, which is of order $|h_\pi|^2$, and d_π, which, according to the Weyl dimension formula, is of degree $CardQ^+$ at the most, and by the way that $\hat{G}$ is parametrized.

5. Hölder continuity

Theorem 5.1. *Let W be a homogeneous and inverse invariant $\mathcal{H}_p$-valued white noise and let $T > 0$. Suppose that for an $\varepsilon > 0$, Q satisfies*

(f) *For any $\gamma \in (0,\varepsilon)$,*

$$\sum_{\pi\in\hat{G}} \frac{q_\pi d_\pi}{(1+\lambda_\pi)^{1-\gamma}} < \infty.$$

Assume also that u_0 is γ-Hölder continuous on G for any $\gamma < \varepsilon$. Then there exists a version of the solution $u = \{u(t,x);\,(t,x) \in [0,T]\times G\}$ to (9) which is $(\gamma/2,\gamma)$-Hölder continuous on $[0,T]\times G$, for any exponent $\gamma \in (0,\varepsilon)$.

Proof. Invoking the Kolmogorov criterion, one only needs to show that for any $p > 2$, and any $(t_1, x_1), (t_2, x_2) \in [0, T] \times G$, we have, for a constant $c_p > 0$, and for all $\gamma < \varepsilon$

$$E[|u(t_1, x_1) - u(t_2, x_2)|^p] \leq c_p(|t_2 - t_1|^{\frac{p\gamma}{2}} + \rho^{p\gamma}(x_2, x_1)).$$

Indeed, since G is locally diffeomorphic to $\mathbb{R}^d$, the Kolmogorov criterion (e.g. [10], Problem 2.2.9) would imply that u has a version which is (α, β)-Hölder continuous in (t, x) for all $\alpha < (p\gamma/2 - (d+1))/p$ and for all $\beta < (p\gamma - (d+1))/p$. Since p is arbitrarily large and γ is arbitrarily close to ε, the conclusion of the theorem would follow. The details of the calculation are omitted. The main ingredients are Burkholder's inequality for W (see [23]), estimates of the heat kernel given above, calculations similar to the proof of Theorem 4.1, and the estimates on the characters χ_π and the eigenvalues λ_π in Proposition 2.1 and in (2.2). □

6. Necessity of the existence and Hölder-continuity conditions

6.1. Results

Note that for any function $f = \sum_{\pi\in\hat{G}} \sum_{i,j=1}^{d_\pi} f_{\pi,i,j}\pi_{i,j}$ in H_{-p}, the operator $(I - \Delta)^{-1/2}$ applied to f can be written as

$$(I - \Delta)^{-1/2} f = \sum_{\pi\in\hat{G}} \sum_{i,j=1}^{d_\pi} f_{\pi,i,j} (1 + \lambda_\pi)^{-1/2} \pi_{i,j}.$$

Proposition 6.1. *Let W be given by (13). The equivalent conditions (a) - (e) for existence and uniqueness of an adapted solution to (9) are equivalent to requiring that the Gaussian field $Y = (I - \Delta)^{-1/2} W$ be function-valued, i.e. that*

$$Y(t, x) = \sum_{\pi\in\hat{G}} \sqrt{q_\pi} (1 + \lambda_\pi)^{-1/2} \sum_{i,j=1}^{d_\pi} \pi_{i,j}(x) W_{k,l}^\pi(t)$$

is an a.s. finite random variable.

Proof. Since $Y(t, x)$ is a centered Gaussian random variable, it will exist if and only if its variance is finite. This variance calculates out to be $t \sum_{\pi\in\hat{G}} q_\pi d_\pi (1 + \lambda_\pi)^{-1}$, and the finiteness of this is exactly condition (a).

This proposition shows that the conditions for existence and uniqueness can be expressed intrinsically as a spatial regularity condition on W, and more specifically, as the existence of a functional "antiderivative" Y for W. We now turn to a characterization of condition (f) for Hölder-continuity, which is also expressed intrinsically in terms of Y. □

Theorem 6.2. *Let W be given by* (13)*. Condition (f) for $(\gamma/2, \gamma)$-Hölder-continuity of the solution to* (9) *for any $\gamma < \varepsilon$, implies each of the following two conditions:*

(g) *the Gaussian field $Y = (I - \Delta)^{-1/2} W$ is almost-surely γ-Hölder-continuous in the space variable for all $\gamma < \varepsilon$.*

(h) *assuming* u_0 *is* γ*-Hölder-continuous in* G *for all* $\gamma < \varepsilon$*, the solution* u *to equation (9) is a.s.* γ*-Hölder-continuous in the space variable for all* $\gamma < \varepsilon$.

The converse also is true, i.e. both (g)⇒(f) and (h)⇒(f), assuming the following structural hypotheses on the group G:

[G] *There is an integer* K_0 *and constants* $c_0, c_1 \in (0,1)$ *such that for any* $\pi \in \hat{G}$ *satisfying* $|h_\pi| \geq K_0$,

$$Card\left\{k : k = 1, \cdots, d_\pi; \left|\mu_\pi^k\right| > c_0 \left|h_\pi\right|\right\} \geq c_1 d_\pi.$$

[H] *Let* $\{h(n)\}_{n=0}^{\infty}$ *be a numbering of* $\{h_\pi\}_{\pi\in\hat{G}}$ *in order of increasing* $|h_\pi|$*. Then* $\{|h(n+1)| / |h(n)|\}_{n=1}^{\infty}$ *is bounded.*

Remark 6.3. We can rephrase the sesqui-equivalence of (f) and (g) as follows. Let Y be a homogeneous and inverse-invariant Gaussian field on G with covariance $R = \sum_{\pi\in\hat{G}} r_\pi \chi_\pi$. If $\sum_{\pi\in\hat{G}} r_\pi d_\pi (1+\lambda_\pi)^\gamma$ converges for all $\gamma < \varepsilon$, then Y is almost-surely γ-Hölder-continuous for any $\gamma < \varepsilon$. The converse is true if [G] and [H] hold.

Condition [G] is a way to require that for each representation π, there is a non-negligeable proportion of its weights whose norm is close to the maximal weight's $|h_\pi|$. Condition [H] just requires that the $|h_\pi|$'s grow no faster than exponentially. These conditions are satisfied for the classical compact and connected Lie groups (see [22]). We now check that they are satisfied in the examples of Section 3.

For the circle S^1, this is trivial since there is only one weight per representation $n \in \mathbb{N}$, and n grows linearly. To study the other examples, we do not need to know the weights explicitly for each π. Indeed, it is known that all the weights of π which are also elements of the dominant chamber D can be obtained from the highest weight h by the formula $h - \sum_{\alpha\in Q^+} \alpha k_\alpha$ for some vector $(k_\alpha) \in \mathbb{N}^{CardQ^+}$.

For $SU(2)$, since $D \cap P = \mathbb{N}(1/2; -1/2)$ and $Q^+ = \{(1,-1)\}$, for $h_m = (-m/2; m/2)$ the number of weights of π_m half as big as h_m is at least $m/4$, and since $d_m = m+1$, the proportion of these weights is large enough. Moreover, the highest weights' norms increase linearly.

For $SU(3)$, we have

$$\begin{aligned} D \cap P &= \left\{h = (x,y,z) \in 1/3\mathbb{Z}^3 : 3y = 3z \bmod 3; x = -y - z; z \leq y \leq -z/2\right\} \\ Q^+ &= \{(1;-1;0);(1;0;-1);(0;1;-1)\}. \end{aligned}$$

The magnitude of $h = (x, y, z_0)$ is the magnitude of z_0. Indeed, $|y| \leq |z_0|$ and $|x| \leq 2|z_0|$. Each negative integer value of z_0 is allowable, so that the norms increase no more than linearly. The number of vectors in D of the form $(x, y, z_0) - \sum_{\alpha\in Q^+} \alpha k_\alpha$ and whose z-coordinate is $> |z_0|/2$, is of order the volume of the planar sector $z \leq y \leq -z/2$ with radii between $|z_0|/2$ and $|z_0|$, which is on the order of $|z_0|^2$. However the dimension d_h is of order $|z_0|^3$. To make up this discrepancy, one needs to realize that each weight may occur multiple times in the list of weights. The Freudenthal multiplicity formula (see [9]) indicates a way to calculate the weights' multiplicity, to make up for the missing factor of $|z_0|$. See [22] for details.

For $SO(4)$, we have $D \cap P = \{(x,y) : x, y \in \mathbb{N}, x \geq y\}$ while Q^+ is formed of the two vectors $\{(1,1);(1,-1)\}$ and $d_{(x,y)} = 1 + 2x + (x-y)(x+y)$. For $\pi \in \hat{G}$ let $h = (x_0, y_0)$ be its highest weight.

Consider first the case $x_0 = y_0$. Then for large $|h|$ the set of integer points on the line $x = y$ with x-coordinates between $x_0/2$ and x_0 are all weights of π, have norm of the order of $|h|/2$ and there are on the order of $d_h/4$ of them. Next if $x_0/2 \leq y_0 \leq x_0 - 1$, consider the parallelogram B delimited by $x = x_0/3$, $x = y_0$, $x = y$, and $x - x_0 = y - y_0$. It has area $(y_0 - x_0/3)(x_0 - y_0) \geq x_0(x_0 - y_0)/6$, and all the non-negative-integer-valued points in it are weights of π. The portion of B that may dip below the x-axis is at most a triangle with area $(x_0 - y_0 - x_0/3)^2/2 \leq x_0(x_0 - y_0)/12$. Therefore we have identified a region A with area $\geq x_0(x_0 - y_0)/12 \geq d_{(x_0,y_0)}/36$ so that for large $|h|$ it contains more than $d_\pi/36$ weights of π. Moreover each such weight μ has x-coordinate greater than $x_0/3$, so that $|\mu| > |h|/6$. The last case needed to verify condition [G] is $y_0 < x_0/2$. Here we consider the triangular region T with vertices $(x_0, y_0), (y_0, y_0)$ and $((x_0 + y_0)/2, (x_0 + y_0)/2)$. Again, all integer-valued points in T are weights of π. T has area $(x_0 - y_0)^2/4$. Let U be the triangular subset of T with x-coordinates $\geq x_0/2$. U has area $(x_0/2 - y_0)^2/2$. $A = T \backslash U$ has area $(x_0^2/2 - y_0^2)/4 \geq x_0^2/16$. We conclude as in the previous case, and condition [G] is established. Condition [H] is clear since the norms of the h's, which are of the order of magnitude of their x-coordinate, increase no more than linearly.

We now outline how the equivalence of (f), (g), and (h) is established. Details are in [22]. That (f) is sufficient for (h), even without Hypotheses [G] or [H], is the result of Theorem 5.1. The converse (h)⇒(f) follows from implication (g)⇒(f). Indeed, assume the latter is true.

The solution to (9) is given explicitly as the sum of a deterministic Hölder-continuous function and of the following Gaussian random field:

$$U(t,x) = \int_0^t W(ds, H_{t-s}(x,\cdot)),$$

whose covariance is calculated as in the proof of Theorem 4.1:

$$E[U(t,x)U(s,z)] = q_1 (s \wedge t) + \sum_{\pi \neq 1} \frac{q_\pi}{2\lambda_\pi} \left[e^{-|t-s|\lambda_\pi} - e^{-(t+s)\lambda_\pi} \right] \chi_\pi (xz^{-1}).$$

showing that $x \mapsto U(t,x)$ is an inverse-invariant homogeneous Gaussian field on G.

Under (g), it is γ-Hölder-continuous in x. With $R = \sum_{\pi \in \hat{G}} r_\pi \chi_\pi$ its homogeneous covariance function, Remark 6.3 shows that for all $\gamma < \varepsilon$,

$$\sum_{\pi \in \hat{G}} r_\pi d_\pi (1 + \lambda_\pi)^\gamma < \infty.$$

However, we have

$$r_\pi = \frac{q_\pi}{2\lambda_\pi} [1 - \exp(-2\lambda_\pi)]; \ r_1 = q_1.$$

Since Δ has a spectral gap, the infimum of all eigenvalues λ_π for $\pi \neq \mathbf{1}$ is a value $c > 0$, and we get

$$\begin{aligned}\infty &> \sum_{\pi\in\hat{G}-\{\mathbf{1}\}} \frac{q_\pi d_\pi}{2\lambda_\pi}\left[1-\exp\left(-2\lambda_\pi\right)\right]\left(1+\lambda_\pi\right)^\gamma \\ &\geq \left(1-\exp\left(-2c\right)\right)\sum_{\pi\neq\mathbf{1}} \frac{q_\pi d_\pi}{\left(1+\lambda_\pi\right)^{1-\gamma}},\end{aligned}$$

proving (f) holds for Q.

The proof of (f)⇒(g) is a simple application of Kolmogorov's lemma as in Theorem 5.1. The converse is more difficult, and invokes the theory of regularity of Gaussian processes of Fernique and Talagrand, which translates as the general proposition below, proved in [22].

Proposition 6.4. *Let $\{Y(x) : x \in G\}$ be a homogeneous Gaussian process indexed by a compact Lie group G. The metric on G is denoted by ρ. The* canonical metric *of Y is the pseudo-metric defined by*

$$\delta(x,y) = E^{1/2}\left[\left(Y(x)-Y(y)\right)^2\right].$$

Then Y is almost-surely γ-Hölder-continuous for all $\gamma < \varepsilon$ if and only if

$$\forall\gamma<\varepsilon, \exists\alpha_0>0 : \forall x\in G, \rho(x,e)<\alpha_0 \Rightarrow \delta(x,e)\leq\rho(x,e)^\gamma. \tag{19}$$

With such a result in hand, a detailed study of the tail of $\sum_\pi r_\pi d_\pi$, and conditions [G] and [H], allow to finish the proof (see [22] for details).

6.2. Analysis of the Hölder-continuity condition for the examples of Section 3

Let W be as in (13). Let $Y = (I-\Delta)^{-1/2} W$. Thus the spatial covariance of Y is $R = \sum r_\pi \chi_\pi$ with $r_\pi = q_\pi (1+\lambda_\pi)^{-1}$. In the last section we showed that the solution u is Hölder-continuous iff Y is. In this section, we indicate the amount of regularity which is needed of R, in the H_p scale, to guarantee Hölder-continuity of Y (and thus of u). This question is relevant for practical purposes, as it is normally the covariance function R which is given in order to define a potential W, as via the representation formula (13). Recall that u exists iff Y exists iff $\sum_\pi r_\pi d_\pi$ converges. Also γ-Hölder continuity for all $\gamma < \varepsilon$ are equivalent for u and Y, and occur iff $\sum_\pi r_\pi d_\pi (1+\lambda_\pi)^\gamma$ converges. And by definition, R is in H_p if $\sum_\pi r_\pi^2 (1+\lambda_\pi)^p$.

1. For S^1, $R \in H_p$ means $\sum_{n\in\mathbb{N}} r_n^2 \left(1+n^2\right)^p$ converges, so

$$r_n = o\left(n^{-(p+1/2)}\right).$$

Thus $r_n d_n (1+\lambda_n)^\gamma = r_n \left(1+n^2\right)^\gamma = o\left(n^{p+1/2-2\gamma}\right)^{-1}$. Therefore, by the equivalence of the conditions for Hölder continuity, the solution is γ-Hölder continuous if and only if $R \in H_p$ for some $p > 2\gamma + 1/2$.

2. For $SU(2)$, $\hat{G}$ is in bijection with $\mathbb{N}$, d_n is equivalent to n and λ_n to n^2. Thus $R \in H_p$ is equivalent to the convergence of the series

$$\sum_n r_n^2 \left(1+n^2\right)^p,$$

so $r_n d_n (1+\lambda_n)^\gamma = o\left(n^{p-1/2-2\gamma}\right)$. Therefore, the solution is γ-Hölder if and only if $R \in H_p$ for some $p > 2\gamma + 3/2$.

3. For $SU(3)$, the calculation is slightly more involved since the set of highest weights is a wedge in $\frac{1}{3}\mathbb{Z}^3$ which in polar coordinates is

$$\frac{1}{3}\mathbb{Z}^3 \cap \{(x,\rho,\theta) : x = -\rho(\sin\theta + \cos\theta); \rho \geq 0; -3\pi/4 \leq \theta \leq -\pi/8\}.$$

Since $\lambda_{x,\rho,\theta}$ is of the order of ρ^2, the convergence of the series for $R \in H_p$ should be equivalent to the convergence of $\int_0^\infty \int_{-3\pi/4}^{-\pi/8} r_{\rho,\theta}^2 \rho^{2p} \rho d\rho d\theta$. If r is isotropic (not dependent on θ), then we would have $r_\rho = o\left(\rho^{-1-p}\right)$. Since moreover $d_{x,\rho,\theta}$ is of the order of ρ^3, then

$$r_\rho d_{\rho,\theta} (1+\lambda_{\rho,\theta})^\gamma = o\left(\rho^{-(p-3-2\gamma)}\right),$$

so that in the isotropic case, the solution should be γ-Hölder if and only if $R \in H_p$ for some $p > 2\gamma + 4$.

To make this argument more rigorous, we assume isotropy, i.e. for $\pi = (\rho,\theta)$, $r_\pi = r(\rho)$, and we assume that asymptotically, the function ρ has relatively moderate oscillations over any interval of length one: there exists $c \in (0,1)$ and ρ_0 such that if $\rho > \rho_0$ and $|\rho - \rho'| \leq 1$ then

$$|r(\rho') - r(\rho)| \leq c r(\rho). \tag{20}$$

For $k \in \mathbb{N}$, let $A_k = \{\pi = (\rho,\theta) \in \hat{G} : [\rho] = k\}$. The condition $R \in H^p$ is then equivalent to

$$\sum_{k=1}^\infty k^{2p} \sum_{\pi=(\rho,\theta)\in A_k} r(\rho) < \infty \quad \Leftrightarrow \quad \sum_{k=1}^\infty k^{2p} r(k)\, Card(A_k) < \infty.$$

$Card(A_k)$ is the number of points of $3^{-1}\mathbb{Z}^2$ in a sector of radii k and $k+1$, and of angular length $5\pi/8$, which for large A_k is approximately equal to $9k \cdot 5\pi/8$. Therefore

$$R \in H^p \quad \Leftrightarrow \quad \sum_{k=1}^\infty k^{2p+1} r(k) < \infty. \tag{21}$$

This implies that $r(k) = o\left(k^{-1-p}\right)$. A similar argument applied to the series $\sum_\pi r_\pi d_\pi (1+\lambda_\pi)^\gamma$ now allows to conclude that if r is isotropic and satisfies (20), then indeed u is γ-Hölder if and only if $R \in H_p$ for some $p > 2\gamma + 4$.

4. For $SO(4)$, the set of highest weights is the cone

$$\mathbb{Z}^2 \cap \{(\rho,\theta) : \rho \geq 0; 0 \leq \theta \leq \pi/4\}.$$

As above, $\lambda_{\rho,\theta}$ is of the order of ρ^2. Thus as above, if r is isotropic and satisfies (20) the equivalence (21) still holds. Now $d_{\rho,\theta}$ is not of any constant order in ρ: $d_{\rho,\theta} = 4 + 2\rho\cos\theta + \rho^2 \sin 2\theta$, so we must be more careful. Let us rewrite the equivalent condition for γ-Hölder-continuity of u.

$$\sum_{\pi} r_\pi d_\pi (1+\lambda_\pi)^\gamma < \infty$$

$$\Leftrightarrow \sum_{k=1}^{\infty} k^{2\gamma} r(k) \sum_{\pi=(\rho,\theta)\in A_k} d_\pi < \infty$$

$$\Leftrightarrow \sum_{k=1}^{\infty} k^{2\gamma} r(k) \sum_{\pi=(\rho,\theta)\in A_k} \left(\rho\cos\theta + \rho^2 \sin 2\theta\right) < \infty$$

$$\Leftrightarrow \sum_{k=1}^{\infty} k^{2\gamma} r(k) \left(k \left[\sum_{\pi=(\rho,\theta)\in A_k} \cos\theta \right] + k^2 \left[\sum_{\pi=(\rho,\theta)\in A_k} \sin 2\theta \right] \right) < \infty.$$

In the term $\sum_{\pi=(\rho,\theta)\in A_k} \sin 2\theta$, if we sum only over those π such that $\theta > \pi/12$, we obtain, for large k, a lower bound that is approximately equal to $\sin(\pi/6)(\pi/2 - \pi/12)k$. Of course we also have the upper bound $k\pi/2$. This proves

$$\sum_{\pi} r_\pi d_\pi (1+\lambda_\pi)^\gamma < \infty \Leftrightarrow \sum_{k=1}^{\infty} k^{2\gamma+3} r(k) < \infty.$$

Since for $R \in H^p$, $r(k)\, k^{2\gamma+3} = o\left(k^{-(1+p-2\gamma-3)}\right)$ we see that γ-Hölder-continuity holds iff $R \in H^p$ for some $p > 2\gamma + 3$.

Note that one should expect p to be twice as large as γ since p measures the regularity of a homogeneous covariance of Y, not of Y itself. However, these examples do show that there is a loss of regularity from R to Y or u. In the scale of Y, regularity typically (in the case of isotropic coefficients) drops by 1/4, 3/4, 2, and 3/2 respectively for S^1, $SU(2)$, $SU(3)$, and $SO(4)$.

7. (Non-Hölder) continuity

We now undertake the task of providing conditions guaranteeing that u almost-surely has a given modulus of continuity on G. These conditions follow the model of those for Hölder-continuity in Section 6. We begin with a general result regarding the canonical metric.

Proposition 7.1. *Let $\{Y(x) : x \in G\}$ be a homogeneous Gaussian process indexed by a compact Lie group G with Riemannian metric ρ. The canonical metric δ of Y is as in Proposition 6.4. Let f be a positive strictly increasing differentiable function defined on a neighborhood of 0 in $\mathbb{R}_+ - \{0\}$, such that $\lim_0 f = 0$. Define the function*

$$w(\varepsilon) = \int_0^\varepsilon f'(x)\sqrt{-\log x}\,dx$$

for all $\varepsilon > 0$. Then, up to multiplication by a deterministic constant, w is almost-surely a modulus of continuity for Y at any point in G if the following two conditions are met:

(k) $\exists \varepsilon_0 > 0 : w(\varepsilon_0) < \infty$

(l) *For any g in a neighborhood of e in G, $\delta(g,e) \leq f(\rho(g))$.*

Proof. Assume (k) and (l) hold. Let $\omega(\varepsilon) = \int_0^\varepsilon \sqrt{-\log|B_{\delta,e}(x)|}dx$. Fernique's and Talagrand's results characterizing continuity for homogeneous Gaussian fields imply that $\omega(\varepsilon)$ is (up to a universal multiplicative constant) a global almost-sure modulus of continuity for Y on G as long $\omega(\varepsilon)$ is finite for some $\varepsilon > 0$, if this modulus is measured with respect to δ (see Corollary 4.7 in [1]). Here $|B_{\delta,e}(x)|$ is the Haar measure of the ball of radius x in the metric δ centered at the identity e. For x small enough we have

$$\begin{aligned} B_{\delta,e}(x) &= \{g \in G : \delta(g,e) < x\}, \\ &\supset \{g \in G : f(\rho(g,e)) < x\}, \\ &= \left\{g \in G : \rho(g,e) < f^{-1}(x)\right\}, \end{aligned}$$

where we used condition (l) and the fact that f is increasing. Using estimate (3) we get

$$|B_{\delta,e}(x)| \geq c_1 f^{-1}(x)^d.$$

We can now estimate, using again condition (l), and the Fernique-Talagrand expression for ω, for all g, g' in a neighborhood of e,

$$\begin{aligned} \sup_{\rho(g,g')<\varepsilon} |Y(g) - Y(g')| &= \sup_{f(\rho(g,g'))<f(\varepsilon)} |Y(g) - Y(g')|, \\ &\leq \sup_{\delta(g,g')<f(\varepsilon)} |Y(g) - Y(g')|, \\ &\leq K_u \int_0^{f(\varepsilon)} \sqrt{-\log|B_{\delta,e}(x)|}dx, \\ &\leq K_u\sqrt{d} \int_0^{f(\varepsilon)} \sqrt{-\log f^{-1}(x)\, c_1^{1/d}}dx \end{aligned}$$

almost surely, where K_u is a universal constant. By assuming moreover that $\varepsilon < c_1^{1/d}$, we get in the above integral $f^{-1}(x) \leq \varepsilon < c_1^{1/d}$ yielding

$$\begin{aligned} &\sup_{\rho(g,g')<\varepsilon} |Y(g) - Y(g')|, \\ \leq\ & K_u\sqrt{2d} \int_0^{f(\varepsilon)} \sqrt{-\log f^{-1}(x)}dx, \\ =\ & K_u\sqrt{2d} \int_0^{\varepsilon} f'(y)\sqrt{-\log y}\,dy = K_u\sqrt{2d}\,w(\varepsilon), \end{aligned}$$

where we used the change of variable $x = f(y)$, establishing the proposition. □

Remark 7.2. A partial converse for this proposition holds. If Y is almost-surely continuous, Fernique's results imply that the entropy integral $\omega(\varepsilon)$ is finite for all ε small enough. Assume now that (l) holds with the direction of the inequality reversed. Condition (k) then follows by the same arguments as above, using the upper bound in (3) instead of the lower bound. We will not use this "converse" in the sequel.

Corollary 7.3. *Let Y, f, and w be as in Proposition* 7.1, *and assume (k) and (l) hold. Let $\hat{w}(\varepsilon) = f(\varepsilon)\sqrt{-\log \varepsilon}$ and*

$$\tilde{w}(\varepsilon) = 2^{-1}\int_0^{\varepsilon} f(x)\, x^{-1}(-\log x)^{-1/2}\, dx.$$

Then $w(\varepsilon) = \hat{w}(\varepsilon) + \tilde{w}(\varepsilon)$ and two cases occur.

(Case 1) *Either $\tilde{w}(\varepsilon)/\hat{w}(\varepsilon)$ is bounded for small ε, in which case $\hat{w}(\varepsilon)$ is a modulus of continuity for Y*

(Case 2) *Or $\tilde{w}(\varepsilon)/\hat{w}(\varepsilon)$ tends to infinity for small ε, in which case $\tilde{w}(\varepsilon)$ is a modulus of continuity for Y.*

Proof. Integration by parts shows that $w(\varepsilon) = \hat{w}(\varepsilon) + \tilde{w}(\varepsilon)$. The other results then follow by the previous proposition. □

Remark 7.4. Case 1 is typically satisfied if $f(x) < (-\log x)^{-r}$ for some $r > 1/2$. In this sufficiently regular situation, the interpretation is that the loss of regularity from δ to Y is exactly equal to the factor $\sqrt{-\log x}$. Case 2 is a more irregular situation, and is typically satisfied for f in the following iterated logarithm scale:

$$f(x) = (-\log x)^{-1/2}\left\{\log_2\left(x^{-1}\right)\cdots\log_m\left(x^{-1}\right)\left[\log_{m+1}\left(x^{-1}\right)\right]^{1+\alpha}\right\}^{-1},$$

where $\log_m$ is the composition of m logarithm functions, and $\alpha > 0$. In this case, the regularity of Y is computed to be $\log_{m+1}\left(x^{-1}\right)$. The interpretation here is not as simple as in Case 1, and the loss of regularity is slightly greater. These losses of regularity are not visible in the Hölder scale, as Theorem 5.1 shows.

As a last step to finding a condition for the solution u to have a specific function as its almost-sure spatial modulus of continuity, the following proposition shows how to produce a canonical metric δ satisfying condition (l).

Proposition 7.5. *Let Y be as above and let $Q = \sum r_\pi \chi_\pi$ be its homogeneous convariance function. Suppose that $\frac{f(x)}{x}$ is decreasing for sufficiently small x. Then, replacing f by a constant multiple of f if necessary, the following condition implies condition (l):*

(m) $\sum_{\pi\in\hat{G}} r_\pi d_\pi / f(1/|h_\pi|)^2$ *converges.*

Proof. One easily calculates that $\delta(x,y)^2 = \sum_{\pi\in\hat{G}} r_\pi \left(d_\pi - \chi_\pi\left(xy^{-1}\right)\right)$. Proposition 2.1 implies that $d_\pi - \chi_\pi(z) \leq 2d_\pi \left(1 \wedge \left[|h_\pi|^2 \rho(z)^2\right]\right)$. Therefore,

$$\delta(z,e)^2 \leq 2f(\rho(z))^2 \sum_{\pi:|h_\pi|^2\rho(z)^2>1} r_\pi d_\pi \frac{1}{f(1/|h_\pi|)^2} + 2 \sum_{\pi:|h_\pi|^2\rho(z)^2\leq 1} r_\pi d_\pi |h_\pi|^2 \rho(z)^2 .$$

The first term yields the required estimate $\delta(z,e) \leq Kf(\rho(z))$ with K^2 given as the sum of the converging series of condition (m). Thus we only need to study the second term. The function $f(x)/x$ is decreasing for small x by assumption. Therefore,

$$\begin{aligned} r_\pi d_\pi |h_\pi|^2 \rho(z)^2 &= r_\pi d_\pi \frac{1}{f(1/|h_\pi|)^2} \rho(z)^2 |h_\pi|^2 f(1/|h_\pi|)^2 \\ &\leq r_\pi d_\pi \frac{1}{f(1/|h_\pi|)^2} \rho(z)^2 f(\rho(z))^2 / \rho(z)^2 \\ &= f(\rho(z))^2 r_\pi d_\pi \frac{1}{f(1/|h_\pi|)^2}, \end{aligned}$$

which is the same upper bound as for the first term, concluding the proof. □

Theorem 7.6. *Let W be a homogeneous and inverse invariant H_{-p}-valued white noise. Let $Q = \sum_{\pi\in\hat{G}} q_\pi \chi_\pi$ be its homogeneous covariance. Let f be a function as in Proposition 7.1. Assume that f is a uniform modulus of continuity for u_0 in G. Let $\overline{f}(x)$ be the larger of $f(x)(-\log x)^{-1/2}$ and $x(-\log x)^{1/2} f'(x)$. Assume $\overline{f}(x)/x$ is decreasing for small x. Then a sufficient condition for the solution u to (9) to have f as an almost-sure uniform modulus of continuity is*

$$\sum_{\pi\in\hat{G}} \frac{q_\pi d_\pi}{(1+\lambda_\pi)\overline{f}\left(\lambda_\pi^{-1/2}\right)^2} < \infty.$$

Proof. As in the calculations of Section 6, it is equivalent to find a condition guaranteeing that f is a modulus of continuity for the random field $U(x) = \int_0^1 W(ds, H_{1-s}(x,\cdot))$ whose covariance function $S = \sum_{\pi\in\hat{G}} s_\pi \chi_\pi$ satisfies $s_\pi = 2^{-1} q_\pi \lambda_\pi^{-1}(1 - \exp -2\lambda_\pi)$ for $\pi \neq \mathbf{1}$. By Condition (m) of Proposition 7.5, Proposition 7.1 and Proposition 2.2, it is sufficient to have convergence of the series

$$\sum_{\pi\in\hat{G}-\{\mathbf{1}\}} \frac{q_\pi\left(1-e^{-2\lambda_\pi}\right) d_\pi}{2\lambda_\pi \check{f}\left(\lambda_\pi^{-1/2}\right)},$$

where $\mathring{f}$ is chosen so that $w_{\mathring{f}} := \int_0^\varepsilon \left(\mathring{f}\right)'(x)\sqrt{-\log x}\,dx$ is equal to f, namely,

$$\mathring{f}(\varepsilon) = \int_0^\varepsilon f'(x)(-\log x)^{-1/2}\,dx,$$

as long as $\mathring{f}(x)/x$ is decreasing near 0. According to Corollary 7.3, the convergence of the series is unchanged if we replace $\mathring{f}(x)$ by $\overline{f}(x)$, the larger of $f(x)(-\log x)^{-1/2}$ and $x(-\log x)^{1/2} f'(x)$ for small x; the theorem follows as long as $\overline{f}(x)/x$ is decreasing for small x. □

Our methods clearly show that the above condition is sharp. We will in fact prove that it is necessary in a future communication.

Remark 7.7. The condition $\overline{f}(x)/x$ is satisfied in the examples of Remark 7.4. More generally, assume that we are in Case 1. Then $\overline{f}(x)/x = f(x)/(x\sqrt{-\log x})$. We see that $\overline{f}(x)/x$ is decreasing as soon as $f(x)/x$ is. This condition is natural if we assume that $\lim_0 f' = +\infty$, since it corresponds to saying that this limit is attained monotonically. The case of $\lim_0 f' < +\infty$ is the case of differentiable Y and u, which can be treated by considering the derivatives of these fields, reducing the study to the case of $\lim_0 f' = +\infty$.

Consider Case 2. We have $\overline{f}(x)/x = f'(x)(-\log x)^{1/2}$. This is decreasing as soon as $f'(x)$ is decreasing. This concave condition on f is a natural assumption in Case 2, in which we always have $\lim_0 f' = +\infty$.

Remark 7.8. The condition $\overline{f}(x)/x$ decreasing can be relaxed considerably to include such oscillating moduli of continuity as $\overline{f}(x) = x^{1/4} + x^{1/2}\sin x^{-1}$. Indeed, such a function satisfies the following condition:

(n) There exists a constant $L > 1$ such that for all x and y in a neighborhood of 0, $x > y \Longrightarrow \overline{f}(x)/x < L\overline{f}(y)/y$.

Using this condition in the proof of Proposition 7.5 only changes the estimate on the series of condition (m) by the factor L, and the result thus remains. It is worth noting however that for practical purposes, one often only requires knowing a modulus of continuity up to a constant factor, in which case an oscillating modulus such as the example above can be replaced by its dominating term which does satisfy the stronger decrease condition.

References

[1] R. Adler, *An introduction to continuity, extrema, and related topics for general Gaussian processes*, Inst. Math. Stat., Hayward, CA, **1990**.

[2] P. L. Chow,*On solutions of the stochastic wave equation*, Communication at the Fourth Joint AMS-SMM meeting. UNT, May 1999.

[3] T. Coulhon, L. Saloff-Coste, N. Varopoulos, *Analysis and Geometry on Groups*, Cambridge University Press, **1992**.

[4] R. C. Dalang and C. Robert, *Extending the martingale measure stochastic integral with applications to spatially homogeneous*, s.p.d.e.'s. Electron. J. Probab., **6** (1999).

[5] R. C. Dalang and N. Frangos, *The stochastic wave equation in two spatial dimensions*, Ann. Probab., **26** (1998), 187–212.

[6] G. DaPrato and J. Zabczyk, *Stochastic equations in infinite dimensions*, Cambridge Univ. Press, **1992**.

[7] H. Fegan, *Introduction to compact Lie groups*, Prentice Hall, **1991**.

[8] G. Folland, *A course in abstract harmonic analysis*, Studies in Advanced Mathematics, CRC Press, Boca Raton, FL, **1995**.

[9] W. Fulton and J. Harris, *Representation Theory*, Springer, **1991**.

[10] I. Karatzas and S. Shreve, *Brownian motion and stochastic calculus*, 2nd ed., Springer V., **1991**.

[11] A. Karczewska and J. Zabczyk, *A note on stochastic wave equation*, Preprint.

[12] P. Kotelenez, *A class of quasilinear stochastic partial differential equations of McKean-Vlasov type with mass conservation*, Probab. Theory Rel. Fields, **102** (1995), 159–188.

[13] N. V. Krylov, *An analytic approach to SPDEs, Stochastic partial differential equations: six perspectives*, Math. Surveys Monogr., **64**, Amer. Math. Soc., Providence, RI, (1999), 185–242.

[14] H. Kunita, *Stochastic flows and stochastic differential equations*, Cambridge Univ. Press, **1990**.

[15] T. Kurtz and P. Protter, *Weak convergence of stochastic integrals and differential equations II: Infinite dimensional case*, Lecture Notes, 1996 CIME school in probab.

[16] M. Marcus and G. Pisier, *Random Fourier series with applications to harmonic analysis*, Princeton UP, **1981**.

[17] A. Millet and M. Sanz-Sole, *A stochastic wave equation in two space dimensions: smoothness of the law*, Ann. Probab., **27** (1999), 803–844.

[18] S. Peszat and J. Zabczyk, *Stochastic evolution equations with a spatially homogeneous Wiener process*, Stochastic Processes Appl., **72 (2)** (1997), 187–204.

[19] S. Peszat and J. Zabczyk, *Nonlinear stochastic wave and heat equation*, Probab. Theory Related Fields., **116** (2000), 421–443.

[20] M. Sanz and M. Sarra, *Hölder continuity for the stochastic heat equation with spatially correlated noise*, This volume.

[21] M. Sugiura, *Fourier series of smooth functions on compact Lie groups*, Osaka J. Math., **8** (1971), 33–47.

[22] S. Tindel and F. Viens, *On space-time regularity for the stochastic heat equation on Lie groups*, J. Func. Analy., **169 (2)** (1999), 559–603.

[23] J. B. Walsh, *An introduction to stochastic partial differential equations*, In: cole d'té de Probabilités de Saint Flour XIV, Lecture Notes in Math., **1180** (1986), 265–438.

UMR CNRS 7539 – Dpartement de Mathmatiques
Universit Paris 13
93430 Villetaneuse
France

Department of Mathematics
University of North Texas
P.O. Box 311430
Denton, TX 76203
United States

Progress in Probability, Vol. 52, 293–302

Forward Integrals and Stochastic Differential Equations

M. Zähle

Abstract. We show that an anticipating stochastic forward integral introduced in [9] by means of fractional calculus is an extension of other forward integrals known from the literature. The latter provide important classes of integrable processes. In particular, we investigate the deterministic case for integrands and integrators from optimal Besov spaces. Here the forward integral agrees with the continuous extension of the Lebesgue-Stieltjes integral to these function spaces.

Generalized quadratic variation processes are defined in a similar manner. A survey on applications to anticipating stochastic differential equations with driving processes $Z^0, Z^1, \ldots, Z^m$ is given, where Z^0 has a generalized bracket $[Z^0]$ and $[Z^0], Z^1, \ldots, Z^m$ are smooth of fractional order greater than 1/2.

1. Anticipating stochastic differential equations

We consider SDE in $\mathbb{R}^n$ of the following type:

$$dX(t) = \sum_{i=0}^{m} a_i\big(X(t),t\big)\, dZ^i(t) + b\big(X(t),t\big)\, dt \ , \ t \in [0,T] \ , \ X(0) = X_0 \, . \tag{1}$$

$a_0(x,t), \ldots, a_m(x,t)$ are random C^1-vector fields whose partial derivatives are locally Lipschitz in x and $b(x,t)$ is continuous and locally Lipschitz in x. X_0 is an arbitrary random vector and the driving processes are as follows: Z^0 is continuous and possesses a generalized bracket $[Z^0](t)$ (see Section 2) and $[Z^0], Z^1, \ldots, Z^m$ have sample paths in the Sobolev-Slobodeckij space W_2^β (which agrees with the Besov space $B_{2,2}^\beta$, see, e.g., [7]) for some $\beta > 1/2$. We do not suppose any adaptedness or independence. The stochastic integrals in (1) are forward integrals and will be introduced in Section 2. An important example with applications in mathematical finance is the case where Z^0 agrees with the Wiener process and $Z^1, \ldots, Z^m$ are fractional Brownian motions with Hurst exponents greater than 1/2.

2000 *Mathematics Subject Classification.* 60H10, 60H05, 34F05.
Key words and phrases. Forward integral, Besov space, quadratic variation process, anticipating stochastic differential equation.

In [10] it is proved that equation (1) has a unique local pathwise solution with generalized bracket satisfying the rules of Itô calculus. It may be determined by the following procedure (which extends the Doss-Sussman approach known from martingale theory for $m = 0$):

$$X(t) = h\big(Y(t), Z^0(t), t\big) \tag{2}$$

where $h(y, z, t)$ is a pathwise local C^1-solution of the differential equation

$$\begin{aligned} \frac{\partial h}{\partial z}\,(y, z, t) &= a_0\big(h(y, z, t), t\big) \\ h\big(Y(0), Z(0), 0\big) &= X_0 \end{aligned} \tag{3}$$

being Lipschitz in y such that

$$\det\left(\frac{\partial h}{\partial y}\,(y, z, t)\right) \neq 0.$$

Given $Y(0)$ and h the stochastic process Y in $\mathbb{R}^n$ is locally determined by the SDE

$$\begin{aligned} dY(t) = \left(\frac{\partial h}{\partial y}\,\big(Y(t), Z^0(t), t\big)\right)^{-1} &\left[\sum_{j=1}^{m} a_j\Big(h\big(Y(t), Z^0(t), t\big), t\Big)\, dZ^j(t)\right. \\ &+ \left(b\Big(h\big(Y(t), Z^0(t), t\big), t\Big) - \frac{\partial h}{\partial t}\,\big(Y(t), Z^0(t), t\big)\right) dt \\ &\left. - \frac{1}{2}\,\frac{\partial a_0}{\partial x}\,\Big(h\big(Y(t), Z^0(t), t\big), t\Big)\, a_0\Big(h\big(Y(t), z^0(t), t\big), t\Big)\, d[Z^0](t)\right]. \end{aligned} \tag{4}$$

This equation has a better analytical behavior than (1), in particular since the processes $Z^1, \ldots, Z^m$ have zero quadratic variations. In [10] a solution procedure by means of a contraction principle in Besov spaces is demonstrated.

Note that because of equation (3) the representation (2) of the unique solution $X(t)$ is, in general, not unique.

2. Stochastic forward integrals and generalized quadratic variations

By means of fractional calculus we are led in [9] to the following version of **stochastic forward integral**: Let X be a càglàd process and Z a càdlàg process on $[0, T]$. Denote $Z_{t-}(s) := \big(Z(s) - Z(t-)\big)\, 1_{[0,t)}(s)$.

Definition 2.1. *X is integrable w.r.t Z on $[0, T]$ if the following limit exists (uniformly in t) in probability:*

$$\int_0^{t-} X\, dZ := \lim_{\varepsilon \to 0} \int_0^1 \varepsilon u^{\varepsilon - 1} \int_0^t X(s)\, \frac{Z_{t-}(s+u) - Z_{t-}(s)}{u}\, ds\, du\,. \tag{5}$$

This is an **extension** of the forward integral

$$\lim_{u \searrow 0} \int_0^t X(s) \, \frac{Z_{t-}(s+u) - Z_{t-}(s)}{u} \, ds \tag{6}$$

with (uniform) convergence in probability introduced in Russo and Vallois [4] for the continuous case (and in a modified version for the discontinuous case).

In order to check this, note first that the kernel $\varepsilon u^{\varepsilon-1}$ acts as the δ-function as $\varepsilon \to 0$. Further, let $Y_t(u)$ be a family of càdlàg processes on (0,1) converging to $Y_t(0)$ in probability as $u \searrow 0$ (uniformly in the parameter t). Suppose that $\sup |Y_t(u)| < \infty$ w.p.1, where the supremum is taken over u (and t, resp.). Then we have

$$\begin{aligned}
\int_0^1 \varepsilon u^{\varepsilon-1} \, Y_t(u) \, du &= Y(0) + \int_0^1 \varepsilon u^{\varepsilon-1} \, \big(Y_t(u) - Y_t(0)\big) \, du \\
&= Y(0) + \int_0^1 \varepsilon u^{\varepsilon-1} \, 1\{u : |Y_t(u) - Y_t(0)| > \delta\}\big(Y_t(u) - Y_t(0)\big) \, du \\
&\quad + \int_0^1 \varepsilon u^{\varepsilon-1} \, 1\{u : |Y_t(u) - Y_t(0)| \le \delta\}\big(Y_t(u) - Y_t(0)\big) \, du \, .
\end{aligned}$$

The last integral does not exceed δ. The first integral on the right-hand side converges to zero in probability as $\varepsilon \to 0$ (uniformly in t), since $|Y_t(u) - Y_t(0)|$ is (uniformly) bounded w.p.1 and

$$\begin{aligned}
&\lim_{\varepsilon \to 0} \big(\sup_t\big) \, \mathbb{E} \int_0^1 \varepsilon u^{\varepsilon-1} \, 1\Big\{u : \big|Y_t(u) - Y_t(0)\big| > \delta\Big\} \, du \\
&\le \lim_{\varepsilon \to 0} \int_0^1 \varepsilon u^{\varepsilon-1} \, \mathbb{P}\Big\{\big(\sup_t\big)\big|Y_t(u) - Y_t(0)\big| > \delta\Big\} \, du \\
&= \lim_{u \searrow 0} \mathbb{P}\Big\{\big(\sup_t\big)\big|Y_t(u) - Y_t(0)\big| > \delta\Big\} = 0
\end{aligned}$$

by assumption. Hence,

$$\lim_{\varepsilon \to 0} \int_0^1 \varepsilon u^{\varepsilon-1} \, Y_t(u) \, du = Y_t(0)$$

with (uniform) convergence in probability. Taking

$$Y_t(u) := \int_0^t X(s) \, \frac{Z_{t-}(s+u) - Z_{t-}(s)}{u} \, ds$$

we obtain the assertion.

In [9] it is proved that uniform convergence in probability of the Riemann-Stieltjes sums implies that in (5). Therefore our integral agrees with the Itô integral in semimartingale theory. Moreover, it also coincides with the forward integral in Malliavin calculus when X has the required regularity and Z is Brownian motion and with Young's approach for functions with finite p-variations (see [9]). In the next section we will show that other types of forward integrals known from the literature also fit into our model. This provides important classes of processes where the integral exists.

Extending Russo's and Vallois' notion [5] we have introduced in [9] the **quadratic variation** of a càdlàg process Z by

$$[Z](t) := \lim_{\varepsilon \to 0} \int_0^1 \varepsilon u^{\varepsilon-1} \int_0^t \frac{\left(Z_{t-}(s+u) - Z_{t-}(s)\right)^2}{u} \, ds \, du + \left(Z(t) - Z(t-)\right)^2 \quad (7)$$

whenever the limit exists (uniformly) in probability. Note that for semimartingales this agrees with the bracket used in the literature.

The covariation process of two such processes is defined similarly. (See [5], [6] and [9] for some Itô calculus.) Any process with generalized bracket is an element of the function space $W_2^{1/2-}$ w.p.1 (see [9]).

Remark 2.2. The averaging procedure in (7) as $u \to 0$ is adapted to Definition (5) via associated Itô formulae. Below this will be related to fractional calculus.

3. Special cases of (stochastic) forward integrals

The background from fractional calculus for our notion (5) is the following representation:

$$\frac{1}{\Gamma(\varepsilon)} \int_0^\infty u^{\varepsilon-1} \int_0^t X(s) \, \frac{Z_{t-}(s+u) - Z_{t-}(s)}{u} \, ds \, du = \int_0^t I_{0+}^{\varepsilon} \, X(s) \, dZ(s)$$

for (random) functions X and Z possessing fractional derivatives in L_p, resp. L_q, of all orders less than α, resp. $1-\alpha$, where $\frac{1}{p} + \frac{1}{q} = 1$ and $0 \le \alpha \le 1$ (see [8], [9]). The integral on the right-hand side is determined by means of such derivatives:

$$\begin{aligned} \int_0^t I_{0+}^{\varepsilon} \, X(s) \, dZ(s) := (-1)^{\alpha-\varepsilon/2} \int_0^t D_{0+}^{\alpha-\varepsilon/2} X_{0+}(u) \, D_{t-}^{1-\alpha-\varepsilon/2} Z_{t-}(u) \, du \\ + X(0+) \left(Z(t-) - Z(0)\right). \end{aligned} \quad (8)$$

(I_{0+}^{ε} stands for the fractional Riemann-Liouville integral of order ε.)

Instead of the limit $\lim_{\varepsilon \to 0} \frac{1}{\Gamma(\varepsilon)} \int_0^\infty u^{\varepsilon-1} \, \ldots \, du$ it is convenient to consider $\lim_{\varepsilon \to 0} \varepsilon \int_0^1 u^{\varepsilon-1} \, \ldots \, du$ as above. If the marginal values α and $1-\alpha$ are included we get the function spaces $I^{\alpha}(L_p)$ and $I^{1-\alpha}(L_q)$ where the limit exists and agrees with (8) for $\varepsilon = 0$.

In particular, the stochastic integrals in equations (1) and (4) for $j = 1, \dots, m$ may be interpreted pathwise in that sense. The stochastic integral w.r.t the process Z^0 in (1) is defined by (5), where $[Z^0]$ is given according to (7) (continuous case).

A further advantage of the averaging procedure in the limit for small u consists in extending the following procedure: In [2] Bedford and Kamae considered the **Cesàro averages** (in the sense of ergodic theory)

$$\int_0^t X\,dZ := \lim_{T\to\infty} \frac{1}{T}\int_0^T\int_0^t X(s)\,\frac{Z_{t-}(s+e^{-v}) - Z_{t-}(s)}{e^{-v}}\,ds\,dv \tag{9}$$

(for continuous X and Z) and proved existence of the (uniform) limit for certain X and self-affine functions Z. Replacing deterministic convergence by (uniform) **convergence in probability** one obtains the stochastic version for càglàd X and càdlàg Z.

Proposition 3.1. *If the stochastic integral in* (9) *is determined then it agrees with that in our Definition* (5).

Proof. The right-hand side of (9) is equal to

$$\lim_{\delta\to 0} \frac{1}{|\ln\delta|}\int_\delta^1 \frac{1}{u}\int_0^t X(s)\,\frac{Z_{t-}(s+u) - Z_{t-}(s)}{u}\,ds\,du \tag{10}$$

so that the difference between (9) and(5) consists only in the averaging procedure as $u \searrow 0$. The kernel $\varepsilon u^{\varepsilon-1}$ in (5) arises from the last kernel $\frac{1}{|\ln\delta|}\,1_{(\delta,1)}(u)\,\frac{1}{u}$ by means of a second averaging kernel $k_\varepsilon(v) := \varepsilon^2 v^{\varepsilon-1}|\ln v|$ on (0,1) as follows: Denote the above inner integral by $Y_t(u)$ and use Fubini in order to get

$$\begin{aligned}\int_0^1 k_\varepsilon(v)\,\frac{1}{|\ln v|}\int_v^1 \frac{1}{u}\,Y_t(u)\,du\,dv &= \int_0^1 \frac{1}{u}\int_0^u \frac{1}{|\ln v|}\,k_\varepsilon(v)\,dv\,Y_t(u)\,du \\ &= \int_0^1 \varepsilon u^{\varepsilon-1}\,Y_t(u)\,du\,.\end{aligned}$$

Since $\lim_{v\searrow 0}\frac{1}{|\ln v|}\int_v^1 \frac{1}{u}\,Y_t(u)\,du = \int_0^t X\,dZ$ in the sense of Bedford and Kamae (stochastic version) and $k_\varepsilon(v)$ acts as the δ-function as $\varepsilon \to 0$ we infer similarly as in Section 2:

$$\lim_{\varepsilon\to 0}\int_0^1 k_\varepsilon(v)\,\frac{1}{|\ln v|}\int_v^1 \frac{1}{u}\,Y_t\,du\,dv = \int_0^t X\,dZ$$

(uniformly) in probability. In view of the above equations this agrees with

$$\lim_{\varepsilon\to 0}\int_0^1 \varepsilon u^{\varepsilon-1}\, Y_t(u)\, du\,.$$

□

The last procedure may be applied to statistically self-similar random processes.

A further **specification** is a modification of a forward integral introduced by Klingenhöfer (personal communication):

$$\int_0^t X\, dZ := \lim_{\delta\to 0}\frac{1}{\ln 2}\int_\delta^{2\delta}\frac{1}{u}\int_0^t X(s)\,\frac{Z_{t-}(s+u)-Z_{t-}(s)}{u}\,ds\,du. \tag{11}$$

(Klingenhöfer considered $\lim_{n\to\infty} 2^{2(n+1)}\int_{2^{-(n+1)}}^{2^{-n}}\int_0^t X(s)\,\big(Z(s+u)-Z(s)\big)\,ds\,du$ in the deterministic case and proved existence for optimal Besov spaces). We here again take (uniform) convergence in probability, càglàd X and càdlàg Z.

Proposition 3.2. *If the integral in* (11) *exists then it agrees with that in the Cesàro averaging* (10).

Proof. We have for

$$Y_t(u) := \int_0^t X(s)\,\frac{Z_{t-}(s+u)-Z_{t-}(s)}{u}\,ds$$

$$\frac{1}{|\ln\delta|}\int_\delta^1\frac{1}{u}\,Y_t(u)\,du = \frac{1}{|\ln\delta|}\sum_{k=1}^{n-1}\frac{1}{\ln 2}\int_{2^k\delta}^{2^{k+1}\delta}\frac{1}{u}\,Y_t(u)\,du$$

$$+\frac{1}{|\log\delta|}\,\frac{1}{\ln 2}\int_{2^n\delta}^{1}\frac{1}{u}\,Y_t(u)\,du$$

where $n=n(\delta)$ is the integral part of $|\ln\delta|/\ln 2$. The last summand tends to 0 as $\delta\to 0$ uniformly in t. The first summand up to the factor $\frac{n}{|\ln\delta|}$ equals

$$\frac{1}{n}\sum_{k=1}^{n}\frac{1}{\ln 2}\int_{2^{-k}2^n\delta}^{2\cdot 2^{-k}2^n\delta}\frac{1}{u}\,Y_t(u)\,du\,.$$

Denote the k-th summand of the last sum by $A_t^{n,\delta}(k)$. Using (11) we infer

$$\lim_{k\to\infty} A_t^{n,\delta}(k) = \int_0^t X\, dZ$$

uniformly (in t and) in n, δ such that $n > k,\ 2^n\delta < 1$ in probability. This implies

$$\lim_{\delta\to 0} \frac{1}{n(\delta)} \sum_{k=1}^{n(\delta)} A_t^{n(\delta),\delta}(k) = \int_0^t X\, dZ$$

(uniformly in t) in probability. Hence, the last integral agrees with the (uniform) limit in probability

$$\lim_{\delta\to 0} \frac{1}{|\ln \delta|} \int_\delta^1 \frac{1}{u}\, Y_t(u)\, du\,.$$

□

Thus, we have obtained the following **implications of integrability** in the above forward integrals:

$$(6) \Rightarrow (11) \Rightarrow (9) \Rightarrow (5).$$

Using a duality result of Triebel [7] we now will provide **optimal classes of Besov spaces** for which the deterministic variant of Russo's and Vallois' integral (6) makes sense. (Recall that the norm on the Besov space $B^\alpha_{p,q}$ on $\mathbb{R}$ for the classical case $0 < \alpha < 1$ and $1 \le p \le \infty$ may be introduced by

$$\|f\|^\alpha_{p,q} := \|f\|_{L_p} + \left(\int_{-1}^{1} \frac{1}{|h|}\, \frac{1}{|h|^{\alpha q}} \left(\|f((\cdot)+h) - f\|_{L_p} \right)^q dh \right)^{1/q}$$

if $1 \le q < \infty$ and

$$\|f\|^\alpha_{p,\infty} := \|f\|_{L_p} + \sup_{|h|\le 1} \frac{1}{|h|^\alpha} \left\| f((\cdot)+h) - f \right\|_{L_p}.$$

For the quasi-normed distribution spaces $B^\alpha_{p,q}$ with general $\alpha \in \mathbb{R},\ p > 0,\ q > 0$ see [7].)

Theorem 3.3. *Suppose* $f \in B^\alpha_{p,q},\ g \in B^{1-\alpha}_{p',q'},\ 1 \le p < \infty,\ 1 \le q < \infty,\ 1/p + 1/p' = 1,\ 1/q + 1/q' = 1,\ 0 \le \alpha \le 1$. *Then the limit*

$$\int f\, dg := \lim_{u \searrow 0} \int f(s)\, \frac{g(s+u) - g(s)}{u}\, ds$$

exists and we have for some constant $c > 0$,

$$\left| \int f\, dg \right| \le c\, \|f\|^\alpha_{p,q}\, \|g\|^{1-\alpha}_{p',q'}\,.$$

Proof. Note that the Schwartz space $\mathcal{S}$ of rapidly decreasing C^∞-functions of $\mathbb{R}$ is dense in $B^{1-\alpha}_{p',q'}$. If $g \in \mathcal{S}$ and g' denotes the derivative we obtain

$$\int f(s)\, \frac{g(s+u) - g(s)}{u}\, ds = \int f(s) \int_0^1 g'(s+ru)\, dr\, ds$$

so that for $0 < \tilde{u} < u$,

$$\left| \int f(s) \, \frac{g(s+u) - g(s)}{u} \, ds - \int f(s) \, \frac{g(s+\tilde{u}) - g(s)}{\tilde{u}} \, ds \right|$$

$$\leq \int_0^1 \int |f(s)| \, |g'(s+ru) - g'(s+r\tilde{u})| \, ds \, dr.$$

According to [7, 2.11.2], the inner integral may be estimated from above by

$$c' \, \|f\|^{\alpha}_{p,q} \, \|g'((\cdot)+ru) - g'((\cdot)+r\tilde{u})\|^{-\alpha}_{p',q'}$$

for some constant c'. (g' is here interpreted as a distribution.)

Further, $\|\phi'\|^{-\alpha}_{p',q'} \leq c'' \, \|\phi\|^{1-\alpha}_{p',q'}$, $\phi \in \mathcal{S}$, for some constant c'' (see, e.g., [7, 2.3.8]). Hence, the above difference does not exceed

$$c \, \|f\|^{\alpha}_{p,q} \int_0^1 \|g((\cdot)+ru) - g((\cdot)+r\tilde{u})\|^{1-\alpha}_{p',q'} \, dr$$

where $c = c'c''$. This estimation extends to arbitrary $g \in B^{1-\alpha}_{p',q'}$ via approximation by functions from $\mathcal{S}$. Similarly,

$$\left| \int f(s) \, \frac{g(s+u) - g(s)}{u} \, ds \right| \leq c \, \|f\|^{\alpha}_{p,q} \int_0^1 \|g((\cdot)+ru)\|^{1-\alpha}_{p',q'} \, dr \, .$$

Thus, it remains to show that for $1 \leq p, q \leq \infty$,

$$\lim_{\Delta \to 0} \, \|\phi((\cdot)+\Delta) - \phi\|^{\alpha}_{p,q} = 0$$

provided $\phi \in B^{\alpha}_{p,q}$. This can be seen as follows. Given $\varepsilon > 0$ choose $\tilde{\phi} \in \mathcal{S}$ such that $\|\phi - \tilde{\phi}\| < \varepsilon$. Then we obtain by the invariance of the norm under translations

$$\begin{aligned} \|\phi((\cdot)+\Delta) - \phi\|^{\alpha}_{p,q} &\leq \|\phi((\cdot)+\Delta) - \tilde{\phi}((\cdot)+\Delta)\|^{\alpha}_{p,q} \\ &\quad + \|\tilde{\phi}((\cdot)+\Delta) - \tilde{\phi}\|^{\alpha}_{p,q} + \|\tilde{\phi} - \phi\|^{\alpha}_{p,q} \\ &< \|\tilde{\phi}((\cdot)+\Delta) - \tilde{\phi}\|^{\alpha}_{p,q} + 2\varepsilon \, . \end{aligned}$$

Hence, it suffices to prove that the last norm tends to zero as $\Delta \to 0$. For the L_p-part of the norm this is obvious. The remaining part (up to modification for

$p = \infty$ or $q = \infty$ or $\alpha = 0, 1)$ is equal to

$$\left(\int_{-1}^{1} \frac{1}{|h|^{\alpha q+1}} \left(\left\| \int_{0}^{\Delta} (\tilde{\phi}'(x+h+r) - \tilde{\phi}'(x+r)) \, dr \right\|_{L_p} \right)^q dh \right)^{1/q}$$

$$\leq \|\tilde{\phi}''\|_{L_\infty} \, \Delta \left(\int_{-1}^{1} |h|^{q(1-\alpha)-1} \, dh \right)^{1/q}$$

$$= \|\tilde{\phi}''\|_{L_\infty} \, \mathrm{const}(\alpha, q) \, \Delta \, .$$

□

Remark 3.4.

1. The last theorem extends a partial result of [3, V. 3], for $q = 1$ and some p and α. Our conditions on the Besov spaces are best possible since we have used implicitly the duality $(B^{\alpha}_{p,q})' = B^{-\alpha}_{p',q'}$. By the norm estimators the forward integral in this case agrees with the continuous extension of the Lebesgue-Stieltjes integral to these function spaces.
2. The ideas of proof remain valid for the stochastic forward integral (5) if the Besov spaces are replaced by some stochastic variants. This will be shown elsewhere. For the method of fractional derivatives introduced in [8, 9] and fractional Brownian motion as integrator such a combination of analysis and stochastics is used in [1], Section 5.

References

[1] E. Alòs, O. Mazet and D. Nualart, *Stochastic calculus with respect to fractional Brownian motion with Hurst parameter less than 1/2,* Stoch. Process. Appl., **86** (2000), 121–139.

[2] T. Bedford and T. Kamae, *Stieltjes integration and stochastic calculus with respect to self-affine functions,* Japan J. Indust. Appl. Math., **8** (1991), 445–459.

[3] Z. Ciesielski, G. Kerkyacharian and R. Roynette, *Quelques espaces fonctionelles associés à des processus gaussiens,* Studia Math., **107** (1993), 71–204.

[4] F. Russo and P. Vallois, *Forward, backward and symmetric stochastic integration,* Probab. Th. Relat. Fields, **97** (1993), 403–421.

[5] F. Russo and P. Vallois, *The generalized covariation process and Itô formula,* Stoch. Processes Appl., **59** (1995), 81–104.

[6] F. Russo and P. Vallois, *Stochastic calculus with respect to continuous finite quadratic variation processes,* Stoch. and Stoch. Reports., **70** (2000), 1–40.

[7] H. Triebel, *Theory of function spaces,* Geest & Portig, Leipzig, **1983** and Birkhäuser, Basel, **1983**.

[8] M. Zähle, *Integration with respect to fractal functions and stochastic calculus I,* Probab. Th. Relat. Fields, **111** (1998), 333–374.

[9] M. Zähle, *Integration with respect to fractal functions and stochastic calculus II*, Math. Nachr., 225 (2001), 145–183.

[10] M. Zähle, *Pathwise solution of anticipating SDE with mixed driving*, preprint.

Department of Mathematics
University of Jena
D-07740 Jena
Germany
E-mail address: zaehle@minet.uni-jena.de

MIX
Papier aus verantwortungsvollen Quellen
Paper from responsible sources
FSC® C105338

If you have any concerns about our products,
you can contact us on
ProductSafety@springernature.com

In case Publisher is established outside the EU,
the EU authorized representative is:
Springer Nature Customer Service Center GmbH
Europaplatz 3, 69115 Heidelberg, Germany

Printed by Libri Plureos GmbH
in Hamburg, Germany